Environmental Health Literacy

Symma Finn • Liam R. O'Fallon
Editors

# Environmental Health Literacy

Springer

*Editors*
Symma Finn
Population Health Branch (PHB)
National Institute of Environmental Health Sciences
Durham, NC, USA

Liam R. O'Fallon
Population Health Branch (PHB)
National Institute of Environmental Health Sciences
Durham, NC, USA

ISBN 978-3-030-06798-4 ISBN 978-3-319-94108-0 (eBook)
https://doi.org/10.1007/978-3-319-94108-0


Softcover re-print of the Hardcover 1st edition 2019


This Springer imprint is published by the registered company Springer Nature Switzerland AG
The registered company address is: Gewerbestrasse 11, 6330 Cham, Switzerland

# Dustjacket Reviews

*"Environmental Health Literacy is vital for making informed decisions aimed at preventing illness and disease. This well-organized book offers readers an accessible introduction to environmental health literacy. It details the emergence of EHL, provides a range of case studies, and describes approaches being used to build EHL. It's inspiring and a must-read for anyone interested in environmental health communication and community engaged research."*

– Linda S. Birnbaum, Ph.D., D.A.B.T., A.T.S., Director, National Institute of Environmental Health Sciences and National Toxicology Program

*"As concerns about the effect of environmental change on population health continue to grow, so too does our need to understand and address gaps in environmental health literacy. You hold in your hands a comprehensive foray into the why, how, where, and when of increasing engagement with health effects of the environment. All stakeholders in this space – from researchers to risk communicators to the lay public – will benefit by having this volume at the top of reading pile."*

– William M. P. Klein, Ph.D., Associate Director, Behavioral Research Program, Division of Cancer Control and Population Sciences, National Cancer Institute

*"In this book, Finn and O'Fallon provide a solid foundation to help readers understand key elements and challenges of environmental health literacy. They also provide recommendations to improve communications efforts so that the science translates to information the public can understand and use to improve and protect their health."*

– Patrick Breysse, Ph.D., CIH, Director, National Center for Environmental Health (NCEH) | Agency for Toxic Substances and Disease Registry (ATSDR), Centers for Disease Control and Prevention

*"Our ancestors were experts at living in harmony with the land. Through research and discovery, we must once again become experts if we wish to live in harmony with ourselves. This book touches upon the importance of culture as part of envi-*

*ronmental health literacy to ensure respectful interactions among all partners in the research process. It is a good resource for anyone working with tribal communities."*

– David R. Wilson, Ph.D. (Dine), Director, Tribal Health Research Office, Division of Program Coordination, Planning, and Strategic Initiatives, Office of the Director, NIH

*"Environmental Health Literacy recognizes that public health problems are heavily impacted by environment and are more likely to be solved in the community than in the clinic. This timely collection provides evidence that social and health sciences in partnership with an informed, engaged public whose intelligence, expertise, and right-to-know are acknowledged and respected are key to finding those solutions."*

– Andrew M Geller, Ph.D., Deputy National Program Director, Sustainable and Healthy Communities Research Program, US Environmental Protection Agency

# Introduction: Environmental Health Literacy

**Abstract** Environmental Health Literacy (EHL) is a recently emerged, distinct subfield of literacy that draws equally upon key principles and methodological approaches from the fields of risk communication, health literacy, environmental health sciences (EHS), communications research, and education (Finn and O'Fallon 2017a; Gray 2018). Although, EHL has obvious roots in health literacy, it is very distinct in its purpose and outcomes. For EHL focuses primarily on improving the understanding of affected communities about risks related to environmental factors rather than solely on an individual's risk of disease, and on prevention rather than disease management and medical adherence. This focus on public health rather than disease management is consistent with the goals of EHS research to prevent environmentally induced disease. However, 'affected communities' are not monolithic in their demographics, and EHL has emerged during an era when there is greater recognition of the importance of cultural sensitivity in community-engaged research and the need for targeted messaging for specific subpopulations and audiences (Brugge et al. 2018; LePrevost et al. 2013; Lesch et al. 2009).

As an emerging field, EHL is being variously defined, however the majority of these definitions refer to the progressive nature of EHL that begins with an individual's understanding of the connection between environmental exposures and human health and proceeds to the ability to create effective public health messages and/or act on this understanding (Gray 2018; Guidotti 2013; Nutbeam 2009). For, similar to the stages of learning that have been delineated in the field of education, EHL is not a static achievement, but an evolutionary process (Bloom 1956; Finn and O'Fallon 2017a).

This book explores the emergence of EHL, provides numerous examples of the specific audiences who benefit from EHL, and describes the various communication and visualization modalities that are being employed to raise EHL among specific subpopulations and audiences.

Part I situates the book within the conceptual constructs related to health literacy, risk communication and public health education; describes EHLs relevance to environmental health sciences research; and delineates the challenges and successes to date in measuring improvements in EHL. Part II provides an overview of the types of stakeholders and audiences who benefit from efforts to raise their EHL; and Part III explores the wide range of communication and visualization modalities that are used to promote EHL.

## Background

The roots of environmental health literacy first began emerging in response to an increased public and media interest in the environment in the 1960s and 1970s, the subsequent development of the scientific discipline of environmental health sciences and the increasing need to communicate the results of EHS research findings to affected communities (Brody et al. 2014; Ramirez-Andreotta et al. 2016a). Concurrent social developments during these decades included empowerment movements, such as patient empowerment that informed early health literacy efforts, and the rise of social health movements that reflected a shift away from a professional dominance of the research enterprise, to more collaborative approaches that engaged affected community members and grassroots community organizations in addressing environmentally-related health concerns (Brown et al. 2004; English et al. 2017; Finn and Collman 2016). The emergence of EHL can additionally be linked to the rapid increase in computer usage among the public in the 1980s and 1990s, the development of digitized methods for visualizing risk, such as Geographic Information Systems (Lahr and Kooistra 2010; Severtson 2013), and the availability of remote imagery that helped to promote understanding of the concept of the Earth as an ecosystem linking people and nature. Together these social, technological and scientific developments led to greater involvement of the public in EHS research and greater understanding of the impact that pollution had on human health, an initial stage of environmental health literacy (Finn and O'Fallon 2017a; D Fitzpatrick-Lewis et al. 2010). This understanding, in turn, led to the recognition and confirmation that the poorest, most politically disenfranchised neighborhoods, often among communities of color, faced disproportionate exposures to harmful toxicants (Eggers et al. 2018). To this end, the emergence of environmental health literacy can be linked with the Environmental Justice movement, increased interest in community based participatory research (CBPR), and later with Citizen Science efforts (Minkler et al. 2008; Ramirez-Andreotta et al. 2016a; Stokes et al. 2010).

**The Evolution of Risk Communication:** The communication of environmental health research findings has evolved over time from generic risk messages aimed at the general public (e.g., Public Service Announcements), to risk messaging, educational modules, online games and mobile apps, and websites that are targeted to specific subpopulations, or segments of society, and delivered in a variety of modalities (Branco et al. 2015; Guidotti 2013; Korfmacher and Garrison 2014). In recent years, the delivery of environmental health risk communications has utilized the reach of social media and mobile based platforms to deliver information to a wider audience (G. L. Kreps 2014b; Neuhauser and Kreps 2003). In addition, environmental health risk communications have taken advantage of advances in visual technologies, such as GIS mapping, to deliver complex messages in an accessible manner that promotes environmental health literacy among the public (Kuchinskaya 2018; Lahr and Kooistra 2010).

There now exists a wealth of information on environmental health risks available to the public and the scientific community. Unfortunately, public information may or may not be evidence-based, there are no criteria for assessing sources of information or the appropriateness of content, nor whether the delivery (dissemination) of such information is understandable, effective and/or actually leads to prevention or mitigation of environmental health risks. Efforts to raise EHL and to measure improvements in EHL therefore benefits from utilization of health literacy, dissemination and implementation, and behavioral research approaches that evaluate the effectiveness of existing environmental health risk messaging and guide development of new messaging for specific audiences (Chinn 2011; Fung et al. 2018).

**Environmental Education and the Achievement Gap:** Environmental education has also informed the growth of environmental health literacy. Classroom and informal education settings have long-served as positive venues for increasing understanding about the environment around us (Hursh and Martina 2011). Through the 1990s there was a growing awareness of toxicants in our environment, and yet there was a lack of education in classrooms about the effect of such exposures on human health (Cunningham and Stubbs 2003; Hursh and Martina 2011). Concurrently, there was a nation-wide realization that USA students were falling behind in scientific understanding compared to students in other countries (Moreno and Tharp 1999; Morrone 2001). As such, there was a need to develop materials to fill these gaps. Inter-disciplinary approaches were established to convey complex scientific information and engage students in hands-on, inquiry-based activities. These approaches have increased interest in STEM subjects and made environmental issues accessible, understandable and actionable (Korfmacher and Garrison 2014; Read et al. 2016). More recently, game-based learning has also been explored as a viable means for increasing critical thinking, creative problem solving and teamwork to address environmental sustainability and management issues (Branco et al. 2015; Madani et al. 2017). These STEM efforts have contributed to our understanding of EHL and how such approaches can serve to inform and inspire youth.

## Book Chapters

The chapter authors represent a range of expertise related to communications research, dissemination and implementation, environmental health sciences, health education, and social scientific research. The majority of the authors are investigators funded by the National Institute of Environmental Health Sciences (NIEHS) who have developed methods or adapted existing modalities to raise the EHL of affected community members, health professionals, policy makers, and/or teachers associated with community-based EHS research studies.

This book explores various aspects of environmental health literacy (EHL) from the perspective of investigators working in this emerging field and their community partners in research

- Part I explores the foundational aspects of EHL – its evolving definition and early attempts to adapt existing measures to track educational attainment in this field – and provides examples of the theoretical frameworks that underlie research to promote or improve EHL. Chapters include an overview and definition of EHL and its relation to health literacy and communications research; and a proposed model for measuring EHL and the use of this model among American Indian research partners. This section also includes a chapter with background information on the increased use of communication research approaches in environmental health science research using a case study of ways to improve the EHL of stakeholders related to risks for breast cancer from environmental factors.
- Part II provides examples of the specific needs of target audiences and how EHL is being promoted to these distinct populations. This includes chapters exploring health professional needs, the benefits of EHL to empower members of environmental vulnerable communities, and the importance of raising the EHL and scientific literacy of students (i.e., STEM efforts). This section also includes a chapter describing research on the return of individual EHS results to affected community members. This exploration of the empowering aspects of EHL is a common theme in all of the chapters in this section.
- Part III describes the types of communication styles that have been used to promote and raise EHL. These chapters delineate the value of culturally appropriate methods that have been successfully employed in specific subpopulations (e.g., Native Americans, Hispanics), and the increased use of social media to disseminate and collaboratively explore environmental health information. Part III chapters explore the broad range of communication modalities that are being used in environmental health sciences research, including written and computer based formats, visual presentations, artistic approaches and the application of a tactile approach that employ hands-on methods to teach complex environmental health topics such as epigenetics to students and the general public. These chapters describe these various methodologies used to deliver messages to distinct audiences and the challenges and successes of traditional written communications vs. the challenges of disseminating non-verbal visual, tactile and artistic representations of risk.

Each chapter concludes with recommendations for research to improve communication efforts in the EHS and future directions for EHL. In addition, the book includes an afterword by the editors with overarching recommendations for the field of EHL based on the individual approaches described in the chapters and on the strategic goals of the NIEHS related to communications and education. Through these combined academic and governmental efforts, we hope to see EHL embraced as an essential component for empowering the public, and affected stakeholders, and for improving environmental health outcomes through this increased understanding.

S. Finn
L. O'Fallon

## References


Bloom, B. S. (1956). *Taxonomy of educational objectives; The classification of educational goals.* 1st ed. New York: Longmans, Green.

Branco, M. A., Weyermuller, A. R., Muller, E. F., Schneider, G. T., Hupffer, H. M., Delgado, J., et al. (2015). Games in the environmental context and their strategic use for environmental education. *Brazilian Journal of Biology, 75*:114–121.

Brody, J. G., Dunagan, S. C., Morello-Frosch, R., Brown, P., Patton, S., Rudel, R. A. (2014). Reporting individual results for biomonitoring and environmental exposures: Lessons learned from environmental communication case studies. *Environmental Health : A Global Access Science Source 13*(1), 40.

Brown, P., Zavestoski, S., McCormick, S., Mayer, B., Morello-Frosch, R., Gasior Altman R. (2004). Embodied health movements: New approaches to social movements in health. Sociology of Health & Illness, *26*, 50–80.

Brugge, D., Tracy, M., Thayer, K., Thayer, A., Dayer, B., Figueroa, N., et al. (2018). The role of environmental health literacy when developing traffic pollution fact sheets for puerto rican adults. *Environmental Justice, 11*, 40–46.

Chinn, D. (2011). Critical health literacy: A review and critical analysis. *Social Science & Medicine, 73*, 60–67.

Cunningham, W. P., Stubbs, H. S. (2003). Information needs related to teaching about air quality. *Environment International, 29*, 331–336.

Eggers, M. J., Doyle, J. T., Lefthand, M. J., Young, S. L., Moore-Nall, A. L., Kindness, L., et al. (2018). Community engaged cumulative risk assessment of exposure to inorganic well water contaminants, crow reservation, montana. *International Journal of Environmental Research and Public Health*, 15.

English, P. B., Olmedo, L., Bejarano, E., Lugo, H., Murillo, E., Seto, E., et al., (2017). The imperial county community air monitoring network: A model for community-based environmental monitoring for public health action. *Environmental Health Perspectives 125,* 074501.

Finn, S., Collman, G., (2016). The pivotal role of the social sciences in environmental health sciences research. *New Solutions*.

Finn, S., O'Fallon, L., (2017). The emergence of environmental health literacy-from its roots to its future potential. *Environmental Health Perspectives, 125,* 495–501.

Fitzpatrick-Lewis, D., Yost, J., Ciliska, D., Krishnaratne, S., (2010). Communication about environmental health risks: A systematic review. *Environmental Health : A Global Access Science Source 9*, 67.

Fung, T. K. F., Griffin, R. J., Dunwoody, S., (2018). Testing links among uncertainty, affect, and attitude toward a health behavior. *Science Communication, 40*, 33–62.

Gray, K. M., (2018). From content knowledge to community change: A review of representations of environmental health literacy. *International Journal of Environmental Research and Public Health,* 15.

Guidotti, T., (2013). Communication models in environmental health. *Journal of Health Communication, 18,* 1166–1179.

Hursh, D., Martina, C. A. (2011). *Teaching environmental health to children: An interdisciplinary approach*. Dordrecht: Springer Science and Business.

Korfmacher, K. S., Garrison, V., (2014). Partnering to reduce environmental hazards through a community-based "healthy home museum": Education for action. *Environ Justice 7*, 158–165.

Kreps, G. L., (2014). Achieving the promise of digital health information systems. *Journal Public Health Research*, *3*, 471.

Kuchinskaya, O., (2018). Connecting the dots: Public engagement with environmental data. *Environmental Communication*, *12*, 495–506.

Lahr, J., Kooistra, L., (2010). Environmental risk mapping of pollutants: State of the art and communication aspects. *The Science of the Total Environment, 408*, 3899–3907.

LePrevost, C., Storm, J., Blanchard, M., Asuaje, C., Cope, W. (2013). Engaging latino farmworkers in the development of symbols to improve pesticide safety and health education and risk communication. *Journal of Immigrant and Minority Health/Center for Minority Public Health, 15*, 975–981.

Lesch, M. F., Rau, P. L., Zhao, Z., Liu, C., (2009). A cross-cultural comparison of perceived hazard in response to warning components and configurations: *US vs. China. Applied Ergonomics, 40*, 953–961.

Madani, K., Pierce, T. W., Mirchi, A. (2017). Serious games on environmental management. *Sustainable Cities and Society, 29*, 1–11.

Minkler, M., Vasquez, V., Tajik, M., Petersen, D. (2008). Promoting environmental justice through community-based participatory research: The role of community and partnership capacity. *Health Education & Behavior : The Official Publication of the Society for Public Health Education, 35*, 119–137.

Moreno, N., Tharp, B. (1999). An interdisciplinary national program developed at baylor to make science exciting for all k-5 students. *Academic Medicine : Journal of the Association of American Medical Colleges, 74*, 345–347.

Morrone, M. (2001). Primary- and secondary-school environmental health science education and the education crisis: A survey of science teachers in ohio. *Journal of Environmental Health, 63*, 26–30.

Neuhauser, L., Kreps, G. L. (2003). Rethinking communication in the e-health era. *Journal of health psychology, 8,* 7–23.

Nutbeam, D. (2009). Defining and measuring health literacy: What can we learn from literacy studies?. *International Journal of Public Health, 54*, 303–305.

Ramirez-Andreotta, M. D., Brody, J. G., Lothrop, N., Loh, M., Beamer, P. I., Brown, P. (2016). Improving environmental health literacy and justice through environmental exposure results communication. *International Journal of Environmental Research and Public Health*, 13.

Read, C. Y., Ricciardi, C. E., Gruhl, A., Williams, L., Vandiver, K. M. (2016). Building genetic competence through partnerships and interactive models. *The Journal of Nursing Education, 55*, 300–303.

Severtson, D. (2013). The influence of environmental hazard maps on risk beliefs, emotion, and health-related behavioral intentions. *Research in Nursing & Health 36*, 330–348.

Stokes, S., Hood, D., Zokovitch, J., Close, F. (2010). Blueprint for communicating risk and preventing environmental injustice. *Journal of Health Care for the Poor and Underserved, 21*, 35–52.

# Contents

# Part I
# Foundations of Environmental Health Literacy

# Chapter 1
# Defining Environmental Health Literacy

**Anna Goodman Hoover**

**Abstract** *Environmental health literacy* (EHL) is an emerging area of study that incorporates content and strategies from environmental, health, and social sciences to promote understanding of the ways environmental contaminants affect health. The basic knowledge and skills needed for comprehending environmental health risks and for devising, assessing, implementing and evaluating potential solutions form the foundations of EHL. EHL strives to improve understanding of how individuals and communities make sense of and act on health-related information about environmental hazards. In this chapter, the concept of EHL and its background are introduced, and foundational contributions from health, risk, and participatory communication are discussed.

**Keywords** Environmental health literacy · Health literacy · Health communication · Dissemination · Risk communication · Participatory research · Environmental health sciences · Community engagement · Research translation · Risk perception

## Introduction

Human beings have long been exposed to substances in the environment that can make them sick, but understanding the relationship between pollutants and human health can be challenging. Many relevant issues beyond the chemistry of a specific contaminant drive what is known about the severity and scale of that contaminant's potential health impacts. For example, how are people exposed to the substance? Do they inhale it when they breathe? Absorb it through their skin? Drink or eat it? Did the contaminant enter food through the soil in which it grew or the vegetation on which it fed? What kinds of solutions can help reduce the contaminant's prevalence in the environment? And how can people best protect their health against a substance after they have been exposed?

A. G. Hoover (✉)
Department of Preventive Medicine and Environmental Health, University of Kentucky
College of Public Health, Lexington, KY, USA
e-mail: Anna.Hoover@uky.edu



S. Finn, L. R. O'Fallon (eds.), *Environmental Health Literacy*,
https://doi.org/10.1007/978-3-319-94108-0_1

These questions are not new, though contaminants of concern and exposure pathways change over time and across settings. From human and animal waste in streets and streams to smoke stacks and chemical spills, societies have learned how to minimize or even prevent some exposures and reduce illnesses related to others using a variety of strategies, including sewage treatment, occupational health and safety planning, and hazardous waste site remediation. New research regularly contributes additional evidence about the best strategies for protecting health in the face of evolving environmental challenges, rendering a grasp of complex environmental health concepts critical as individuals and communities attempt to identify and address concerns.

Implicit in many definitions of EHL is recognition that understanding scientific evidence can spur individual and/or collective actions to prevent unnecessary exposures and minimize negative health impacts related to unavoidable exposures. Because individual and community environmental health decisions are rooted within complex historical, political, regulatory, and community contexts, moving from initial levels of EHL to action often requires additional knowledge and skills that extend beyond comprehension of environmental health content to incorporate familiarity with decision-making structures, assessment of the feasibility of potential solutions, recognition of the inherent scientific uncertainties underlying these potential solutions, and identification of relevant local contextual factors that could affect the success of a chosen solution. In other words, EHL competencies also may include broader understanding of roles and responsibilities in environmental health decision processes, the universe of feasible choices, the uncertainties inherent in those choices, and the available strategies for taking action.

## Background

Both the integrative nature and action orientation of EHL were recognized in an early definition developed by the Society of Public Health Education (2007), which noted that EHL brings together environmental and health concepts to help people "seek out, comprehend, evaluate, and use environmental health information to make informed choices, reduce health risks, improve quality of life and protect the environment." More recently, Finn and O'Fallon (2017) described EHL as occurring along a continuum that "begins with an understanding of the connection between environmental exposures and human health"; such understandings can better inform how individuals and communities identify and address risk. Although Finn and O'Fallon distinguished between basic levels of understanding and more advanced action-focused EHL, they share with the SOPHE definition *explicit* recognition of EHL's multidisciplinarity and *implicit* recognition of action as a likely outcome.

Building on this convergence, environmental health literacy can be defined as *an emerging and evolving multidisciplinary field that seeks to better understand how individuals and communities make sense of and act on health-related information about environmental hazards*. This definition makes clear that EHL requires basic scientific knowledge about contaminants, human health, and exposure pathways; however, such knowledge may not be sufficient for action. Rather, people at more advanced levels of EHL also need to understand the complicated roles, responsibilities, and uncertainties related to environmental health decision and solution implementation processes.

The U.S. Superfund regulatory framework provides just one example of the legislative, political, and bureaucratic complexity that underlies environmental health governance and decisions, creating potential challenges for both understanding and action. The Comprehensive Environmental Response, Compensation, and Liability Act of 1980, or CERCLA, established liability of potentially responsible parties – called PRPs – and authorized the U.S. Environmental Protection Agency (US EPA), along with federal, state, and tribal governments, to recover damages caused by hazardous substances. In addition, the original CERCLA legislation established the Agency for Toxic Substances and Disease Registry, or ATSDR, at the Centers for Disease Control and Prevention (CDC). CERCLA was followed six years later by the Superfund Amendments and Reauthorization Act of 1986, or SARA, which increased state involvement in Superfund and reinforced the program's human health focus, establishing the Superfund Research Program – SRP – within the National Institutes of Health's National Institute of Environmental Health Sciences (NIEHS). Simultaneously, the Emergency Planning and Community Right-to-Know-Act (EPCRA) was signed into law, requiring local and state involvement in emergency planning and establishing the right of public access to information about chemical hazards and risks in communities. As a result, Local Emergency Planning Committees (LEPCs) and Citizen or Stakeholder Advisory Boards (CABs or SABs) began playing important roles in site discussions, alongside state Departments of Environmental Quality (DEQs), state and local health departments (SHDs and LHDs), and other state and local government entities.

Together, this interwoven set of regulations, organizations, and relationships has resulted in an acronym-driven shorthand that often appears as unintelligible "alphabet soup", even to highly educated communities of practice directly concerned with potential exposures from hazardous waste sites (Hu 1994; Theodore and Kiraz 2005). Basic understanding of contaminants and health risks without practical knowledge of how to navigate these existing structures would make taking protective action much more difficult. Therefore, progressing from a basic understanding of environmental health risks to more advanced understanding requires the incorporation of policy knowledge as well. Putting these and many other multidisciplinary pieces together in ways that make content clear and actionable, however, requires integration of social scientific – and particularly communication – theories and strategies into EHL research and practice.

## Theoretical Frameworks

Although the fields of toxicology, public health, medicine, engineering, geography, and policy all contribute to the scientific and technical knowledge base on environmental exposures and human health, communication provides the theoretical and methodological crucible within which these diverse disciplines combine. From the field of health communication, EHL inherits both health literacy's focus on information and action, and health communication campaign strategies for changing attitudes and behavior. From risk communication, EHL draws important frameworks for understanding how technical, emotional, historical, and interpersonal factors jointly influence how people make sense of and evaluate messages about hazards. Finally, from participatory communication, EHL derives principles and practices through which scientific and contextual knowledge can be converted to health-protective actions.

## Health Communication

Improving environmental health requires individual and community capacity to read and understand evidence-based information, as well as "expert" capacity to develop and deploy clear, relevant, and actionable messages about protective strategies. These twin needs require EHL to incorporate foundational theories and constructs from the field of health communication, "the study and use of methods to inform and influence individual and community decisions that enhance health" (Freimuth and Quinn 2004). While the utility of health communication theory and practice will be explored in greater depth in subsequent chapters (e.g., Chap. 3), the process of defining EHL requires some discussion of specific contributions from two key areas: health literacy and theory-driven health communication campaigns.

## Health Literacy

Health literacy is commonly understood as "the degree to which individuals have the capacity to obtain, process, and understand basic health information and services needed to make appropriate health decisions" (Ratzan and Parker 2000; IOM 2004). Like its descendant EHL, health literacy is an action-oriented concept that recognizes knowledge should lead to "good health-care decisions" (Benjamin 2010). Also like EHL, health literacy has seen its definition evolve over time, with no fewer than 17 distinct but related definitions emerging within a single decade (Sørensen et al. 2012). Because almost all definitions include explicit or implied focus on the ability to access and understand written and oral *information* to support health-related decisions, health literacy is strongly connected to health communication research.

Although it has generated interest since the mid-1970s, health literacy rose in prominence as a tool for improving health and well-being when it was designated as the first objective in the *Health Communication and Health Information Technology* domain of Healthy People 2020 (2017). Because of this placement as a key policy benchmark, it is essential that health literacy be examined within a web of existing hierarchical structures to ensure that its use does not further marginalize the already vulnerable. Consistent associations among low health literacy and age, education, and ethnicity have led to suggestions that health literacy itself be considered among the social determinants of health (Paasche-Orlow et al. 2005; Baur 2010).

It is no accident, therefore, that critics have noted that traditional health literacy competencies often are themselves rooted in privilege. Some skills and knowledge needed to be deemed "health literate" rely on high levels of formal education in science, statistics, and information-seeking that simply are not available to everyone (Chinn 2011). Furthermore, communication and information scholars have pointed out that the development of health literacy has been a top-down phenomenon led by academics and government officials – an approach that has left little room for balancing the concerns and lived contexts of the very populations that the field strives to assist (Huber et al. 2012). However, a more recent movement toward critical health literacy has begun recognizing that the social contexts of target audiences must be incorporated into messages and materials (Nutbeam 2008).

A related concern for health literacy is its reliance on the presumed neutrality and benevolence of biomedical research (Chinn 2011). Models of health literacy often are situated within a positivistic framework that assumes a "right answer" can be derived from technical or scientific expertise. This approach can miss contextual elements that are vitally important for developing and implementing action. Furthermore, some traditional learning models on which health literacy is based assume that simply *knowing* about a topic leads to *solving* topic-related challenges (Colucci-Gray et al. 2006); however, evidence indicates that movement from knowledge to action is mediated by a number of constructs, including attitudes, self-efficacy, communication channels, and perceived message relevance (Bettinghaus 1986; Rimal 2000; Noar et al. 2007; Webb et al. 2010).

A key difference between traditional health literacy and EHL lies in their units of analysis. The earliest conceptualizations of health literacy focused primarily on the measurement and attainment of individual skills; recognition of health literacy as a community phenomenon only followed much later (Lurie and Parker 2007; Berkman et al. 2010). In contrast, recognition of the community unit is foundational to EHL, which has at its core a desire to assess and mitigate the threat of exposures on a community level or regional scale. Often such mitigation involves taking protective statutory or regulatory action on contaminants that threaten human health. Thus, entire communities and communities of practice must come to a common understanding in order to act on evidence-based information to prevent exposures and minimize illness related to unavoidable exposures.

By foregrounding community contexts, EHL makes room for multiple types of knowledge, recognizing that making "an informed decision" does not necessarily equate to making the decision that scientists, regulatory officials, or other technical experts might deem best. In recognizing the community as a unit of EHL action, it is essential to recognize that decisions intended to improve environmental health outcomes sometimes affect economic and socio-cultural dimensions of life. Balancing these concerns in the decision process can escalate complexity to levels that simply do not exist in laboratory settings.

In other words, EHL promotes the ability to interpret scientific information *not* as an end unto itself, but as a means of helping individuals and communities consider technical recommendations and guidelines *in balance* with their own needs and values. EHL further recognizes that fostering knowledge of socioeconomic, political, and other local contextual drivers among scientists, officials, and other experts is key for building trust among constituencies involved in the decision process (Hoover 2013). In essence, EHL builds on its health literacy foundations by fostering solid scientific ground on which to exchange information about health and hazards, while supporting collaborative decisions that incorporate context into solutions.

## Health Communication Campaigns

Health literacy pioneer Nutbeam (2000) explicitly recognized the field's link with health communication, calling health literacy "a composite term to describe a range of outcomes to health education and communication activities." In fact, health education and promotion theories have long provided critical foundations for the development and dissemination of the evidence-informed messages that are embedded in health communication campaigns. Often focused on individual-level health behavior change, such campaigns strive to educate at-risk people on issues from seat belt safety to the dangers of illegal drug use, with a goal of catalyzing protective behaviors.

Using health promotion models and constructs, health communication campaigns have helped build health literacy on targeted issues by both increasing knowledge *and* promoting action. Early models, such as the Health Belief Model (HBM), deployed health-related information primarily to change perceptions of an individual's susceptibility to a poor health outcome, the severity of that outcome, the benefits of taking protective action, and the barriers to taking that action (Rosenstock 1974). However, it soon became apparent that additional constructs were needed to strengthen the outcomes of such informational campaigns. Health communicators and their health promotion partners subsequently developed new models to account for additional drivers of action.

Efficacy is one critical construct found across many health communication models. Popularized through Bandura's (1977) Social Learning and Cognitive Theory, *self-efficacy*, or "the conviction that one can successfully execute the behavior required to produce the [desired] outcomes," began appearing in various forms across both new and existing health promotion and communication models (Rosenstock et al. 1988). A very similar construct, *perceived behavioral control*, was added to the Theory of Reasoned Action to produce the Theory of Planned Behavior (Fishbein 1979; Ajzen 1991).

Models that incorporate self-efficacy posit that people will act on health-related information only when they are confident that they have the ability to take the desired action. For example, people who believe they do not have time to prepare healthful meals will be less likely to undertake weight loss programs designed around cooking new recipes. Therefore, addressing beliefs about self-efficacy, particularly among vulnerable populations who might face significant barriers to action, is an important step in crafting messages that promote protective action.

On the other hand, people may feel confident that they *can* take a particular action but are not confident that the action itself will achieve the touted outcome. For example, people may feel that they have the time to cook new recipes but that these meals will not actually help them lose weight – a belief that also reduces the likelihood of their undertaking the weight loss program. In her Extended Parallel Process Model, or EPPM, Witte (1992) addressed this challenge by incorporating yet another efficacy construct, *response efficacy*.

Response efficacy, in short, is the belief that taking a specific action will lead to the desirable outcome. EPPM considers response efficacy an important mediator for how people respond to fear-inducing messages. Considered in the EHL context, a person living near a hazardous waste site might have adequate time and resources to attend a U.S. EPA-hosted community meeting but also might feel that attending the meeting will change nothing, thereby rendering that person less likely to participate. However, if the person has sufficient knowledge of the structures and mechanisms of environmental regulation and if promotional materials for the meeting include very specific messages about how attendance can help prioritize local concerns, response efficacy – and consequently the individual's likelihood of participating – could be increased.

Communication campaigns use the models and constructs described above – and others like them – to generate and disseminate awareness-raising messages to large audiences in hopes of spurring specific outcomes (Rogers and Storey 1987). In recent years, dissemination and implementation science – or D&I – researchers and their practice partners increasingly have deployed lessons learned from health communication campaigns to help shrink the research-to-practice and research-to-policy gaps. As scientists strive to ensure the adoption of evidence-informed policies and practices to improve health, they increasingly borrow and adapt audience-specific health communication strategies to ensure carefully crafted messages are delivered by trusted sources through the most appropriate channels (Brownson et al. 2018). By doing so, D&I scientists hope to optimize the real-world impact of emerging knowledge for very distinct audiences.

Today, EHL is similarly positioned to draw on the cumulative lessons of decades of campaign-related implementation and evaluation. Meta-analyses of health communication campaigns, for example, provide vital information. Some underscore the prevalence of small campaign effect sizes (Snyder et al. 2004). Others illustrate that messages strong in both fear appeals and efficacy are more likely to generate behavior change (Witte and Allen 2000). Still others recognize the importance of audience targeting for achieving campaign goals (Noar 2006). As EHL grows and matures, it will be important for researchers and practitioners to learn from such findings.

## Risk Communication

Developing strategies for advancing EHL also requires some understanding of risk communication history and practice. Risk communication coalesced as a field in the 1970s and 80s, primarily in response to a series of crises, including revelations about the excessive exposures along Love Canal, the Bhopal chemical disaster, and the Three Mile Island and Chernobyl nuclear accidents. With each event, government officials, scientists, and industry leaders became increasingly aware of the need to inform at-risk communities about existing and emerging environmental and health risks (Perrow 1984; Weick 1988; NRC 1989; Heath and O'Hair, 2010).

The earliest iterations of risk communication thus were driven by a "deficit model" that implied disagreements about risk were driven by lack of scientific or technical knowledge among lay audiences rather than potential differences in priorities and/or values across stakeholder groups (Horlick-Jones and Farre 2010). However, the mere provision of information was not very successful in changing either attitudes or behaviors. The unidirectional approach had fallen victim to what Sellnow and Sellnow (2010) call "the general tenet of communication studies": information does not equate to understanding.

The National Research Council (1989) subsequently called for risk communication approaches to include "democratic dialogue," to be "interactive," and to incorporate stakeholder "concerns, opinions, or reaction to risk messages." Responding to – and in some cases anticipating – these needs, risk communicators in government offices, private industry, and academia have worked to develop, test, implement, evaluate, and refine best practices for the field.

In many ways, the best practices approach began with US EPA's *Seven Cardinal Rules of Risk Communication* (Covello and Allen 1994). The *Cardinal Rules* and later approaches, such as CDC's Crisis and Emergency Risk Communication and the Best Practices for Crisis Communication, draw from the premise that people understand risk as a function of both technical hazard and community outrage (Sandman 1993; Covello and Sandman 2001; Reynolds and Seeger 2005; Seeger 2006). In this view, information based in technical hazard – that is, the likelihood of a harmful event occurring and the severity of the consequences – privileges scientific expertise; in contrast, the at-risk parties experience outrage that draws from their lived experiences, fears, and concerns. Messages that address technical issues but ignore emotional issues and motivations to act will not be effective. As Sandman (1993) puts it, "[a]s long as the outrage goes unmanaged, the public is unlikely to notice that the hazard is well-managed."

While early risk communicators assumed that the mere act of sharing information was a positive step, subsequent risk communication theory and practice make clear that how people make sense of risks is central to how they respond to hazard-related messages. To draw from Weick (1988), the act of communicating *can* facilitate mutual understanding, but communicative action also can have detrimental effects that play out over time (Hoover 2017). This is the *enactment* perspective, which describes an ongoing, iterative process through which people both shape and are constrained by their environments. These constraints subsequently influence how people make sense of information.

The enactment perspective explicitly recognizes three potential mediators that affect sensemaking processes. *Commitment* to a particular position can cause the formation of "blind spots" and the "tenacious justification" of that position in the face of contrary evidence; *capacity* limits enactment through the homogeneity of perspectives among potential sensemakers; and *expectations* can lead to assumptions that become "self-fulfilling prophecies" (Weick 1988). In the often emotionally charged push-pull of risk communication, delays in information sharing or contradictory messages can lead to additional confusion among stakeholders about what to believe and whom to trust, further exacerbating challenges for collaborative action (Sellnow et al. 2009).

EHL researchers and practitioners must recognize that diverse stakeholders often have competing perspectives about contaminants and health. These people might have received conflicting or erroneous information, and they almost certainly have differing baseline levels of technical knowledge and emotional investment in environmental health issues. In other words, EHL cannot be divorced from the broader cultural and sociopolitical landscape. Rather, risk communication's sensemaking framework recognizes that an individual's understanding is formulated through dialogue with others. By extension, it is possible that individuals with higher EHL levels can, through dialogue with other stakeholders, positively affect the ability of the whole to collectively understand, diagnose, and solve environmental health problems. As Chinn (2011) notes, such collective action relies on "critical health literacy competencies [that] include knowledge of the local community and organizations, [and] skills in working in groups." These vital competencies can be supported by leveraging strategies from participatory communication.

## Participatory Communication

In recent years, risk communication has seen a marked turn from its unidirectional, information-driven beginnings to a partnership development approach (Fischhoff 1995). This evolution has been driven in part by critical pedagogy that recognizes students as co-creators of knowledge (Freire 1996; Colucci-Gray et al. 2006), as well as by advances in community-based participatory research (CBPR) and community-engaged research (CEnR) methods. As these influences have been brought to bear on both risk communication and environmental health, approaches have become more focused on equity and representation.

For two decades, federal agencies, scientists and others have recognized the need to incorporate communities and communities of practice as partners in the development and conduct of research (Israel et al. 1998; Israel et al. 2001; O'Fallon and Dearry 2002; Winterbauer et al. 2016). These participatory and engaged paradigms have arisen in reaction to the traditional extractive research model in which data is gathered and analyzed solely by outsiders, such as urban university researchers conducting research in rural communities without sharing subsequent findings with that community. In addition to supporting inclusion of local knowledge in studies themselves, participatory strategies also have become vital for the translation of research findings into action. Wallerstein and Duran (2010) in particular have argued that incorporating indigenous knowledge and culture into complex systems-level approaches can help bridge jargon-driven communicative disconnects and promote trust – both of which are common challenges for communicating environmental health research findings.

With the development of participatory communication methods, the communication discipline has been a leader in this important area. With its foundations in action research, participatory communication builds on an iterative social scientific approach that includes both researchers and participant-subjects (Lewin 1946), recognizing that stakeholders' local knowledge has inherent value for solving health-related challenges. Thus, participatory communication focuses on horizontal information exchange and mutual learning that can build dialogue and trust among often disparate stakeholders.

Participatory communication endeavors encourage multi-directional communication – deploying visualizations, interviews, and group work to facilitate dialogue and collaborative decision-making among all stakeholders (Anyaegbunam et al. 2010). By involving community partners, participatory communication encourages the development, implementation, and evaluation of culturally-appropriate decisions and policies, assisting participants in developing mutual understandings that can help reduce conflict (Ting-Toomey and Oetzel 2002; Anyaegbunam et al. 2004).

Because participatory communication directly addresses the social construction of risk, its approaches can build EHL capacity and support action that is situated within a particular social context. With its unit of analysis often set at the community level, EHL requires such collaborative activities to help communities define, identify, and solve health problems through action (Beeker et al. 1998). In short, the integration of participatory communication into EHL fulfills the field's promise of promoting culturally appropriate, multidirectional information-sharing that is more likely to move knowledge into action.

## Methods

As an emerging field that draws its theoretical underpinnings from many disciplines, EHL also integrates both quantitative and qualitative methods from these disciplines. For example, health literacy's focus on knowledge and skill attainment requires the development and validation of surveys and questionnaires to assess competencies (Altin et al. 2014); such instruments also are essential for assessing EHL. In contrast, health communication campaigns strive to change behaviors in target audiences, which requires extensive formative research, including audience segmentation and analysis, identification of optimal communication channels, and development of appropriate message content (Snyder 2007); all of these activities are key for increasing EHL. Building on this health communication groundwork and complementing it with additional approaches from the marketing field, dissemination and implementation science researchers have created frameworks and tools for audience-to-channel matching, as well as evaluative tools to help determine the reach and effectiveness of their efforts (Becker 2015; Neta et al. 2015; Brownson et al. 2018). Risk communication research uses a variety of methodological tools, including case studies examining the social construction of risk, media studies that evaluate the interplay of mass communication and risk perceptions, and mental models that support message development (McComas 2006); because the ways in which people make sense of risk affects their subsequent interpretation of environmental health information, each of these tools has utility for EHL. Finally, participatory communication approaches that privilege context and promote collaborative decisions are vital for achieving EHL's potential of promoting action that can improve health.

## Research/Engagement Examples

Many foundational efforts for the emerging EHL field have arisen through efforts to translate environmental health research findings and engage communities to solve problems. At times, these activities have deployed social scientific tools to ensure the cultural relevance of research (Hoover et al. 2015; Ramirez-Andreotta et al. 2014, 2015), thereby encouraging the exchange of both technical and contextual information. For example, efforts to incorporate indigenous knowledge into data collection (Rock et al. 2007) builds participants' knowledge of technical processes while also raising scientists' awareness of cultural context.

In some cases, researchers have worked to build capacity for understanding and acting on technical information to solve very specific local problems, such as determining acceptable and unacceptable future uses of a hazardous waste site (KRCEE 2011). In other cases, researchers have documented the importance of community trust for the interpretation and actionability of scientific evidence (Scammell et al. 2009). Some studies have even focused on identifying visualizations and other communication tools that can promote understanding of hazards and potential health impacts (Cleary et al. 2017). Whether formally acknowledged as EHL or not, all of these activities have contributed immensely to the field's development.

## Future Directions

As EHL continues to evolve, both research and practice will benefit from its formal recognition as a cohesive, multidisciplinary field. Ongoing efforts to capture and characterize EHL, such as those described in this book, will support future classification of various approaches, consolidation of research findings across sectors and domains, and translation of these findings to communities, practice, and policy. The continuous engagement of new communities, contexts, and investigators with diverse expertise will help elucidate the relationship between EHL and the social determinants of health.

Importantly, in helping individuals and communities improve understanding and make informed decisions, EHL must strive to recognize the value of even basic levels of understanding as well as avoid stigmatizing communities and individuals who do not yet have the resources, formal education, or lived experience to interpret and/or act on technical information as readily as others. Further, EHL must ensure that environmental health scientists, practitioners, and other "technical experts" also are seeking and gaining knowledge about the contexts in which lay audiences are interpreting and acting on their messages. Understanding local priorities and values will not only generate better messages delivered through more relevant channels, but also will enable exposure-reducing, health-promoting actions that are feasible, collaborative, sustainable and resonate with affected populations.

## Conclusion

As an emerging field, EHL will continue to see its definition evolve and become refined in coming years. New evidence about how individuals and communities make sense of the relationship between environmental hazards and human health will contribute vital information for this evolution, as will improved approaches for incorporating community contexts into health protective recommendations and evidence-informed action. Inevitably these new understandings will arise from studies that build on the theoretical and methodological foundations discussed in this chapter. Lessons from health literacy can help EHL researchers create tools to identify and assess critical skills and competencies. Health communication and dissemination science approaches can help guide formative work to develop and target EHL information for key audiences. Risk communication can help EHL practitioners incorporate more holistic understandings of sensemaking into their capacity-building processes. Finally, participatory communication strategies can help minimize further marginalization of vulnerable populations, creating a multidirectional, horizontal EHL framework. In the often chaotic world of competing risk perceptions and values, assessing and improving EHL can unite practitioners, researchers, advocates, industries, and communities to raise understanding and promote actions that minimize exposures and mitigate exposure-related health threats.

## References


Ajzen, I. (1991). The theory of planned behavior. *Organizational Behavior and Human Decision Processes, 50*(2), 179–211.

Altin, S. V., Finke, I., Kautz-Freimuth, S., & Stock, S. (2014). The evolution of health literacy assessment tools: A systematic review. *BMC Public Health, 14*(1), 1207.

Anyaegbunam, C., Mefalopulos, P., & Moetsabi, T. (2004). *Participatory rural communication appraisal: A handbook for rural development practitioners* (2nd ed.). Rome: UN FAO.

Anyaegbunam, C., Hoover, A. G., & Schwartz, M. (2010). *Use of community-based participatory communication to identify community values at a Superfund site*. Providence: American Society of Civil Engineers Annual Meeting.

Bandura, A. (1977). Self-efficacy: Toward a unifying theory of behavioral change. *Psychological Review, 84*, 191–215.

Baur, C. (2010). New directions in research on public health and health literacy. *Journal of Health Communication, 15*(S2), 42–50.

Becker, S. J. (2015). Direct-to-consumer marketing: A complementary approach to traditional dissemination and implementation efforts for mental health and substance abuse interventions. *Clinical Psychology: Science and Practice, 22*(1), 85–100.

Beeker, C., Guenther-Grey, C., & Raj, A. (1998). Community empowerment paradigm drift and the primary prevention of HIV/AIDS. *Social Science & Medicine, 46*(7), 831–842.

Benjamin, R. M. (2010). Improving health by improving health literacy. *Public Health Reports, 125*(6), 784.

Berkman, N. D., Davis, T. C., & McCormack, L. (2010). Health literacy: What is it? *Journal of Health Communication, 15*(S2), 9–19.

Bettinghaus, E. P. (1986). Health promotion and the knowledge-attitude-behavior continuum. *Preventive Medicine, 15*(5), 475–491.

Brownson, R. C., Eyler, A. A., Harris, J. K., Moore, J. B., & Tabak, R. G. (2018). Getting the word out: New approaches for disseminating public health science. *Journal of Public Health Management and Practice, 24*(2), 102–111.

Chinn, D. (2011). Critical health literacy: A review and critical analysis. *Social Science & Medicine, 73*(1), 60–67.

Cleary, E. G., Patton, A. P., Wu, H. C., Xie, A., Stubblefield, J., Mass, W., Grinstein, G., Koch-Weser, S., Brugge, D., & Wong, C. (2017). Making air pollution visible: A tool for promoting environmental health literacy. *JMIR Public Health and Surveillance, 3*(2), e16.

Colucci-Gray, L., Camino, E., Barbiero, G., & Gray, D. (2006). From scientific literacy to sustainability literacy: An ecological framework for education. *Science Education, 90*(2), 227–252.

Covello, V. T., & Allen, F. W. (1994). *Seven cardinal rules of risk communication*. Washington, D.C: US Environmental Protection Agency.

Covello, V., & Sandman, P. M. (2001). Risk communication: Evolution and revolution. In *Solutions to an environment in peril* (pp. 164–178).

Finn, S., & O'Fallon, L. (2017). The emergence of environmental health literacy—From its roots to its future potential. *Environmental Health Perspectives, 125*(4), 495–501. https://doi.org/10.1289/ehp.1409337.

Fischhoff, B. (1995). Risk perception and communication unplugged: Twenty years of process. *Risk Analysis, 15*(2), 137–145.

Fishbein, M. (1979). A theory of reasoned action: Some applications and implications. In *Nebraska symposium on motivation* (Vol. 27, pp. 65–116).

Freimuth, V. S., & Quinn, S. C. (2004). The contributions of health communication to eliminating health disparities. *American Journal of Public Health, 94*(12), 2053–2055.

Freire, P. (1996). *Pedagogy of the oppressed* (Rev ed.). New York. Continuum.

Healthy People 2020 [Internet]. Washington, DC: U.S. Department of Health and Human Services, Office of Disease Prevention and Health Promotion [cited 17 July

2017]. Available from: https://www.healthypeople.gov/2020/topics-objectives/topic/health-communication-and-health-information-technology
Heath, R. L., & O'Hair, H. D. (2010). The significance of risk and crisis communication. In R. L. Heath & H. D. O'Hair (Eds.), *The handbook of risk and crisis communication*. New York: Routledge.
Hoover, A. G. (2013). *Communication at superfund sites and the reification of division: Toward a convergence-building model of risk communication*. University of Kentucky.
Hoover, A. G. (2017). Sensemaking, stakeholder discord, and long-term risk communication at a US superfund site. *Reviews on Environmental Health, 32*(1–2), 165–169.
Hoover, E., Renauld, M., Edelstein, M. R., & Brown, P. (2015). Social science collaboration with environmental health. *Environmental Health Perspectives, 123*(11), 1100–1106.
Horlick-Jones, T., & Farre, J. (2010). On the communicative constitution of risk objects in mediated times. *Catalan Journal of Communication & Cultural Studies, 2*(2), 131–143.
Hu, H. (1994). Decision-making in human health impact assessments: A clinician's perspective. *Environmental Impact Assessment Review, 14*(5–6), 439–450.
Huber, J. T., Shapiro, R. M., & Gillaspy, M. L. (2012). Top down versus bottom up: The social construction of the health literacy movement. *The Library Quarterly, 82*(4), 429–451.
Institute of Medicine. (2004). In D. A. Kindig, A. M. Panzer, & L. Nielsen-Bohlman (Eds.), *Health literacy: a prescription to end confusion*. Washington, DC: National Academies Press.
Israel, B. A., Schulz, A. J., Parker, E. A., & Becker, A. B. (1998). Review of community-based research: Assessing partnership approaches to improve public health. *Annual Review of Public Health, 19*(1), 173–202.
Israel, B. A., Schulz, A. J., Parker, E. A., & Becker, A. B. (2001). Community-based participatory research: Policy recommendations for promoting a partnership approach in health research. *Education for Health (Abingdon, England), 14*(2), 182–197.
Kentucky Research Consortium for Energy and the Environment., 2011. *Community visions for the Paducah Gaseous Diffusion Plant site* (No. UK/KRCEE doc#: 27.1 2011). Kentucky Research Consortium for Energy and Environment, University of Kentucky, Lexington, KY.
Lewin, K. (1946). Action research and minority problems. *Journal of Social Issues, 2*, 34–46.
Lurie, N., & Parker, R. (2007). Moving health literacy from the individual to the community. *American Journal of Health Behavior, 31*(1), S6–S7.
McComas, K. A. (2006). Defining moments in risk communication research: 1996–2005. *Journal of Health Communication, 11*(1), 75–91.
National Research Council. (1989). *Improving risk communication*. Washington, DC: National Academies Press.
Neta, G., Glasgow, R. E., Carpenter, C. R., Grimshaw, J. M., Rabin, B. A., Fernandez, M. E., & Brownson, R. C. (2015). A framework for enhancing the value of research for dissemination and implementation. *American Journal of Public Health, 105*(1), 49–57.
Noar, S. M. (2006). A 10-year retrospective of research in health mass media campaigns: Where do we go from here? *Journal of Health Communication, 11*(1), 21–42.
Noar, S. M., Benac, C. N., & Harris, M. S. (2007). Does tailoring matter? Meta-analytic review of tailored print health behavior change interventions. *Psychological Bulletin, 133*(4), 673–693.
Nutbeam, D. (2000). Health literacy as a public health goal: A challenge for contemporary health education and communication strategies into the 21st century. *Health Promotion International, 15*(3), 259–267.
Nutbeam, D. (2008). The evolving concept of health literacy. *Social Science & Medicine, 67*(12), 2072–2078.
O'Fallon, L. R., & Dearry, A. (2002). Community-based participatory research as a tool to advance environmental health sciences. *Environmental Health Perspectives, 110*(Suppl 2), 155–159.
Paasche-Orlow, M. K., Parker, R. M., Gazmararian, J. A., Nielsen-Bohlman, L. T., & Rudd, R. R. (2005). The prevalence of limited health literacy. *Journal of General Internal Medicine, 20*(2), 175–184.
Perrow, C. (1984). *Normal accidents*. New York: Basic Books.

Ramirez-Andreotta, M. D., Brusseau, M. L., Artiola, J. F., Maier, R. M., & Gandolfi, A. J. (2014). Environmental research translation: Enhancing interactions with communities at contaminated sites. *Science of the Total Environment, 497*, 651–664.

Ramirez-Andreotta, M. D., Brusseau, M. L., Artiola, J., Maier, R. M., & Gandolfi, A. J. (2015). Building a co-created citizen science program with gardeners neighboring a superfund site: The Gardenroots case study. *International Public Health Journal, 7*(1).

Ratzan, S.C. and Parker, R.M., (2000). Introduction. In: *National library of medicine current bibliographies in medicine: Health literacy.* NLM Pub. No. CBM 2000-1. Selden CR, Zorn M, Ratzan SC, Parker RM, Eds. Bethesda, MD: National Institutes of Health/ U.S. Department of Health and Human Services.

Reynolds, B. and W. Seeger, M., (2005). Crisis and emergency risk communication as an integrative model. Journal of Health Communication, 10(1), pp.43–55.

Rimal, R. N. (2000). Closing the knowledge-behavior gap in health promotion: The mediating role of self-efficacy. *Health Communication, 12*(3), 219–237.

Rock, T., Brugge, D., Slagowski, N., Manning, T., & Lewis, J. (2007). Lessons from the Navajo: Assistance with environmental data collection ensures cultural humility and data relevance. *Progress in Community Health Partnerships: Research, Education, and Action, 1*(4), 321.

Rogers, E. M., & Storey, J. D. (1987). Communication campaigns. In C. R. Berger & S. H. Chaffee (Eds.), *Handbook of communication science* (pp. 817–846). Beverly Hills: Sage.

Rosenstock, I. M. (1974). Historical origins of the health belief model. *Health Education Monographs, 2*, 328–335.

Rosenstock, I. M., Strecher, V. J., & Becker, M. H. (1988). Social learning theory and the health belief model. *Health Education Quarterly, 15*(2), 175–183.

Sandman, P. M. (1993). *Responding to community outrage: Strategies for effective risk communication*. Fairfax: American Industrial Hygiene Association.

Scammell, M. K., Senier, L., Darrah-Okike, J., Brown, P., & Santos, S. (2009). Tangible evidence, trust and power: Public perceptions of community environmental health studies. *Social Science & Medicine, 68*(1), 143–153.

Seeger, M. W. (2006). Best practices in crisis communication: An expert panel process. *Journal of Applied Communication Research, 34*(3), 232–244.

Sellnow, T. L., & Sellnow, D. (2010). The instructional dynamic of risk and crisis communication: Distinguishing instructional messages from dialogue. *The Review of Communication, 10*(2), 112–126.

Sellnow, T. L., Ulmer, R. R., Seeger, M. W., & Littlefield, R. S. (2009). *Effective risk communication: A message-centered approach*. New York: Springer Science+Business Media, LLC.

Snyder, L. B. (2007). Health communication campaigns and their impact on behavior. *Journal of Nutrition Education and Behavior, 39*(2), S32–S40.

Snyder, L. B., Hamilton, M. A., Mitchell, E. W., Kiwanuka-Tondo, J., Fleming-Milici, F., & Proctor, D. (2004). A meta-analysis of the effect of mediated health communication campaigns on behavior change in the United States. *Journal of Health Communication, 9*(S1), 71–96.

Society of Public Health Education. (2007). *National health education planning guide*. Washington, DC: SOPHE Available at: http://citeseerx.ist.psu.edu/viewdoc/download?doi=10.1.1.179.3181&rep=rep1&type=pdf.

Sørensen, K., Van den Broucke, S., Fullam, J., Doyle, G., Pelikan, J., Slonska, Z., & Brand, H. (2012). Health literacy and public health: A systematic review and integration of definitions and models. *BMC Public Health, 12*(1), 80.

Theodore, L., & Kiraz, R. G. (2005). A review of nanotechnology. *Nanotechnology Law & Business*, 305.

Ting-Toomey, S., & Oetzel, J. G. (2002). Cross-cultural face concerns and conflict styles. *Handbook of International and Intercultural Communication, 2*, 143–164.

Wallerstein, N., & Duran, B. (2010). Community-based participatory research contributions to intervention research: The intersection of science and practice to improve health equity. *American Journal of Public Health, 100*(S1), S40–S46.

Webb, T. L., Joseph, J., Yardley, L., & Michie, S. (2010). Using the internet to promote health behavior change: A systematic review and meta-analysis of the impact of theoretical basis, use of behavior change techniques, and mode of delivery on efficacy. *Journal of Medical Internet Research*, (1), 12.
Weick, K. E. (1988). Enacted sensemaking in crisis situations. *Journal of Management Studies, 25*(4), 305–317.
Winterbauer, N. L., Bekemeier, B., VanRaemdonck, L., & Hoover, A. G. (2016). Applying community-based participatory research partnership principles to public health practice-based research networks. *SAGE Open, 6*(4).
Witte, K. (1992). Putting the fear back into fear appeals: The extended parallel process model. *Communications Monographs, 59*(4), 329–349.
Witte, K., & Allen, M. (2000). A meta-analysis of fear appeals: Implications for effective public health campaigns. *Health Education & Behavior, 27*(5), 591–561.

# Chapter 2
# Measuring Environmental Health Literacy

**Kathleen M. Gray and Marti Lindsey**

**Abstract** Environmental health literacy (EHL) is an emerging framework that defines the knowledge and skills that prepare people to make health-protective decisions using available environmental data (Finn S, O'Fallon L. Environ Health Perspect 125(4):495–501 (2017)). According to the Society for Public Health Education (Society for Public Health Education. What is environmental health literacy? Retrieved from: http://www.sophe.org/environmentalhealth/key_ehl.asp (2014)), application of knowledge and skills is essential to achieving EHL, as evidenced in its definition of EHL as "the wide range of skills and competencies that people need in order to seek out, comprehend, evaluate, and use environmental health information to make informed choices, reduce health risks, improve quality of life and protect the environment." Early research on EHL has focused on how people understand connections between environmental exposures and health and, to a more limited extent, on the ways that improving literacy can lead to policies or infrastructure developments that reduce environmental exposures (Gray KM. Int J Environ Res Public Health 15(3):466 (2018)).

The perspectives that inform this chapter include adult literacy, sociocultural theory, and the principles of community-based participatory research (CBPR). Understanding the elements of EHL and how they might be measured is bolstered by familiarity with a range of literacies, including science literacy, health / public health literacy, and environmental literacy; and the range or continuum of skills mastery embodied in EHL parallels the widely accepted representation of adult literacy (National Center for Education Statistics. What is NAAL? In: National Assessment of Adult Literacy (NAAL). A nationally representative and continuing assessment of English language literary skills of American Adults. Retrieved from http://nces.ed.gov/naal/ (2003)). Additionally, the lens of sociocultural theory and constructivism, which has been applied nationally and internationally in educational research and practice (Amineh RJ, Asl HD. J Soc Sci, Lit Lang 1(1):9–16

K. M. Gray (✉) · M. Lindsey
UNC Chapel Hill Institute for the Environment, Chapel Hill, NC, USA
e-mail: kgray@unc.edu



S. Finn, L. R. O'Fallon (eds.), *Environmental Health Literacy*,
https://doi.org/10.1007/978-3-319-94108-0_2

(2015); Jacobson SK, McDuff MD, Monroe MC. Conservation education and outreach techniques. Oxford University Press, Oxford (2015); Liu CH, Matthews R. Intl Edu J 6(3):386–399 (2005)), provides a context for understanding the development of EHL. Its emphasis on learning through social interaction with more skilled instructors, who can be peers, and the influence of cultural context in learning, underscores the importance of culturally relevant instruction and materials and non-standard forms of assessment. Further, CBPR promotes active community involvement in the design and conduct of research and educational interventions; and its principles are grounded in commitment to active collaboration and participation from all partners at every stage of research (O'Fallon L, Dearry A. Environ Health Perspect 110:155–159 (2002)). Taken together, these perspectives suggest that EHL goes beyond understanding a specific environmental exposure to a broader understanding of the varied ways that the environment affects human health in community contexts.

This chapter begins with a description of five, critical literacies – adult, science, health, public health, and environmental – and approaches to measuring them, all of which inform the conception and measurement of EHL as discussed in this chapter. Research relevant to measuring EHL is described, including a section on culturally relevant measurement. The chapter concludes with implications for EHL measurement approaches going forward.

**Keywords** Environmental health literacy · Measurement · Assessment · Skills · Knowledge · Health literacy · Adult literacy · Environmental literacy · Community-based

## Conceptualization and Measurement of Literacy Levels

### *Adult Literacy*

Literacy is broadly understood to mean that a person can make and communicate meaning from and use a variety of symbols, such as words, letters, and numbers (UNESCO 2002). The literacy movement in the United States arose after World War II, when it became national policy to raise basic literacy in order to increase national security and global competitiveness (Brandt 2004). Yet even 70 years later, the problem of low literacy is common in the United States (DeWalt et al. 2004), as shown by the following national literacy assessment efforts: National Adult Literacy Survey (1993), the National Assessment of Adult Literacy (2005), and the Program for the International Assessment of Adult Competencies (2016).

**Measuring Adult Literacy** In 1992, the National Center for Education Statistics (NCES) surveyed a nationally representative sample of nearly 13,600 adults in the United States, age 16 and older, using the National Adult Literacy Survey (NALS) (Kirsch et al. 1993). The NALS established a mechanism for assessing three types of adult literacy: (a) prose literacy – the knowledge and skills needed to search,

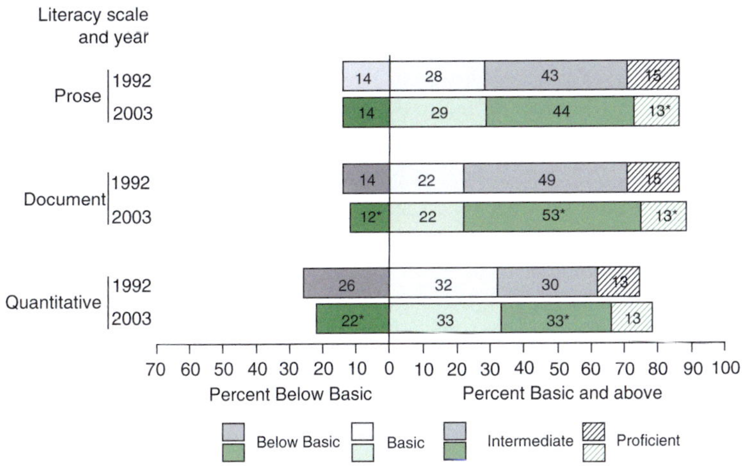


**Fig. 2.1** Percent of U.S. Adults in Each Literacy Level: 1992 & 2003. Source: NCES, 2003

comprehend, and use *continuous* texts (such as editorials, news stories, and instructional materials), (b) document literacy – knowledge and skills needed to search, comprehend, and use *non-continuous* texts (such as job applications, payroll forms, transportation schedules, and drug/food labels), and (c) quantitative literacy – knowledge and skills required to identify and perform computations, either alone or sequentially, using numbers embedded in printed materials (such as balancing a checkbook, figuring out a tip, and completing an order form).

In 2003, NCES conducted a second survey, the National Assessment of Adult Literacy, or NAAL, with over 19,000 adults in the United States (National Center for Education Statistics 2003). This analysis led to the use of four skills-based performance levels to represent an individual's ability to deal with increasing complexity, as follows:

- Below Basic: no more than the most simple and concrete literacy skills
- Basic: can perform simple and everyday literacy activities
- Intermediate: can perform moderately challenging literacy activities
- Proficient: can perform complex and challenging literacy activities

In comparison to 1992, the 2003 NAAL showed little progress towards the national goal of eliminating illiteracy (Fig. 2.1, below). Yet in highlighting the importance of adult literacy for policy and programs, the NAAL represented an early attempt to define and understand the nature and implications of health literacy, specifically low health literacy. This assessment documented that almost 50 percent of Americans (or 89 million adults) experienced difficulty with health information and navigating the health system (Kutner et al. 2006). These findings informed the development of the concept of health literacy, which is discussed later in this chapter.

In the third large study of adult literacy, the Program for the International Assessment of Adult Competencies (PIAAC) (NCES 2016), adults (ages 16–65) in 35 countries were surveyed between 2012 and 2014. When compared to the international average distribution of literacy skills, a larger percentage of U.S. adults performed at both the top and bottom of the distribution. Thirteen percent of U.S. adults performed at the highest proficiency level (4/5) on the PIAAC, compared to 12 percent of the international participants. Eighteen percent of U.S. adults performed at the lowest level of the PIAAC literacy scale (at or below Level 1), versus the international average of 16 percent. The Organization for Economic Cooperation and Development (OECD) plans to build on these assessments to further define literacy and numeracy, incorporate problem-solving to emphasize skills used in technology-rich environments, and improve understanding of the reading skills of individuals with low levels of literacy.

## *Science Literacy*

Following national efforts to measure adult literacy, science literacy became a goal of primary and secondary science education as early as the 1980s (Snow and Dibner 2016). *Science for All Americans* (AAAS 1990) defined a science-literate person as someone who:

> *…is aware that science, mathematics and technology are interdependent human enterprises with strengths and limitations; understands key concepts and principles of science; is familiar with the natural world and recognizes both its diversity and unity; and uses scientific knowledge and scientific ways of thinking for individual and social purposes.* (Introduction, para. 19)

This framework is widely recognized as a first step towards setting national standards in science for all students and a major influence on science education reform efforts. Miller (1998) refined this definition of science literacy to focus on three dimensions: (1) an understanding of key scientific terms and concepts that would enable a reader to make sense of popular press coverage, (2) an understanding of the process or nature of scientific inquiry; and (3) some level of understanding of the impact of science and technology on society.

**Measuring Science Literacy** Despite identifying three dimensions, Miller (1998) asserted that scientific literacy was best conceptualized as a two-dimensional measure, including a vocabulary dimension associated with basic scientific constructs and a process or inquiry dimension. According to Laugksch (2000), there has been continued debate about Miller's third dimension, which emphasizes a problem-solving approach in science teaching and a focus on science-technology-society issues, particularly because the sociocultural context varies substantially when measured across nations. In the decades since, researchers have used large-scale samples, standardized questions, and survey techniques to describe and compare trends associated with gains in science content knowledge, attitudes toward science, and support for science among the public. A majority of this research has focused on science literacy at the individual rather than societal level (Snow and Dibner 2016).

## *Health Literacy and Public Health Literacy*

**Health literacy** is defined as the "degree to which individuals have the capacity to obtain, process, and understand basic information and services needed to make appropriate decisions regarding their health" (Ratzan and Parker 2000 as cited in IOM 2004, Executive Summary, p. 2). As described above, the National Assessment of Adult Literacy (NAAL) was part of early efforts to define health literacy (NCES 2003; Nutbeam 2000). Specifically, the NAAL identified relationships between health literacy and variables such as educational attainment, age, race/ethnicity, sources of information about health issues, and health insurance coverage. It defined health literacy to include prose, document, and quantitative literacy and suggested that health information should be tailored to the capacities of target audiences.

Initially, health literacy focused on strengthening patient–provider relationships, specifically improving communication between the two so that individuals and families could make better health decisions and better adhere to medical regimens (Nutbeam 2008). The Institute of Medicine (IOM) definition focused on individuals, because research had shown that people with low health literacy had difficulty understanding health information, received less preventive care and paid more for care (IOM 2004; Nutbeam 2008). Recent discussions about health literacy have highlighted the importance of moving beyond an individual focus and considering health literacy as an interaction between the skills of individuals (and their families/caregivers) and the demands of health systems.

**Public Health Literacy** Freedman et al. (2009) asserted that the traditional framing of health literacy, with its focus on individual capacity and competency, resulted in lower literacy levels being associated with individual deficiencies rather than socio-cultural dynamics. Thus, efforts to improve literacy focused on improving interpersonal communication and facilitating information exchange between patients and providers. The limitations of this approach led to the conceptualization of public health literacy (PHL) as "the degree to which individuals and groups can obtain, process, understand, evaluate, and act upon information needed to make public health decisions that benefit the community" (Freedman et al. 2009, p. 448). This concept differs from health literacy in several ways. First, PHL emphasizes the ability to evaluate the quality of information and includes action, which assumes that individuals and groups have the power to accomplish public health goals through civic engagement. Further, although individual skills related to accessing, interpreting, and using health information contribute to PHL; they do not ensure societal-level outcomes. Finally, PHL recognizes that individuals exist in environmental and social contexts rather than viewing them only as part of a patient-provider dyad.

**Measuring Health Literacy** Low health literacy has been described by the IOM as limited health knowledge, health behaviors, and health outcomes (IOM 2004). Even a person with adequate reading comprehension skills may not be able to understand printed health materials, especially if they lack the preparation to understand the technical information contained in those materials (Baker 2006). In the United States, low health literacy is a pervasive and an under recognized problem (Williams et al. 2002). As noted above, the 2003 NAAL estimated that more than 89 million American adults had limited health literacy skills (Kutner et al. 2006). Specifically, using the performance levels mentioned above, NAAL reported that a majority of adults (53%) had Intermediate health literacy; and roughly one-third had Basic (22%) or Below Basic (14%) health literacy, meaning they were likely to have difficulties with health-related documents. Just 11% had a health literacy level of Proficient. Problems with health literacy included trouble with basic reading and numerical skills (Safeer and Keenan 2005). Additionally, Plimpton and Root (1994) identified a gap between the reading level of most adults, approximately eighth grade, and that of most health care materials, most often above the tenth grade. In other words, lower individual literacy levels often correlate with limited health literacy.

At least 112 health literacy instruments measuring a range of skills have been developed, but few have been rigorously validated, nor do they cover the full range of dimensions of health literacy (Snow and Dibner 2016). Further, comprehensive, direct measurement of health literacy is impractical in most clinical settings and in scoping most health education projects (Pleasant and McKinney 2011). However, in a research setting, an individual's general and health reading fluency and vocabulary can be measured using established tools, such as the Newest Vital Sign (NVS), Test of Functional Health Literacy in Adults (TOFHLA), and the Rapid Estimate of Adult Literacy in Medicine (REALM). Each of these has been validated and used in research and clinical settings, and all three are discussed briefly below. Thus, for specific topics, health knowledge can be measured as can the complexity, reading level and difficulty of written materials. In each case, the purpose of the screening tool is to identify patients who might have difficulty reading or understanding health information.

The NVS (Weiss et al. 2005) was developed as a quick screening test for limited literacy in primary health care settings. It measures reading comprehension and numeracy skills using a nutrition label accompanied by six questions. Fewer than four correct responses indicate the possibility of limited literacy. This tool has been shown to be effective with research participants and clinical patients (Rowlands et al. 2013), particularly in identifying limited health literacy in younger adult populations and with older Spanish- and English-speaking patients (Ramirez-Zohfeld et al. 2015).

The TOFHLA and a shorter version (S-TOFHLA) are timed reading comprehension tests in which words in a passage are omitted and replaced with blank spaces (Parker et al. 1995). Patients must select a word for each blank space from four multiple-choice options. These tests have been shown to be valid, reliable indicators of patient ability to read health-related materials, with results indicating that a high proportion of patients cannot perform basic reading tasks (Parker 2000; Baker et al. 1999).

Finally, the REALM, which was developed and validated in the early 1990s, continues to be used in health literacy assessment (Haun et al. 2014). It was designed to assist health care professionals in identifying patients with low literacy levels so that simplified medical instructions and patient education materials could be provided to meet patient needs (Murphy et al. 1993). The REALM tests recognition of medical terms, and a shorter form (REALM-SF) enables quicker assessment (Arozullah et al. 2007). With the REALM-SF, patients read the following words aloud to a healthcare provider: behavior, exercise, menopause, rectal, antibiotics, anemia, and jaundice. If the patient does not recognize a word or takes longer than 5 seconds to read it, the word is marked as unread. Scores are accompanied by recommendations for appropriate types of health education materials. Dumenci et al. (2013) have questioned the value of REALM scores, asserting that the ability of a person to read and pronounce health-related terms does not adequately represent health literacy. They suggest that the ability to read and pronounce printed health-related terms should be viewed as a predictor of low health literacy rather than as a being synonymous with health literacy.

## *Environmental Literacy*

Whereas public health literacy considers decision-making and broad community contexts, environmental literacy identifies knowledge and skills *and* presupposes that knowledge leads to action, both individual and collective. Specifically, the North American Association for Environmental Education (NAAEE) defines an environmentally literate person as:

> *…someone who, both individually and together with others, makes informed decisions concerning the environment; is willing to act on these decisons to improve the well-being of other individuals, societies, and the global environment; and participates in civic life. (North American Association for Environmental Education (NAAEE)* 2011*)*

The full definition describes cognitive, affective, and behavioral components of environmental literacy, noting that these components are both interactive and developmental. Further, NAAEE recognizes that environmental literacy develops along a continuum over time. This definition is consistent with ideas first outlined in the Tbilisi Declaration (United Nations Environment Programme 1977), which defined the objectives of environmental education as awareness, knowledge, attitudes, skills and participation.

**Connecting Environmental Literacy to Environmental Health Education** Chepesiuk (2007) highlighted environmental literacy as a useful framework for environmental health education, given its emphasis on critical thinking skills

**Table 2.1** Skills and actions embodied in literacies that inform EHL

| | Skills | Action/application |
|---|---|---|
| Health literacy | Obtain, process and understand information/services needed to make individual health decisions | Strengthen patient-provider communication |
| | | Improve individual health decisions |
| Public health literacy | Obtain, process and understand information/services needed to make individual *and group* health decisions | Make public health decisions that benefit the community, as well as self |
| Environmental literacy | Identify and analyze environmental issues | Make informed decisions about the environment |
| | Evaluate potential solutions | Improve the well-being of other individuals, societies & global environment |
| | Propose and justify actions to address the issues | Participate in civic life |

and developing active citizens who could apply their knowledge to specific issues. He underscored the wide range of skills needed to understand, assess and use environmental health information, especially when behavior change was a goal. Chepesiuk also pointed to a gulf between health literacy and environmental literacy, noting that, at that time, health literacy initiatives typically did not address the environment.

**Measuring Environmental Literacy** Several approaches have been used to measure environmental literacy, often in school settings. Working with 2004 6th and 8th grade students in 48 counties in the United States, McBeth and Volk (2010) developed the Middle School Environmental Literacy Survey (MSELS). They found moderate to high levels of ecological understanding and moderately positive attitudes associated with willingness to take action supporting the environment; however, students did not report actual behaviors to address environmental conditions. Additionally, results showed that most students were lacking critical thinking and decision-making skills that would be useful in resolving environmental problems. The MSELS also was used to compare the environmental literacy of 739 6th and 8th grade students in schools with school-wide environmental education (EE) programs against students in schools without such programs (Stevenson et al. 2013). Results showed improved environmental literacy when published EE curricula were combined with time outdoors.

As these examples indicate, although adult literacy and science literacy are foundational for EHL, health/public health literacy and environmental literacy may more directly inform current conceptions of EHL. Table 2.1 represents the skills and action-oriented or applied contexts of the latter literacies, which suggest framing for EHL and aspects that could be measured.

### *Knowledge and Skills That Constitute Environmental Health Literacy*

Drawing on some of the approaches described above, Valdez et al. (2014) explored the core competencies needed to be considered environmental health literate. For this research, EHL competency was defined as having the necessary ability, knowledge and skills to use environmental health information successfully for one's self or one's community (SOPHE 2014). The purpose of this research was to: (a) describe common themes regarding the skills and knowledge required to identify, understand, and take action regarding environmental exposures, and (b) to refine these themes through a survey distributed to NIEHS grantees nationwide. This effort was viewed as foundational to understanding varied levels of EHL and developing a continuum of knowledge and skills to explore in subsequent research.

Valdez et al. used a mixed-methods approach, including 28 semi-structured interviews of environmental health professionals in two states (Arizona and New York) and a quantitative survey of 181 environmental health professionals across the United States. The interviews were analyzed with a grounded theory approach, which then informed development of the survey questions. The survey was disseminated online to professionals in academia, government agencies, K-12 settings, and others who shared environmental health information with various public and school audiences. Participants included environmental health educators (41%), researchers (40.5%), and others (18.5%). Through the interviews, a total of 69 knowledge items and 37 skills were identified.

Interestingly, participants were better able to describe and agree on what a person needed to *know* to be considered environmental health literate than they were able to describe what person needed to be able to *do*. Twelve of the knowledge items were considered essential by at least 70% of survey respondents, while just six skill items were identified as being essential by more than 50% of survey respondents. Table 2.2 presents the top six knowledge items and skills identified by survey respondents. These results suggest that there is a knowledge base, or a set of knowledge items agreed upon by a majority of respondents, deemed essential for EHL. At the same time, lower agreement on the importance of specific skills may indicate a higher value placed on knowledge over skills by all participants, or it may mean that the participants do not conceive of EHL as always including skills. It may also indicate limitations in approach, such as not eliciting sufficient skill items during the interviews.

Results also indicated that while EHL encompassed several pre-defined types of literacy, it also included some items unique to EHL. According to participants, the focus of EHL was largely on foreseeing the probability that certain health outcomes may follow various environmental exposures. EHL went beyond understanding how to navigate the health care system, comprehending the essentials of illnesses, or proper care and treatment of existing health conditions and extended to how environmental or social conditions could be changed to reduce exposures. These preliminary results warrant further study.

**Table 2.2** Essential knowledge and skills for EHL

| Knowledge | Skills |
|---|---|
| Understanding that environmental exposures influence health | Ability to determine whether an information source was reliable |
| Understanding that environmental agents can enter the body through three primary routes: ingestion, inhalation and dermal absorption | Ability to identify well-known or established hazards in one's environment |
| Understanding the harmful impacts of specific environmental agents | Ability to find information about hazards in one's microenvironment, such as at home or in the workplace |
| Understanding that environmental agents can be reduced but not always avoided | Ability to find information about regional/community environmental hazards/issues |
| Awareness that reliable information about environmental exposures can be provided through research | Ability to find information about how to reduce environmental health risks in one's life |
| Awareness that research on how environmental exposures influence health takes a long time | Ability to convey one's concerns about environmental health risks to others |

These findings suggest an opportunity to further explore the skills needed to use environmental health information. A more complete assessment of such skills could inform approaches to evaluating attainment of EHL. Additionally, the identification of essential knowledge and skills sets the stage for development of a continuum of measures of EHL, which also could provide insight into the consequences of low EHL.

## *Measuring Environmental Health Literacy*

To date, no protocol exists for measuring EHL, although at least one survey instrument is under development (Ratnapradipa et al. 2015). However, as described below, several studies have represented EHL, by describing individual and community-level outcomes and by reporting the level of environmental health knowledge and awareness of participants. A few studies have even characterized the self-efficacy of participants with respect to certain teaching skills (Butterfield et al. 2011; LePrevost et al. 2014). Gray (2018) reviewed studies that described changes in EHL and found that EHL was measured or represented in the following ways:

- As individual understanding of a connection between environmental exposures and health outcomes
- As a score on a survey of environmental health knowledge or increased content knowledge using pre/post-tests
- As individual choices to change behaviors in response to environmental exposures
- As participation in report-back studies
- As community changes in response to environmental exposures

Gray asserted that these examples demonstrate the continuum of EHL that connects understanding of how environmental exposures influence health with individual and community action to reduce exposures. Further, she called for explicit incorporation of elements of self-efficacy and skills that enable health-protective decisions into the conceptualization of EHL. Below we provide examples of these studies, organized by their focus on either individual level EHL, both individual and community-level EHL or community level EHL.

## Individual Level EHL

**Understanding the Connection between Environmental Exposures and Health** A starting point for representing EHL is demonstrating that people are making connections between environmental exposures and their health. Chan et al. (2015) investigated 72 college women's use of personal care products and their views on potential health effects from exposures during the preconception period to endocrine disrupting chemicals in these products. In this study, participants reported lack of awareness of the potential health effects of environmental toxicants in these products. In another study, 894 women were surveyed about their attitudes toward environmental chemicals in food and personal care products (Barrett et al. 2014). Researchers found that college-educated women were more likely to believe that environmental chemicals were dangerous, and these beliefs were associated with subsequent healthy behavior choices.

Other studies have represented participant identification of potential health risks from environmental exposures as a form of EHL. For example, White et al. (2014) found that residents of a public housing complex ($N = 42$) defined environmental health risk factors broadly, including risks from pollutants as well as physical safety concerns from crime and law enforcement interactions. Schure et al. (2013) explored perceptions of health with 27 members of the Confederated Tribes of the Umatilla Indian Reservation. In this study, participants identified a range of environmental health concerns, underscoring the importance of local contexts to environmental health literacy. Participant concerns included: poor air quality from industrial sources, smoking and mold; water pollution from hazardous waste sites and cattle ranching; and toxic chemical exposure from a nuclear facility, pesticides and methamphetamine labs.

**Measuring EHL with Surveys** A team of researchers has been developing tools to measure EHL with a goal of using these tools to define an EHL baseline. Ratnapradipa et al. (2015) developed an EHL survey in 2009 and recently tested it with 32 participants in four states. Results indicated that participant perceptions of environmental health were based on misinformation or lack of information, which could lead to poor environmental health decision making. With an earlier version of this instrument, Ratnapradipa et al. (2011a) surveyed 395 undergraduate students in introductory health education courses at a Midwestern university in the United States.

Results indicated a general lack of knowledge of environmental health concepts, which corresponded to a lack of protective behaviors when faced with environmental risks. In yet another earlier version, Ratnapradipa et al. (2011b) assessed the EHL of 101 pre-service teachers at a Midwestern university and found deficiencies in basic environmental health knowledge and a statistically significant correlation between knowledge scores and environmental health-related behavior scores.

Another instrument, the Environmental Health Engagement Profile, was used to assess how 433 people experienced environmental health hazards and their associated risks as well as individual and collective responses to those risks (Dixon et al. 2009). Findings suggested that personal actions were more likely than community-level actions in response to environmental health concerns.

**Measuring Post-Intervention Gains in Content Knowledge** Other studies have focused on documenting gains in environmental health knowledge following specific interventions as a means of demonstrating EHL. Ramos et al. (2012) surveyed residents of 498 households along the Texas-Mexico border about their knowledge of environmental health and disease, following an educational intervention led by community health workers. The authors reported gains in content knowledge related to pesticide exposures, water exposures and smoking-related diseases. These gains included knowledge of behaviors that could reduce exposure to environmental contaminants.

In another study, researchers assessed farmworker knowledge of pesticide safety messages following an hour-long lesson conducted by trained educators (LePrevost et al. 2014). Participants ($N = 20$) demonstrated significant increases in knowledge of pesticide safety messages between pre- and post-assessments. Similarly, Gray and Graves (2016) (Fig. 2.2) described the results of a 6-hour training focused on identifying and reducing exposure to environmental health hazards in homes, conducted with over 300 public health professionals. Following the training, pre-/post-tests showed increased knowledge of environmental health hazards, and a subset of participants reported actively working with clients and patients to implement exposure-reduction strategies modeled in the training, related to asthma triggers, mold and pests.

**Changing Behavior in Response to Environmental Exposures** Several studies illustrate the importance of formal and informal educators as facilitators of behavior changes that could represent EHL. For instance, in a study addressing contaminated well water in Bangladesh, elementary school teachers were trained in an arsenic education curriculum as part of a school-based intervention that encouraged residents to switch to low-arsenic wells (Khan et al. 2015). The intervention included an in-depth training for teachers, weekly 15-minute educational sessions, bimonthly fieldwork, and an annual competition among schools. Parental engagement also was emphasized. In this case-control study, 773 children from 14 schools participated in pre-/post-surveys about well safety and groundwater knowledge along with well water testing. Findings showed that families of children who received the intervention were five times more likely to switch from high- to low-arsenic wells. In the northeastern United States, Paul et al. (2015) designed an intervention to address barriers to well water testing, and the program boosted well testing rates. In a similar Canadian study

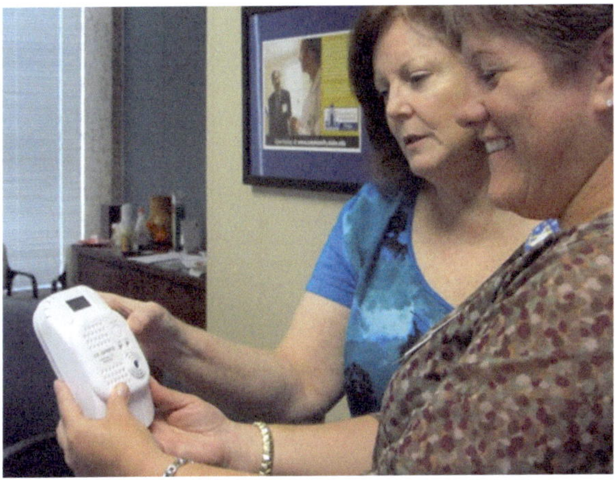

**Fig 2.2** Public health professionals participating in healthy homes training

of well testing behaviors, researchers demonstrated that education alone was not sufficient to ensure that well owners tested their wells annually (Imgrund et al. 2011). The authors reported that improving access to public health departments (which were responsible for testing) was more effective than education alone.

Korfmacher and Kuholski (2008) demonstrated the impact of asking study participants to commit to actions, in writing, following an intervention. The authors were part of a community-based effort to address high levels of lead poisoning in their local community. This effort included a physical *Healthy Home*, in which members of a community-based organization led tours through the house, using low-literacy materials and hands-on displays to show how to reduce environmental health hazards in homes. At the end of the tour, visitors could elect to complete a form describing one or more actions they intended to take to reduce hazards in their own homes. Six-month phone and email follow-up (with over 100 respondents) revealed that almost all reported having completed or partially completed the actions they had identified.

**The Role of Community Health Workers** The effectiveness and value of skilled instructors who are based in communities underscores the relevance of sociocultural theory to EHL measurement. Butterfield et al. (2011) assessed the impact of a home-based intervention on reducing exposures to several environmental health hazards in 235 homes in two northwestern states. Public health nurses led the intervention, which included the collection of environmental samples and biological samples from children in the home, followed by in-home education tailored according to each participant's sample results and presented via an interactive book that highlighted findings in each room of the home. The researchers assessed the impact on parents' cognitive and behavioral outcomes – specifically, their sense of self-efficacy and their adoption of risk-reducing behaviors. Results showed significant improvements on measures of self-efficacy and precaution adoption among the intervention groups.

More recently, Mankikar et al. (2016) evaluated the effectiveness of a home-based environmental health intervention conducted in an urban center in a north-eastern state. Working with 150 low-income families, researchers analyzed the influence of the intervention on residents' knowledge of environmental health hazards, medical visits for children with asthma, and the presence of home health hazards. The intervention consisted of two home visits (led by a community health worker) that included a home assessment, environmental education, and distribution of supplies. Results suggested that the home-based intervention increased participants' knowledge of hazards. Additionally, the presence of home hazards was reduced over the course of the study, and fewer asthma-related medical visits were reported.

## *Spanning Individual and Community-Level EHL*

Report-back protocols that increasingly are being incorporated into biomonitoring studies, in which biological samples are collected in response to an environmental exposure, also have resulted in improvements in EHL. Such studies have demonstrated how study participants use new knowledge to inform action, both at the individual level and collectively. For instance, Ramirez-Andreotta et al. (2016) interviewed parents of children who had participated in a toxic metals exposure study to determine whether they understood study results and how they applied them in their lives. The authors reported that participants understood their children's data and used it to ask new questions. Participants also took action to reduce their family's health risks, using the personal exposure data to inform their choices, which led the authors to assert that the study advanced participants' EHL.

Madrigal et al. (2016) reported increased EHL among youth participants in a 3-year study of young Latinas' exposure to endocrine disrupting compounds (EDCs) in personal care products. In addition to involving the youth in measuring their own exposure, researchers engaged them in developing educational messages for other youth, publicizing the study findings, and developing advocacy activities. Participants' EHL, along with leadership skills, career orientation and self-esteem, were assessed using written reflections, end-of-event questionnaires, interviews, and observation. The researchers used Bloom's Taxonomy to evaluate participants' progression to higher levels of critical thinking, such as occurs when a person goes from being able to identify EDCs in products to being able to share information about EDCs with peers. At the same time, they echoed questions raised by Finn and O'Fallon (2017) about the limits of Bloom's framework when behavior change is a goal of education.

## *Community-Level EHL*

Another approach to documenting EHL has been to show evidence of community change in response to knowledge gains. In a community-based participatory research (CBPR) study with members of the Apsaalooke Tribe, researchers tested 97 wells for uranium and other contaminants and conducted household water use surveys (Eggers et al. 2015). Following tests, well owners received personalized results, a spreadsheet comparing their test results with EPA standards, and an in-person follow-up visit. Results were disseminated through varied activities, ranging from community meetings and displays at local health fairs and health departments to newspaper articles and school presentations. In response to reporting high levels of contaminants in well water and requests from study participants, the tribal water and wastewater authority raised funds for and installed an automated dispensing system in the main Reservation town, which enabled rural residents to purchase filtered municipal water at low cost.

In another CBPR study focused on well water, residents using community wells for drinking water contacted researchers with concerns about contamination of these wells with perfluorooctanoic acid (PFOA) from a local industry. Researchers tested the levels of PFOA in residents' blood and found PFOA levels to be ~80 times higher than in the general population (Emmett et al. 2009). Together they explained the results to the broader community and developed response strategies. Outcomes included provision of free bottled water to residents until the local water utility could upgrade its filtration system.

Citizen science also is an emerging approach for EHL efforts, as demonstrated by Ramirez-Andreotta et al. (2015) (see Chapter 4). In this study, residents of a southwestern community with arsenic contamination learned the skills needed for them to assess uptake of arsenic in garden vegetables and be able to characterize the potential risk from gardening and consuming vegetables from their home gardens. Researchers evaluated individual learning, programmatic outcomes and community-level outcomes. As with the report-back/biomonitoring studies described above, these motivated citizen scientists sought an understanding of the scientific results and implemented behavioral changes to reduce their arsenic exposure as a result of this understanding. Similar to other citizen science efforts, these participants became more involved in resource-related issues, including addressing contamination in the local public water supply.

In summary, biomonitoring and environmental monitoring studies that have incorporated report-back protocols have resulted in increased understanding of environmental exposures among individuals, behavior changes, and community-wide impacts. Many of these studies applied CBPR principles, underscoring connections between EHL and community engagement.

## *Addressing Cultural Context in Measurement of EHL*

Cultural competency was emphasized in the design and implementation of several of the educational interventions described above (Eggers et al. 2015; Ramos et al. 2012), suggesting that culturally sensitive programming more effectively positions communities to address health disparities. As early as 2001, researchers identified the importance of cultural context in fostering EHL. Zarcadoolas et al. (2001) presented a case study of a partnership to produce a community guide to brownfields that included literacy experts, environmental scientists and residents of inner-city communities with brownfields. The authors anticipated modern challenges to EHL, as they described how the combination of declining literacy, increasing cultural diversity, and the complexity of the environmental health sciences made environmental health information inaccessible to the public. Focusing on the use of oral and written language and graphical representations, they facilitated understanding of technical information and the subsequent use of such information to protect health. They termed their process "cooperative composing" (p. 19) and argued that pairing community-produced knowledge with expert knowledge enabled residents to better engage in policy processes. Like the CBPR studies described above, this study was grounded in: commitment to active collaboration and participation from all partners at every stage of research, fostering co-learning, disseminating results in useful terms, and ensuring the cultural competency of the intervention.

## *Promotion of Environmental Health Literacy in Tribal Communities*

A similar convergence of cultural diversity, complexity of environmental health issues and low literacy can be found in many American Indian communities in the United States (NCES 2015). As noted above, a range of environmental health issues have been identified as relevant in American Indian communities, including poor air quality, water pollution, and toxic chemical exposures, among others (Schure et al. 2013). At the same time, students in tribal schools are significantly more likely to live in poverty and attend rural schools, and they are less likely to have parents who graduated college or have a computer at home (Sparks 2017). Additionally, only 11 percent of American Indians have earned bachelor's degrees compared to 27 percent of the total population in the United States (National Education Association 2017). Despite this lower educational attainment, tribal community members often possess intergenerational knowledge about environmental conditions in their communities, which represents a tribal form of EHL.

For over a decade, Lindsey (2017a, b) and colleagues have worked in tribal communities in Arizona, engaging adults and youth in understanding how the environment affects human health. To address underlying issues with low literacy, they adopted *universal precautions*, such as *scaffolding* educational materials, or

Fig. 2.3 Community members participating in asthma information walk

providing materials at progressively more complex levels, to take lower literacy levels into account (Baker 2006).

For instance, the *Information Walk* (Fig. 2.3) incorporates interactive activities for all ages as well as simple evaluation. Distinct *Information Walks* have been co-created with six tribes in Arizona, working closely with the tribal cultural departments to tailor them with appropriate language and images. An *Information Walk* is comprised of multiple stations, with displays and one-on-one activities, similar to a mobile museum exhibit. These stations, which typically are staffed by university personnel and community volunteers, promote face-to-face interactions in a nonthreatening environment and prompt individuals to ask questions; in-depth information is available for those who request it. A simple evaluation tool is used to assess participant learning, identifying the top three things the person learned and questions they still have. Recently, Lindsey's team has used similar tools in concert with science cafés in American Indian communities. Cards and pens are placed on the tables to encourage participants to write down questions if they are reluctant about asking them aloud. For cultural groups where listening is a sign of respect and there are varying degrees of comfort with asking questions, this approach enables more active participation. This work led to the conclusion that assessment of EHL should be done in culturally appropriate ways, including use of verbal rather than written assessment.

Other examples of environmental health education projects developed in partnership with tribal communities have resulted in outcomes that could be interpreted as changes to individual and community EHL. For example, DeWeese et al. (2009) developed and assessed the efficacy of an educational intervention designed to increase awareness of fish consumption advisories in tribal communities in three Midwestern states. They relied on formative research (e.g., risk calculations tailored to tribal consumption patterns and focus groups with tribal members) to develop maps that enabled decisions about reducing mercury exposure while preserving tribal customs. Pre- and post-intervention survey data from fish harvesters and women of childbearing age showed significant increases in awareness and concern about mercury contamination and changes to harvesting and food preparation behaviors to reduce mercury exposure. The authors concluded that community participation in fish consumption advisory development improved advisory acceptance by incorporating critical cultural attitudes and beliefs.

In a similar vein, McOliver et al. (2015) described a tribally-designed maternal biomonitoring program for organohalogens and heavy metals that was implemented alongside sampling of marine fish and sea mammals. When biomonitoring results showed an association between mercury in participants' blood and coastal residence, an educational intervention was designed to raise awareness among subsistence fishermen that younger, smaller sea mammals had lower levels of contaminants, and land mammals had very low levels. Initial results showed maternal blood levels of contaminants decreasing as awareness of contaminants in food sources increased, suggesting that the intervention was effective. These examples underscore the importance of cultural competency and active community engagement in EHL initiatives.

## Implications for Future Measurement of EHL

Over time, the concept of literacy as applied to scientific disciplines has evolved. Early models focused primarily on whether individuals understood specific concepts and processes and how they might encounter these concepts in daily lives (AAAS 1990). Then the idea of literacy in the scientific realm grew to incorporate skills and abilities and, in some cases, application to decision making and community contexts (Freedman et al. 2009; North American Association for Environmental Education (NAAEE) 2011). The literacy conversation also began to include sociocultural dimensions, recognizing that individuals acting alone may not be able to influence community-scale and policy issues (Snow and Dibner 2016). This shift is evident in the emerging representations of EHL as a continuum of knowledge, skills and practice (Finn and O'Fallon 2017; Gray 2018; Hoover 2014; Marsili 2016).

With reference to specific measurement tools, a variety of tools have been developed to assess science, health/public health, and environmental literacy (McBeth and Volk 2010; Pleasant and McKinney 2011; Snow and Dibner 2016). Many of the science and health literacy tools have focused primarily on content knowledge (Snow and Dibner 2016). Yet the results of studies referenced in this chapter have shown that content knowledge does not necessarily correspond with behavior (Ratnapradipa et al. 2011a, b), raising questions about why content knowledge has been deemed a sufficiently robust measure. An additional concern is that achievement on content knowledge measures may reflect prior schooling as much as immediate learning. As Falk (2002) argued, science literacy tests that mimic school-based testing may be misleading, because those who scored well "were likely to be those who had completed the most school courses in science" (p. 64). Further, many measurement tools target literacy at the individual level, meaning they may miss community-scale outcomes. Given a definition of EHL that explicitly incorporates community-scale dimensions, content-focused measurement tools alone do not seem sufficient.

Initial work on EHL measurement tools has not yet grappled with the potential need to assess content mastery and skills development. For instance, Ratnapradipa et al. (2015) found that such instruments may need to be issue-specific, to be understandable to lay audiences, and simplified, given the number of content areas spanned by the environmental health sciences. At the same time, when environmental health concepts are reduced to a brief set of questions, detail and complexity are lost, which may limit one's ability to apply that knowledge in meaningful ways and may limit researchers' ability to measure how such knowledge is applied. Given the need for behavior change, both at individual and community levels, to reduce or eliminate some environmental exposures, tools to measure EHL should include an assessment of skills and actions (or intent to act) associated with content knowledge measures.

As Valdez et al. (2014) noted a working knowledge of environmental health sciences is critical for people to be able to generalize from one environmental exposure to others. Further, the ability to accurately interpret science-related knowledge is an important dimension of EHL and should inform an individual's ability to locate and use new information to understand the impact of additional exposures (Kolstø 2001).

The characterization of adult literacy by performance level (i.e., assigning a value from Below Basic to Proficient based on acquisition of certain skills; NAAL, 2003) offers a potential model for assessing skills development associated with EHL. Using the core knowledge and skills identified by Valdez et al. (2014) and presented in Table 2.2, assessments could incorporate content knowledge about environmental health science, broadly defined, and could be implemented in concert with a skills assessment tailored to the essential skills. Additionally, assessment of content knowledge could be tailored for specific exposures, such as toxic metals in well water or fish, and to be responsive to cultural norms.

Related to skills mastery, measurement of EHL also may need to include some assessment of self-efficacy for health-protective behaviors, especially where behavior change is an identified outcome. As Bandura (1998) noted in his conception of social cognitive theory, individuals choose a specific behavior when they believe in their ability to perform that behavior (self-efficacy) *and* expect it to produce desired outcomes (outcome expectancy). Lacking information about these important variables may obscure opportunities to increase EHL that are not related to content knowledge or existing skills.

Defining EHL comprehensively provides several entry points to measuring EHL, and we recommend that any definition include the following: (a) knowledge of the health effects of a specific environmental exposure underpinned by broader understanding of environmental health science concepts, (b) the ability to seek out and use information, and (c) positive perceptions of self-efficacy related to health-promoting behaviors. This multi-faceted approach also recognizes that EHL is neither a linear progression nor comprehensive knowledge about environmental factors, but rather depends on the specific exposure, individual and community health, and cultural context, among other factors.

## Future Research Opportunities

The research described within this chapter demonstrates a continuum of EHL in which knowledge of environmental exposures and their influences on health is mediated by self-efficacy for health-protective behaviors and skills related to information seeking and processing. Taken together, these elements of EHL can lead to individual and community action to reduce harmful environmental exposures. However, because this scholarship is in an early stage, further research is needed to understand how people move along the EHL continuum. What intrinsic interests or motivation first stimulate awareness and pursuit of environmental health knowledge? How does self-efficacy for specific behaviors promote or inhibit additional learning and action? Many research opportunities lie in the answers to these questions, and they could inform not only the design of educational interventions but also the measurement of their effectiveness and of EHL itself. Given the body of research on self-efficacy and its influence on actions taken in response to information, adaptation of self-efficacy measures used in science education, especially in informal science settings, also could enhance our understanding of EHL.

Future research also could provide insights into the baseline environmental health literacy of target populations, in the context of specific environmental exposures, as well as the influence of community engagement efforts or educational interventions on literacy levels. But to establish a baseline or measure the ways in which an intervention fosters EHL, measurement tools and approaches will be needed. Toward that end, development and validation of multi-faceted tools to formally assess baseline EHL and changes against baseline seems likely. Any such instruments should be responsive to the challenges described above. Several existing literacy measures seem well positioned to be adapted for EHL in research settings. A process-focused approach, similar to Newest Vital Sign, could be developed using an artifact that incorporates environmental health information (such as a household cleaner label). Associated questions could then shed light on the ability of a participant to interpret the available information. This approach combines knowledge and skills, making it more robust than a knowledge-only assessment. Similarly, a case study approach could present a scenario in which participants draw conclusions using environmental health knowledge and other skills. Either of these approaches would ensure EHL would not be reduced to a single number or series of quantitative values, thus presenting a fuller picture of where an individual may be on the EHL continuum and providing greater insight into how to move forward.

As a starting point, future research could revisit the essential knowledge items and skills identified by Valdez et al. (2014) with a goal of refining and building on those items and identifying the combination of knowledge and skills that would constitute varying literacy levels. Similar to the NAAL (NCES 2003), these measures could be used to represent Below Basic, Basic, Intermediate and Proficient literacy levels. Such characterizations could provide environmental public health practitioners and researchers with a better understanding of the abilities of the people they serve, ideally leading to more effective communication about reducing harmful environmental exposures.

Finally, application of the environmental health literacy framework to a range of environmental health issues and impacted populations is needed to understand how the dimensions of EHL interact and what factors ultimately influence lower versus higher levels of literacy. Such research could assist in developing educational interventions to improve environmental health literacy. It also could illuminate the individual and societal implications of a range of literacy levels, which may have actionable implications for environmental health education practices and could inform health-protective decisions.

## References


American Association for the Advancement of Science (AAAS). (1990). *Science for all Americans*. New York: Oxford University Retrieved from http://www.project2061.org/publications/sfaa/online/sfaatoc.htm.

Amineh, R. J., & Asl, H. D. (2015). Review of constructivism and social constructivism. *Journal of Social Sciences, Literature and Languages, 1*(1), 9–16.

Arozullah, A. M., Yarnold, P. R., & Bennett, C. L. (2007). Development and validation of a short-form, rapid estimate of adult literacy in medicine. *Medical Care, 45*(11), 1026–1033 PMID: 18049342.

Baker, D. W., Williams, M. V., Parker, R. M., Gazmararian, J. A., & Nurss, J. (1999). Development of a brief test to measure functional health literacy. *Patient Education and Counseling, 38*(1), 33–42.

Baker, D. W. (2006). The meaning and the measure of health literacy. *Journal of General Internal Medicine, 21*(8), 878–883.

Bandura, A. (1998). Health promotion from the perspective of social cognitive theory. *Psychology and Health, 13*, 623–649.

Barrett, E. S., Sathyanarayana, S., Janssen, S., Redmon, J. B., Nguyen, R. H., Kobrosly, R., & Swan, S. H. (2014). Environmental health attitudes and behaviors: Findings from a large pregnancy cohort study. *European Journal of Obstetrics & Gynecology and Reproductive Biology, 176*, 119–125.

Brandt, D. (2004). Drafting US literacy. *College English, 66*(5), 485–502.

Butterfield, P. G., Hill, W., Postma, J., Butterfield, P. W., & Odom-Maryon, T. (2011). Effectiveness of a household environmental health intervention delivered by rural public health nurses. *American Journal of Public Health, 101*(supplement 1), S262–S270.

Chan, L. M., Chalupka, S. M., & Barrett, R. (2015). Female college student awareness of exposures to environmental toxins in personal care products and their effect on preconception health. *Workplace Health & Safety, 63*(2), 64–70.

Chepesiuk, R. (2007). Environmental literacy: Knowledge for a healthier public. *Environmental Health Perspectives, 115*(10), A494–A499.

DeWeese, A. D., Kmiecik, N. E., Chiriboga, E. D., & Foran, J. A. (2009). Efficacy of risk-based, culturally sensitive Ogaa (walleye) consumption advice for Anishinaabe tribal members in the great lakes region. *Risk Analysis, 29*(5), 729–742.

DeWalt, D. A., Berkman, N. D., Sheridan, S., Lohr, K. N., & Pignone, M. P. (2004). Literacy and health outcomes. *Journal of General Internal Medicine, 19*(12), 1228–1239.

Dixon, J. K., Hendrickson, K. C., Ercolano, E., Quackenbush, R., & Dickson, J. P. (2009). The environmental health engagement profile: What people think and do about environmental health. *Public Health Nursing, 26*(5), 460–473.

Dumenci, L., Matsuyama, R. K., Kuhn, L., Perera, R. A., & Siminoff, L. A. (2013). On the validity of the rapid estimate of adult literacy in medicine (REALM) scale as a measure of health literacy. *Communication Methods and Measures, 7*(2), 134–143.

Eggers, M. J., Moore-Nall, A. L., Doyle, J. T., Lefthand, M. J., Young, S. L., Bends, A. L., et al. (2015). Potential health risks from uranium in home well water: An investigation by the Apsaalooke (Crow) Tribal Research Group. *Geosciences, 5*(1), 67–94.

Emmett, E. A., Zhang, H., Shofer, F. S., Rodway, N., Desai, C., Freeman, D., & Hufford, M. (2009). Development and successful application of a "Community-First" communication model for community-based environmental health research. *Journal of Occupational and Environmental Medicine, 5*(12), 146–156.

Falk, J. H. (2002). The contribution of free-choice learning to public understanding of science. *Interciencia, 27*(2), 62–65.

Finn, S., & O'Fallon, L. (2017). The emergence of environmental health literacy—From its roots to is future potential. *Environmental Health Perspectives, 125*(4), 495–501.

Freedman, D. A., Bess, K. D., Tucker, H. A., Boyd, D. L., Tuchman, A. M., & Wallston, K. A. (2009). Public health literacy defined. *American Journal of Preventive Medicine, 36*(5), 446–451.

Gray, K. M. (2018). From content knowledge to community change: A review of representations of environmental health literacy. *International Journal of Environmental Research and Public Health, 15*(3), 466.

Gray, K., & Graves, N. (Producers). (2016). *Leveraging partnerships to advance lead and healthy homes programming in North Carolina [webinar]*. Retrieved from https://www.niehs.nih.gov/research/supported/translational/peph/webinars/healthy_homes/

Haun, J. N., Valerio, M. A., McCormack, L. A., Sørensen, K., & Paasche-Orlow, M. K. (2014). Health literacy measurement: An inventory and descriptive summary of 51 instruments. *Journal of Health Communication, 19*(sup2), 302–333.

Hoover, A.G. (2014). *Connecting disciplines to inform and develop the emerging field of environmental health literacy. [Webinar]*. Retrieved from: https://www.niehs.nih.gov/research/supported/assets/docs/a_c/connecting_disciplines_to_inform_and_develop_the_emerging_field_of_environmental_health_literacy_508.pdf

Imgrund, K., Kreutzwiser, R., & de Loe, R. (2011). Influences on the water testing behaviors of private well owners. *Journal of Water and Health, 9*(2), 241–252.

Institute of Medicine. (2004). *Health literacy: A prescription to end confusion*. Washington, DC: The National Academies.

Jacobson, S. K., McDuff, M. D., & Monroe, M. C. (2015). *Conservation education and outreach techniques*. Oxford: Oxford University Press.

Khan, K., Ahmed, E., Factor-Litvak, P., Liu, X., Siddique, A. B., Wasserman, G. A., et al. (2015). Evaluation of an elementary school-based educational intervention for reducing arsenic exposure in Bangladesh. *Environmental Health Perspectives, 123*(12), 1331–1336.

Kirsch, I. S., Jungeblut, A., Jenkins, L., & Kolstad, A. (1993). *Adult literacy in America: A first look at the findings of the National Adult Literacy Survey*. Washington, DC: US Department of Education, National Center for Education Statistics.

Kolstø, S. D. (2001). Scientific literacy for citizenship: Tools for dealing with the science dimension of controversial socioscientific issues. *Science Education, 85*(3), 291–310.

Korfmacher, K. S., & Kuholski, K. (2008). Rochester's healthy home: A community-based innovation to promote environmental health action. *Environmental Practice, 10*(1), 94–106.

Kutner, M., Greenburg, E., Jin, Y., & Paulsen, C. (2006). *The health literacy of America's adults: Results from the 2003 National Assessment of Adult Literacy*. NCES 2006–483. Washington, DC: US Department of Education, National Center for Education Statistics.

Laugksch, R. C. (2000). Scientific literacy: A conceptual overview. *Science Education, 84*(1), 71–94.

LePrevost, C. E., Storm, J. F., Asuaje, C. R., Arellano, C., & Cope, W. G. (2014). Assessing the effectiveness of the pesticides and farmworker health toolkit: A curriculum for enhancing farmworkers' understanding of pesticide safety concepts. *Journal of Agromedicine, 19*(2), 96–102.

Lindsey, M. (2017a). *Activities for the general public: Information walks*. Retrieved from http://swehsc.pharmacy.arizona.edu/outreach/activities-general-public
Lindsey, M. (2017b). *Partnerships with tribal communities for environmental public health*. Retrieved from http://swehsc.pharmacy.arizona.edu/outreach/highlights/american_indian
Liu, C. H., & Matthews, R. (2005). Vygotsky's philosophy: Constructivism and its criticisms examined. *International Education Journal, 6*(3), 386–399.
Madrigal, D. S., Minkler, M., Parra, K. L., Mundo, C., Gonzalez, J. E., Jimenez, R., et al. (2016). Improving Latino youths' environmental health literacy and leadership skills through participatory research on chemical exposures in cosmetics: The HERMOSA study. *International Quarterly of Community Health Education, 36*(4), 231–240.
Mankikar, D., Campbell, C., & Greenberg, R. (2016). Evaluation of a home-based environmental and educational intervention to improve health in vulnerable households: Southeastern Pennsylvania lead and healthy homes program. *International Journal of Environmental Research and Public Health, 13*, 900.
Marsili, D. (2016). A cross-disciplinary approach to global environmental health: The case of contaminated sites. *Annali dell'Instituto Superiore di Sanita, 52*(4), 516–523.
McBeth, W., & Volk, T. (2010). The national environmental literacy project: A baseline study of middle grade students in the United States. *The Journal of Environmental Education, 41*(1), 55–67.
McOliver, C. A., Camper, A. K., Doyle, J. T., Eggers, M. J., Ford, T. E., Lila, M. A., et al. (2015). Community-based research as a mechanism to reduce environmental health disparities in American Indian and Alaska Native communities. *International Journal of Environmental Research and Public Health, 12*(4), 4076–4100.
Miller, J. D. (1998). The measurement of civic scientific literacy. *Public Understanding of Science, 7*, 203–223.
Murphy, P. W., Davis, T. C., Long, S. W., Jackson, R. H., & Decker, B. C. (1993). Rapid estimate of adult literacy in medicine (REALM): A quick reading test for patients. *Journal of Reading, 37*(2), 124–130.
National Center for Education Statistics. (2003). What is NAAL? In *National Assessment of Adult Literacy (NAAL). A nationally representative and continuing assessment of English language literary skills of American Adults*. Retrieved from http://nces.ed.gov/naal/
National Center for Education Statistics. (2005). What are the grade levels of NAAL materials? In *National Assessment of Adult Literacy (NAAL), A nationally representative and continuing assessment of English language literary skills of American Adults*. Retrieved from https://nces.ed.gov/NAAL/faq.asp#faq3
National Center for Education Statistics. (2015) *National Indian Education Study 2015*. Retrieved from https://nces.ed.gov/nationsreportcard/pdf/studies/2017161.pdf
National Center for Education Statistics. (2016). *Program for the International Assessment of Adult Competencies (PIAAC)*. Retrieved from https://nces.ed.gov/surveys/piaac/
National Education Association. (2017). *American Indians/Alaska Natives: Education issues*. Accessed online 4 Jan 2017 at http://www.nea.org/home/15596.htm
North American Association for Environmental Education (NAAEE). (2011). *Developing a framework for assessing environmental literacy: Executive summary*. Washington, DC: NAAEE Retrieved from: https://naaee.org/sites/default/files/envliteracyexesummary.pdf.
Nutbeam, D. (2000). Health literacy as a public health goal: A challenge for contemporary health education and communication strategies into the 21st century. *Health Promotion International, 15*, 259–267.
Nutbeam, D. (2008). The evolving concept of health literacy. *Social Science and Medicine, 67*(12), 2072–2078.
O'Fallon, L., & Dearry, A. (2002). Community-based participatory research as a tool to advance environmental health sciences. *Environmental Health Perspectives, 110*, 155–159.
Parker, R. M., Baker, D. W., Williams, M. V., & Nurss, J. R. (1995). The test of functional health literacy in adults. *Journal of General Internal Medicine, 10*(10), 537–541.
Parker, R. (2000). Health literacy: A challenge for American patients and their health care providers. *Health Promotion International, 15*(4), 277–283.

Paul, M. P., Rigrod, P., Wingate, S., & Borsuk, M. E. (2015). A community-driven intervention in Tuftonboro, New Hampshire, succeeds in altering water testing behavior. *Journal of Environmental Health, 78*(5), 30–39.

Pleasant, A., & McKinney, J. (2011). Coming to consensus on health literacy measurement: An online discussion and consensus-gauging process. *Nursing Outlook, 59*(2), 95–106.

Plimpton, S., & Root, J. (1994). Materials and strategies that work in low literacy health communication. *Public Health Reports, 109*(1), 86.

Ramirez-Andreotta, M. D., Brody, J. G., Lothrop, N., Loh, W., Beamer, P. I., & Brown, P. (2016). Improving environmental health literacy and justice through environmental exposure results communication. *International Journal of Environmental Research and Public Health, 13*(7), 690–717.

Ramirez-Andreotta, M. D., Brusseau, M. L., Artiola, J., Maier, R. M., & Gandolfi, A. J. (2015). Building a co-created citizen science program with gardeners neighboring a superfund site: The Gardenroots case study. *International Public Health Journal, 7*(1) Pages 139–153.

Ramirez-Zohfeld, V., Rademaker, A. W., Dolan, N. C., Ferreira, M. R., Eder, M. M., Liu, D., et al. (2015). Comparing the performance of the S-TOFHLA and NVS among and between English and Spanish speakers. *Journal of Health Communication, 20*(12), 1458–1464.

Ramos, I. N., He, O., & Ramos, K. S. (2012). Improvements in environmental health literacy along the Texas-Mexico border following community-wide health education. *Environmental Justice, 5*(1), 32–37.

Ratnapradipa, D., Brown, S. L., Middleton, W. K., & Wodika, A. B. (2011a). Measuring environmental health perception among college students. *The Health Educator, 43*(2), 13–20.

Ratnapradipa, D., Rhodes, D. L., & Brown, S. L. (2011b). Evaluating pre-service teacher workforce: Environmental health knowledge, attitude and behavior. *Online Journal of Workforce Education and Development, 5*(1), 1–18.

Ratnapradipa, D., Wodika, A. B., Brown, S. L., & Preihs, K. (2015). What does the public know about environmental health? A qualitative approach to refining an environmental health awareness instrument. *Journal of Environmental Health, 77*(8), 22–28.

Rowlands, G., Khazaezadeh, N., Oteng-Ntim, E., Seed, P., Barr, S., & Weiss, B. D. (2013). Development and validation of a measure of health literacy in the UK: The newest vital sign. *BMC Public Health, 13*, 116. https://doi.org/10.1186/1471-2458-13-116.

Safeer, R. S., & Keenan, J. (2005). Health literacy: The gap between physicians and patients. *American Family Physician, 72*(3), 463–468.

Schure, M. B., Kile, M. L., Harding, A., Harper, B., Harris, S., Uesugi, S., & Goins, R. T. (2013). Perceptions of the environment and health among members of the confederated tribes of the Umatilla Indian reservation. *Environmental Justice, 6*(3), 115–120.

Snow, C. E., & Dibner, K. A. (Eds.). (2016). *Science literacy: Concepts, contexts, and consequences*. Washington, DC: National Academies Press.

Society for Public Health Education. (2014). *What is environmental health literacy*? Retrieved from: http://www.sophe.org/environmentalhealth/key_ehl.asp

Sparks, S. D. (2017). *Nation's report card shows strengths, shortfalls for Native American students*. Retrieved from http://blogs.edweek.org/edweek/inside-school-research/2017/03/native_american_students_naep.html#

United Nations Environment Programme. (1977). *Tbilisi declaration*. Retrieved from: http://www.eenorthcarolina.org/Documents/tbilisi_declaration.pdf

Stevenson, K. T., Peterson, M. N., Bondell, H. D., Mertig, A. G., & Moore, S. E. (2013). Environmental, institutional, and demographic predictors of environmental literacy among middle school children. *PLoS One, 8*(3), 1–11.

UNESCO. (2002). *Global monitoring report, 2002: Is the world on track?* Retrieved October 1, 2010, from http://www.unesco.org/education/efa/monitoring/monitoring_2002.shtml

Valdez, M. K., Velez, M., Jenkins, I., Lindsey, M., & Chen, S. (2014). *Knowledge and skills associated with environmental health literacy*. Presented at the PEPH Environmental Health Literacy Conference.

Weiss, B. D., Mays, M. Z., Martz, W., Castro, K. M., DeWalt, D. A., Pignone, M. P., et al. (2005). Quick assessment of literacy in primary care: The newest vital sign. *Annals of Family Medicine, 3*(6), 514–522. https://doi.org/10.1370/afm.405.

White, B. M., Hall, E., & Johnson, C. (2014). Environmental health literacy in support of social action: An environmental justice perspective. *Journal of Environmental Health, 77*(1), 24–29.

Williams, M. V., Davis, T., Parker, R. M., & Weiss, B. D. (2002). The role of health literacy in patient-physician communication. *Family Medicine-Kansas City, 34*(5), 383–389.

Zarcadoolas, C., Timm, E., & Bibeault, L. (2001). Brownfields: A case study in partnering with residents to develop an easy-to-read print guide. *Journal of Environmental Health, 64*(1), 15–20.

# Chapter 3
# Communication Research in the Environmental Health Sciences

**Kami J. Silk and Daniel Totzkay**

**Abstract** The need to effectively communicate evidence-based environmental health messages to a range of stakeholders is fundamental for informed decision making about environmental risk factors. There are a range of strategies used to help communicate emerging environmental health science, but there are clear perceived drawbacks when relying on many of these approaches. For example, research findings and their environmental health implications are often communicated by scientists who focus on the accuracy and comprehensive explanations of the research rather than sensitivity to factors that will increase the acceptance of the environmental health messages. Communication efforts also are sometimes led by creative professionals who might be perceived as oversimplifying science so much that the true implications of the research are not appropriately reflected. Science journalists, often a primary source of information for lay audiences, are generally reliable interpreters of research; however, their availability and the outlets for which science journalists typically write are unlikely to maintain an ongoing focus on specific environmental health issues unless a crisis or timely news event is unfolding – and soon after their initial report on an issue, their interest wanes. While scientists, creative professionals, and journalists have an important role to play in communicating about environmental health issues, so do communication researchers.

In this chapter, a communication science approach is proposed to facilitate the creation of high quality environmental health risk messages that are evidence-based, theory-driven, and tailored to the needs of priority audiences. Key topics addressed in the chapter are the importance of formative research for understanding specific audience's preferences and beliefs (audience analysis) and the identification of relevant media formats or channels to disseminate information through (channel selection), theory informed message design, implementation strategies, and evaluation approaches to reach a wide range of stakeholders with maximum impact. Finally, the chapter will

K. J. Silk (✉)
Department of Communication, University of Delaware, Newark, Delaware
e-mail: silkk@msu.edu

D. Totzkay
Department of Communication, Michigan State University, East Lansing, MI, USA



S. Finn, L. R. O'Fallon (eds.), *Environmental Health Literacy*,
https://doi.org/10.1007/978-3-319-94108-0_3

systematically incorporate examples from the Breast Cancer and Environment Research Program (BCERP) to illustrate key topics, processes, and strategies.

**Keywords** Communication · Communication science · Evaluation · Message design · Breast cancer and the environment research program · Extended parallel process model · Heuristic systematic model · Theory of normative social behavior · Diffusion of innovations

## Communication Challenges

The development of effective environmental health messages is rife with challenges. First, there is the question of readiness to communicate about a particular environmental factor; some groups think the precautionary principle should reign, while others are much more cautious and would rather wait to see more evidence before identifying an environmental factor as a risk. The precautionary principle is invoked when evidence is inconclusive, only derived from animal studies or when there is no consensus about the implications of findings. In those cases, risk messages may take a better safe than sorry approach. In the BCERP, for example, there is a natural tension of when to share emerging scientific findings from studies between advocates and scientists due to their differing, yet complementary roles in research. Advocates have specific stakeholders who rely on them to keep them informed about potential risks and their implications as findings emerge from current studies. In contrast, scientists work on long-term programs of research, which often means they are more tentative in making risk reduction recommendations based on their emerging findings.

Whether or not the audience will feel susceptible or perceive a risk as severe enough to warrant their attention is also a challenge, which can undermine any potentially recommended risk reduction activity if perceived susceptibility or severity is low. For example, women with no family history of breast cancer may not feel susceptible to the disease. This perception is problematic because women with no known family history of breast cancer are at risk due to a large range of factors not just family history. Another challenge is the inherent complexity associated with communicating about scientific topics, which may immediately cause individuals to tune out if they perceive the environmental health message as too cognitively complex to process. For example, individuals with limited English proficiency as well as those with low literacy or low scientific literacy may be intimidated by seeing words like perfluorooctanoic acid, endocrine disruptors, and susceptibility in an environmental health message. Finally, there is the production challenge of creating an environmental health message that is engaging, impactful, and effective among diverse audiences. There is indeed a benefit in trying to reach a large audience through media markets due to the potential volume of individuals an environmental health message might reach, but the ability of a more general, less tailored message to meaningfully impact that large audience decreases. In attempting to reach a large audience of different stakeholders, the BCERP currently has online toolkits that provide educational information about breast cancer and the environment. The toolkit includes information for parents and caregivers, health professionals, and outreach organizations, but the effectiveness of the toolkit materials for each of these audiences currently remains unknown. In sum, this range of challenges can be addressed by a focus on message production, processing, and evaluating as fundamental to the development of effective messages.

## Theoretical Frameworks

Prior to explaining the processes for designing effective messages, it is important to understand the theoretical constructs that communication researchers use to guide their development of messages and to understand audience receptivity. Although there are a wide range of theoretical frameworks used in communication research, four theories/frameworks are prioritized in this next section. Their selection was based on their relevance to environmental health message design, implementation, and evaluation, as well as their potential to contribute to research related to environmental health literacy.

**Extended Parallel Process Model** The Extended Parallel Process Model (EPPM; Witte 1992), evolved from Leventhal's danger/fear control work and Rogers' Protection Motivation Theory to explain how individuals process fear appeal messages and how that processing influences danger and fear control processes (Leventhal 1971; Rogers 1995). Fear appeals are persuasive messages designed to scare people by describing through words and/or visuals all of the awful things that will happen if they don't engage in the recommendations of the message (Witte 1992). The EPPM specifies under what conditions fear appeals will work and when they will not. Specifically, the EPPM suggests people respond to threatening messages in different ways. Individuals who perceive a threat and think they have the ability to reduce or eliminate that threat will engage in danger control (adaptive) processes. Contrarily, individuals who perceive a threat, but do not feel capable of taking preventative action will engage in fear control (maladaptive) processes. Finally, individuals who do not perceive any threat will have no response. There is debate over the use of fear appeals as a persuasive strategy as some would perceive that the use of positive affect is a more ethical approach. However, there has been a large body of research to suggest that individuals need to perceive a threat before they will consider taking any action (Maloney et al. 2011). The EPPM acknowledges that a threat needs to be balanced with a strong efficacy component to the messages so that individuals can act on the threat.

In the environmental health context, different environmental exposures have emerged over time as posing threats to human health. For example, once there is enough scientific evidence to deem a chemical as potentially harmful, efforts to communicate about the chemical as a risk factor follow. These efforts often take a fear approach as message designers frame the chemical as a "threat" and then provide the recommendation to avoid the chemical, perhaps providing specific strategies like reading labels, choosing different products, wearing gloves, or removing products from households. Messages that follow this type of pattern may unknowingly be using the fear approach specified in the EPPM, particularly if the threat uses words and visual cues to intensely characterize the threat as immediate, pervasive, and severe for those who are exposed. If the messages are created with a high enough threat and efficacy level, they have the greatest likelihood of success. In the BCERP, messages about specific risk factors did not engage the EPPM because messages pertained to puberty as a window of susceptibility for young girls and also because the threat of breast cancer is not immediate for this audience. Discussing the negative and potentially severe impacts of breast cancer was deemed inappropriate for messages designed for caregivers to share with their daughters; rather than induce fear in caregivers and young girls, focusing on lifelong risk reduction strategies was a more ethical and positive approach. A key point here is that implementing a single theoretical approach to all environmental health message design will be ineffective. A consideration of a full range of possibilities related to the intended audiences and their information needs is necessary to make informed decisions about theoretical message design approaches.

**The Heuristic Systematic Model** The heuristic systematic model (HSM) of persuasion (Chaiken et al. 1989) outlines conditions under which individuals cognitively process environmental health messages either deliberately, called systematic processing, or using surface-level cues unrelated to the content of the message itself, known as heuristic processing. When a message is processed systematically, its information is scrutinized for its validity and relevance to pre-existing knowledge. This makes messages more easily recalled and inspires thoughts relevant to the message's topic (Chaiken and Ledgerwood 2011; Zuckerman and Chaiken 1998). On the other hand, when a message is processed heuristically, only a limited amount of cognitive resources is applied through simple decision rules like if the message's source is expert or if the viewpoint expressed is popular (Eagly and Chaiken 1993). In this case, past experiences aid in making rules that provide fast inference-making to form "good enough" judgements (such as credible sources being believable) (Chaiken et al. 1989).

An environmental health message will be systematically processed when a reader believes they are not confident they can make accurate judgments (Chaiken et al. 1989). So, to motivate individuals to more extensively process a message and retain its information, they must perceive they need more information and be presented with a message that clearly indicates it contains that kind of information. However, the level of knowledge that one perceives is needed is reliant on an individual's motivation and ability. Motivation to process a message generally pertains to the involvement felt in environmental health issues and/or how important environmental health is to oneself (see Cho and Boster 2005). If it is believed the environment plays little role in one's health or someone thinks it is not worth their time to think about environmental health issues, it is unlikely an environmental health message will be closely attended to. Ability to process information operates similarly and is particularly relevant to environmental health literacy. If an individual simply cannot understand an environmental health message, perhaps due to limited time or literacy, they will simply not be able to process the message systematically (Smith et al. 2017). Heuristic cues like the message source's credibility or attractiveness will drive the evaluation of the environmental health message and information will likely not be retained. So, an audience with limited environmental health literacy may automatically rely on surface-level cues of environmental health messages or may not even be motivated to allocate cognitive resources to scrutinize such messages.

It is worth noting that little research on the HSM focuses on the idea of having sufficient information and the explicit role it plays in information processing (Nabi and Moyer-Gusé 2013). According to the original conceptualization of the HSM (Eagly and Chaiken 1993), individuals process environmental health information systematically only until they feel they have enough environmental health information. After "sufficiency" is reached, individuals processing environmental health information would then use less cognitively-taxing processing strategies like relying on expertise and popular viewpoints. Most research bypasses the idea of sufficiency, however, and instead focuses only on the role of motivation and ability on processing strategy. In the future, research on environmental health information processing should refine measurements of information sufficiency and determine more precise relationships between audience or message factors and information processing strategies.

**The Theory of Normative Social Behavior** One way that audience members may make decisions about information processing or in taking cues about how to act is by referencing group norms. The Theory of Normative Social Behavior (TNSB) delineates how perceptions of the thoughts and actions of close, referent others often shape one's own behavior (Rimal and Real 2005), especially in the context of the environment and health. Leveraging the perception of a behavior's frequency/popularity (descriptive norms) and the perceived (dis)approval that comes with enacting a behavior (injunctive norms), social norms research has shown consistent influence of norms on behaviors such as reduction of littering (e.g., Cialdini et al. 1990), eating habits (e.g., Rah et al. 2004), hand washing (e.g., Lapinski et al. 2013), and water conservation (Lapinski et al. 2007).

The TNSB predicts that one's perception of a behavior's descriptive norm directly influences whether one will intend to also perform that behavior. Descriptive norms provide information about what is socially adaptive in a given context (Reno et al. 1993) and act as a heuristic cue (Cialdini et al. 1990). This is the case, however, only when a favorable injunctive norm is perceived, such that people close to an individual are perceived to approve of doing that behavior. So, an individual will be more likely to perform an environmental risk reduction behavior when important people in their social circle are thought to already do it and when those same close others are thought to approve of also doing that behavior. The TNSB also predicts that an environmental health behavior's descriptive norm will exert influence on whether an individual will also perform that behavior when it is perceived to provide the benefits one seeks (outcome expectancies) (Bandura 1986), when the individual feels closely tied to their social network (group identity) (Wilder 1990), and when the environmental health behavior or issue is closely tied to one's sense of self (ego involvement) (Johnson and Eagly 1989). Environmental health messages can leverage social norms by first conducting formative research to understand the behavior's popularity and, then, to communicate social approval information around the behavior (i.e., people who the audience member cares about think it is a good idea for them to adopt the risk reduction behavior). These messages can also explain how the behavior will help an individual achieve what is important to them, and highlight the ways the issue related to that behavior is central to the individual's sense of self.

To date, research has supported the role of outcome expectations on the norm–behavioral intention relationship (e.g., Rimal 2008; Rimal et al. 2011), as has research regarding the role of group identity (e.g., Glynn et al. 2009; Louis et al. 2007). Little research, though, has examined the role of involvement. There is some recent evidence to support the role of the multifaceted construct of involvement (Lapinski et al. 2015), but more research is needed to explicate how norms, characteristics of the behavior, and involvement interact. This is especially the case since involvement is comprised of the centrality of a given issue or behavior to one's self-concept (value-relevant involvement), the importance of enacting a behavior in order to be accepted socially (impression-relevant involvement), and, the original conception of involvement, i.e., how central an issue or behavior is to one's own goal-attainment (outcome relevant involvement) (Cho and Boster 2005).

The consideration of social norms is especially relevant when considering lay individuals with varied levels of environmental health literacy. Given that social norms, especially descriptive norms, can be used as a simple decision cue (Rimal and Real 2005), they lend themselves well to interventions targeting individuals with a low ability to consider nuanced and complex environmental health information, as noted with the HSM, previously. Environmental health message designers can employ the TNSB by identifying the prevalence of the specific behavior, like reducing exposure to an environmental risk, within a target audience while also gauging relative approval and expectations around that behavior. For example, an environmental health message could state that most women support reducing exposure to perfluorooctanoic acid (PFOA) in household products and you and your family would also benefit from doing so too. Emphasizing a behavior's popularity and the social approval that comes with it, as well as how the behavior will help achieve one's goals and be in-line with their sense of self, environmental health message designers can potentially promote environmental health behaviors, regardless of literacy demands.

**Diffusion of Innovations** A final framework that also leverages the influence intrinsic to one's social network is the diffusion of innovations (Rogers 2003), which outlines a variety of factors to explain how information, behaviors, programs, and products spread or get adopted across social groups. This has been applied to many contexts, including HIV prevention (e.g., Collins et al. 2006; Dearing et al. 2013), public policy (e.g., Boushey 2012), tobacco control (Studlar 1999), and physical activity programs (Dearing et al. 2006). Overall, the success of adopting the diffusion approach to environmental health messages and programs is in the segmentation of an audience into different categories of readiness to adopt a new, innovative environmental health behavior. These segments make up a larger social system for the innovation to spread across (Dearing 2008) and choose whether to adopt an innovation for a number of social and individual reasons. Diffusion of innovations also leverages opinion leaders, or certain individuals who assume a role in their social network(s) that comes with increased influence in the spread of a new behavior or product (Dearing 2008). The diffusion paradigm offers environmental health research an efficient framework such that a behavior can be specially targeted to a small subset of potential, influential adopters who will in turn influence a majority within their social system (Dearing 2008), and facilitate wide-spread adoption of the behavior.

An audience differs with regards to how innovative they are and can be segmented based on how early in the diffusion process they will adopt an innovation (Dearing 2008). Audiences are generally segmented into those who are the earliest to adopt an innovation based on its novelty and their own perception of limited risk (innovators), those who carefully appraise the innovation along numerous dimensions before choosing of whether to adopt it (early adopters), and a subsequent large majority, who adopt an innovation gradually as early adopters exert social, normative pressure. As noted, interventions leveraging this paradigm target opinion leaders, who are early adopters who tend to be highly influential and can become conduits of specialized, innovation-related knowledge. But, opinion leaders and other early adopters decide to adopt an innovation only when they positively appraise it along a number of attributes. These include the innovation's complexity, compatibility, observability, and trialability (Wejnert 2002).

Complexity refers to how difficult the behavior is to understand or do, such that if a risk reduction behavior involves many steps or is completely novel, it is likely to be seen as overly complex and will not be adopted. But, if it is simple, like purchasing a readily available product or by changing how often pesticides are applied, the innovation will appear simple and adoption is likely. It is this dimension that most readily applies to environmental health literacy, with a focus on reducing the perceived complexity of information and overcoming cognitive barriers to understanding (Finn & O'Fallon, this volume). Another related attribute that is appraised prior to adoption of the innovation is whether it is compatible with one's lifestyle. If an environmental health risk reduction behavior is atypical or otherwise opposed to an individual's normal, established pattern of behavior, it is not likely to be adopted. Further, an innovation should be observable such that it can easily be seen prior to the decision to adopt is made. This means that if an individual can physically witness an environmental health risk reduction behavior being modeled by others or can readily observe its positive consequences, then it will more likely be adopted. Finally, an innovation is trialable when it can easily be implemented without full commitment on the part of the adopter. This means that a risk reduction behavior that can easily be given up after a short-term "trial period" is one that will be more likely adopted. By being able to test out a behavior prior to a larger commitment, uncertainty and other barriers that interfere with behavioral adoption can be overcame.

Once positive appraisal of the environmental health innovation is established among the early adopters, members of the "late majority" will begin to pay attention to it and seek additional information from opinion leaders. This is to reduce the uncertainty created from learning about some unknown innovation and to learn if it is advantageous to how things are currently done (Dearing 2009; Wejnert 2002). Thus, an innovation is diffused through a process of learning about the innovation (awareness), feeling social pressure either directly from an opinion leader's advocacy or by perceiving a normative pressure to conform to what is popular (persuasion), deciding whether or not to adopt the innovation (decision), actually utilizing the innovation (implementation), and making the innovation a sustainable activity or perception that will be upheld over time (continuation). This means that an environmental health intervention using the diffusion paradigm should provide messaging that will promote the social influence activities of opinion leaders. Opinion leaders will then market the environmental health innovation to be seen as something that inspires curiosity and thus a drive to seek out information on it. Then, the innovation may reach a critical mass within the audience and be adopted via social influence and normative forces. Overall, the environmental health innovation should be easy to understand, should fit into the audience's current lifestyle, should be readily observable, and should be low-commitment in the short-term. Environmental health innovations that are seen as otherwise will likely not be adopted and thus will not diffuse across an audience. As such, the designers of environmental health messages should consider these attributes in the audience of interest and to what extent they should be represented in messages put forth related to environmental health risks.

One key limitation of research on diffusion and a venue for future research lies in the leveraging of opinion leaders. Specifically, diffusion relies on the classic mechanism studied in communication of uncertainty reduction (Dearing 2009), and assumes that the perception of an unfamiliar innovation will inspire uncertainty. Following assumptions outlined in uncertainty reduction approaches to communication (Hogan and Brashers 2009), potential adopters are fundamentally opposed to feeling any type or amount of uncertainty and will seek information to reduce it, particularly from opinion leaders who are seen as being familiar with the innovation. However, prominent communication theories (i.e., Afifi and Weiner 2004; Brashers et al. 2000) suggest that uncertainty is not necessarily maladaptive and that, in certain circumstances, individuals may actually desire to be uncertain. For instance, knowing exactly how many dangerous chemicals one is routinely exposed to and extensive details on the negative impact on one's health that exposure may result in is likely to create discomfort. In fact, one may actually wish to be blissfully ignorant, thus maintaining uncertain about such a topic. In this case, that individual may avoid new information related to those environmental risks or even find disconfirming information to question the certainty of a particular environmental health message. The question remains, then, of how does this impact the diffusion process? More research is needed to determine how these uncertainty management approaches may combine with diffusion and in what cases might they actually undermine the diffusion process.

## Processes for Designing Effective Messages

The aforementioned theories help to increase the likelihood that an environmental health message will be effective by providing predictive empirical relationships between theoretical constructs and a framework for considering the uptake of an innovation. Communication research provides evidence that incorporating relevant theory as well as tailoring messages to audience needs and preferences, and creating messages that are engaging in their content and design will maximize the potential for desired message outcomes. Evaluating the effectiveness of prosocial messages – those messages that aim to improve or benefit others, health outcomes, and/or environmental outcomes – is a crucial component that is rarely assessed. While advertisers and businesses strive to evaluate the effectiveness of their messages through measuring public exposure to messages and product sales, the impact of other types of messages such as environmental health messages remain largely unknown as assessment efforts are minimal to nonexistent. Especially for groups who want to demonstrate the impact of their environmental health messages or campaigns, using a rigorous approach to develop messages will aid their ability to engage in evaluation of effects and effectiveness. Formative research, process evaluation, and summative evaluation techniques are three important components that provide a comprehensive approach to the development and assessment of environmental health messages.

**Formative Research for Message Design** Before creating message concepts, it is imperative for all message designers to engage in formative research so they understand audience characteristics, needs, and preferences. Formative research can take the form of comprehensive literature reviews of relevant research on potential audience segments as well as data collection efforts to better understand barriers, facilitators, and preferences of potential audience segments. Formative research should help to identify preferred channels for audiences as well as credible sources they turn to for information about the topic of interest. The mode of communication selected is important as there is a tendency for groups and organizations to develop websites and then point people to them as a key strategy for communicating about health, environment, and science topics with lay audiences, whether or not the audiences are apt to refer to the websites. This tendency is not ideal because audiences are inundated with information and websites are often static with few updates to maintain their interest in the content. While people do seek environmental health information on websites every day, websites that are not well-managed will not be accessed frequently by users.

Identifying credible sources for the information is important because of distrust felt sometimes within communities toward figures of authority, government sources, and other organizations and individuals who are perceived to be less credible. For example, before developing messages for the BCERP, communication researchers and advocates engaged in focus group research to better understand parents' and caregivers' knowledge, efficacy, intentions, and communication preferences related to environmental risk factors and breast cancer. Focus groups were conducted by community advocates across different regions of the United States so that outreach and engagement personnel could tailor their messages according to local audience needs. Communication preferences asked about information sources, preferred channels for receiving environmental health messages, and other questions about how the information should be visualized. It is also important to note that the questions asked during in the BCERP focus groups were based on theoretical constructs relevant to factors that can make an environmental health message more or less effective. Some of these factors were described earlier in this chapter (see Theoretical Frameworks, pages 47–52) as relevant theoretical frameworks that address how audiences process messages. All of the research activities completed before any message content is written or designed, such as focus groups or other methodologies that capture audience perceptions and communication preferences, is considered pre-production formative research; activities done to create and test messages are considered production research.

Production research is where science and researcher creativity begin to work synergistically for maximizing environmental health message effects. The use of theory in the message design processes is fundamental to a communication science approach. At the production stage, environmental health message designers may decide they want to use a fear appeal approach to increase threat perceptions, or an inoculation approach to help prepare audiences for "attacks" on their belief systems, or some other theoretical framework that provides structure for creating impactful messages. As noted in the previous section, there are multiple theories for consideration when designing messages and for understanding how audiences may process those messages. The production stage is when environmental health message concepts are created, tested, refined, and ideally piloted again to garner audience response data. Response data could again be collected via focus groups, but also more generalizable data could be collected from online experimental message exposures or theatre-style research where an audience is exposed to the would-be message. The goal of the production phase is to finalize environmental health messages with this additional input so they are refined and polished for dissemination. Silk et al. (2014) provides results from collaborative formative research that tests breast cancer risk reduction messages on mothers with daughters. Research from the BCERP (which is available at bcerp.org) has resulted in a range of articles that illustrate communication approaches that can improve the effectiveness of reaching audiences with high quality environmental health messages.

**Process Evaluation During Message Implementation** Once audiences have been defined and environmental health messages have been created, tested, and revised, a communication plan should be developed to disseminate messages to audiences. A communication plan specifies measurable objectives and organizes communication efforts by identifying a timeline for the dissemination of messages. It includes details on what messages will go to what audiences through which channels at what time and with what frequency. When creating measurable objectives, environmental health message designers need to define if the goal of the messages are to increase awareness, knowledge, efficacy, intentions, or to change specific behaviors. If a campaign has unclear or ambiguous objectives, it is difficult to engage in process and summative evaluation procedures. Unfortunately, although a communication plan is a strategic document that is typically used for larger campaign efforts, it is seldom created and used in small-scale community based efforts that involve few messages and channels. The lack of a communication plan in these instances is a mistake because a communication plan is a document that will explicate communication objectives and specifics that provide a baseline for evaluation of communication efforts. Specifically, process evaluation evaluates how closely actual implementation and dissemination efforts have aligned with the communication plan; thus, without a communication plan with clear, measurable objectives it is impossible to identify whether or not dissemination has gone as planned.

Process evaluation is important because it allows for an understanding of why particular communication efforts are effective, and it allows for changes in implementation in the case that efforts are not going as planned – and often, not everything goes as planned. Ultimately, process evaluation is the monitoring of implementation of designated communication activities and it is linked to one's ability to demonstrate message effects and effectiveness.

**Summative Evaluation for Effects and Effectiveness** At the end of communication efforts or campaigns, it is essential to determine whether exposure to environmental health messages has an impact. Summative evaluation generally refers to the determination of intended and unintended effects as well as the determination of effectiveness. Effects refer to any impacts traced back to communication efforts even if those effects are not illustrative of campaign objectives. For example, in designated driver campaigns, messages were successful in increasing the use of designated drivers, which was an intended outcome that also demonstrated effectiveness based on campaign objectives. However, an unintended effect was that those individuals who were not designated drivers actually drank more than they normally would because they knew they did not need to drive; thus, an unintended effect was an increase in total alcohol consumption among those exposed to the campaign.

Methods for capturing effects require research designs that ideally include control conditions, which is sometimes difficult to do. Generally, a pre-test/post-test design where one samples a population before exposure to environmental health messages and then samples the same population after exposure is a minimum approach to be able to attribute any effects to the exposure. Another strategy is to find reliable baseline data prior to the launch of any communication efforts and then compare those data to post-test data collected after the communication efforts. Ideally, a control group in the form of an equivalent community or matched geographic area can be identified as a comparison group. There are more sophisticated research designs to measure effects but these require resources that are usually not designated for the evaluation of environmental health communication efforts.

## Research and Outreach

Ample communication research in the realm of environmental health behavior and environmental health literacy has taken place through the BCERP, a transdisciplinary collaboration funded through the National Institutes of Environmental Health Sciences and the National Cancer Institute. In the BCERP, communication scientists use theories and methodologies such as the ones previously outlined to partner with community advocates, biologists, and epidemiologists to understand the role of environmental factors – such as chemicals like bisphenol-A (BPA) and PFOAin breast cancer development with a particular focus on exposure during the "window of susceptibility" of puberty. As the scientific evidence about these risks grow, BCERP communications researchers work to understand the perceptions and information needs of women across the United States related to breast cancer and the environment in order to create more effective environmental health messages. This formative research then informs strategies for message translation and dissemination for the findings gleaned from projects lead by BCERP biologists and epidemiologists to promote knowledge gain and risk reduction of potentially harmful environmental exposures. A primary target population for BCERP activities and research is parents and caregivers of pre-pubertal daughters in order to motivate the avoidance of environmental risk exposures for said daughters, as parents are identified as these young girls' primary decision-makers. Thus, the issue of environmental health literacy is intrinsic to the research questions examined by BCERP researchers and the challenges associated with the concept have manifested themselves across BCERP studies.

The most salient issue concerning environmental health literacy in communicating to this population is the consequences of literacy on response to environmental health messages. While complicated messages with more demanding literacy levels likely prevent individuals of lower environmental health literacy from comprehending the message and subsequently acting on any recommendations, evidence suggests that tailoring such messages to lower literacy levels helps everyone. In two studies by BCERP researchers, it was found that messages written at a lower health literacy level promoted knowledge gain of various chemical exposures and breast cancer risk, regardless of participant literacy level (Smith et al. 2013; Smith et al. 2017). This finding demonstrates that considerations of message literacy demand can lead to greater retention of information, such that messages written at less cognitively-demanding levels can promote more extensive processing of messages, per the HSM (see above; Chaiken et al. 1989). These studies demonstrate that while greater literacy promoted message processing, tailoring messages to those with lower ability to understand breast cancer and breast cancer risks benefited all individuals, regardless of their individual ability. However, while this seems promising for environmental health literacy research and message implications, other BCERP research complicates message recommendations. In another study examining the effect of message literacy level on women of different ability, contrary results were found such that message literacy level may not facilitate knowledge gain as expected.

Specifically, more accessible messages about a chemical exposure did not increase belief in the chemical being a risk nor attitude toward the chemical (Hitt et al. 2015). This inconsistency in results indicates there is still work needed in the scholarship of environmental health literacy and more care is needed around messaging for even more commonly studied risk exposures.

Yet, there are other issues related to environmental health literacy that have been addressed by BCERP researchers that are not directly related to the processing and response to environmental health messages. Specifically, studies have identified complex cognitions around breast cancer and the environment that complicate message processing and risk reduction behavior, in addition to examining the prevalence of breast cancer and environment information in women's memory and in their greater informational environment. First, evidence from focus groups and surveys suggests that women are, on the whole, uncertain about or confused with information linking environmental exposures to breast cancer (Neuberger et al. 2011; Silk et al. 2006; Volkman and Silk 2008; Yun et al. 2009). This uncertainty and ambivalence regarding the role of the environment in breast cancer risk likely reduces the likelihood of women extensively processing and considering relevant environmental health risk information, per the HSM (Chaiken et al. 1989). A consideration of environmental health literacy therefore must include a consideration of how diminished literacy affects appraisals of uncertainty or ambivalence, as these can, at best, prevent adequate processing of environmental health recommendations or, at worst, undermine them and prevent action.

If this uncertainty is overcome, though, the HSM predicts effortful processing of messages involves relating new information in the message back on to one's knowledge and experiences. One way to assess the information women may refer to in their memory is by analyzing influential messages from one's past that can guide behavior, if recalled, known as "memorable messages" (Smith and Ellis 2001; see Knapp et al. 1981). BCERP researchers examined some of these accessible messages that related to environmental breast cancer risks in terms of the sources of those messages, whether the focus of said messages are on detection or prevention of risk-related morbidity, and other factors that likely affect and are affected by one's EHL. These studies found that the proportion of messages that were prevention-focused were significantly less prominent than awareness-, detection-, or treatment-focused messages (Smith et al. 2009a, b), showing that there remains a strong need for greater focus on prevention/risk reduction activities. Further, negative emotions were frequently evoked from both prevention and detection messages (Smith et al. 2010), and messages from healthcare professionals were associated with increased likelihood of enacting detection rather than prevention behaviors (Smith et al. 2009b).

Given the nature of messages held in women's memory regarding breast cancer and the environment, it is not surprising that BCERP studies suggest action like seeking out information connecting breast cancer to the environment or taking action to reduce one's daughter's exposure to environmental breast cancer risks is not likely. According to the EPPM (see above; Maloney et al. 2011), action would likely be taken if a mother perceives environmental risks to be serious and to have a substantial effect on breast cancer development, as well as to feel like risk reduction is possible and effective. Without a perception of threat, no impetus to even consider action exists, and without appraisals of efficacy, behavioral response will either fail or never happen. BCERP studies suggest that when women consider breast cancer and environment factors, they are pre-occupied with the negative consequences of risk reduction and disease detection behaviors, like the pain involved in mammography (Silk et al. 2006; Silk et al. 2014; Volkman and Silk 2008. Health literacy has been associated with diminished efficacy in the past (i.e., Torres and Marks 2009), so it is reasonable to believe that inaction in the face of environmental risks to breast cancer is in part due to a perceived lack of efficacy for a risk that may occur decades in the future and, in part, as a result of limited environmental health literacy. As a lack of knowledge is perceived in this population (Silk et al. 2006; Volkman and Silk 2008), leveraging environmental health literacy in support of more favorable efficacy appraisals is one way to promote the reduction of exposure to environmental risks.

Assuming ample efficacy to take action does indeed exist within this population and women choose to seek additional information on reducing the risk of breast cancer through avoidance of environmental risks, there is evidence to suggest that information available to them is both unsupportive of immediate actions and at odds with their potentially limited environmental health literacy. For instance, current and past research has documented the extent to which news publications and easily accessible webpages contain information related to breast cancer and its link to the environment. Early studies show a large amount of breast cancer-related information is available, but little strategic behavior change motivators are present. Specifically, prevention-based news stories were found to focus narrowly on the use of pharmaceuticals (Atkin et al. 2008), while websites with breast cancer information were void of strategic behavior change motivators like those outlined in the EPPM (Clark-Hitt et al. 2010; Whitten et al. 2008). Because of this, these presentations are less likely to prompt action among women or their daughters. Further, another analysis found that breast cancer websites also seldom tailored content based on diversity elements or literacy levels, overly relying on statistics and numeric information that is difficult for low literacy individuals to understand (Whitten et al. 2011). While a wealth of information can be found about breast cancer in news media and on the Internet, more attention to motivational cues is needed in conveying that information. In fact, preliminary evidence suggests that this trend of limited motivational content continues for websites even in the contemporary informational landscape (Totzkay et al. 2017). This demonstrates that modern advances in information access has not led to the provision of comprehensive information regarding the environment and breast cancer, and many factors studied by BCERP biologists and epidemiologists for their contribution to breast cancer morbidity are not found on webpages most likely to be accessed by those seeking breast cancer information via popular Internet search engines. This all has a wide variety of implications for those of varied environmental health literacy levels, as the absence of information, both in general and specifically tailored to lay audiences, likely contributes to further misunderstanding of the importance of the environment to health.

Lastly, a variety of studies have measured the effect and perceptions of environmental health messages regarding breast cancer. Evidence suggests that mothers, acting somewhat as opinion leaders per the diffusion paradigm (see above; Rogers 2003), are ambivalent regarding the role of the environment in breast cancer development (Neuberger et al. 2011; Yun et al. 2009), despite feeling capable of providing a healthy lifestyle for their daughters. This is concerning from an environmental health literacy perspective since it implies that knowledge of the effects of environmental factors on one's health is missing, perhaps due to how complex presentations of this information is. It could be that the information either does not exist in one's informational ecosystem (e.g., Totzkay et al. 2017), or it is just not presented in effective ways (Clark-Hitt et al. 2010; Whitten et al. 2008; Whitten et al. 2011). It could also be that the environmental health information around breast cancer is overly vague or presented in abstract language, as BCERP research has indicated that messages presented like this were less likely to prompt action than recommendations using specific and direct language (i.e., Silk et al. 2014). It appears that mothers are not adopting the innovation of concern for environmental risks to breast cancer development due to an unfavorable appraisal of its innovation characteristic (Dearing 2008; Wejnert 2002). In considering environmental health literacy, the attribute being weighed here is likely complexity, in that thinking about the abstract nature of environmental risk exposure and the ways it can promote breast cancer development later in life is difficult to think about and some risk reduction activities may be difficult to implement. As a result, mothers likely do not favorably appraise the innovation of concern for the environment in breast cancer development and, thus, dismiss it. If favorable appraisals do occur, however, there is some evidence that mothers can view this type of environmental health information as useful and may promote its dissemination to their personal networks. In fact, combining much of the research outlined here into understandable, accurate, and easy-to-use magazine-style ads for breast cancer risk and the role of the environment resulted in perceptions of credibility and relevance, in addition to that intention to spread the information (Silk et al. 2014).

## Future Directions

Communication science is novel to many in the environmental health sciences. However, there are strong possibilities for the discipline to help further translational activities. Communication scientists are natural partners for facilitating increased public understanding of science, including environmental health issues, because of their audience-centered approach and understanding of theoretically designed messages. For example, communication scientists are well-situated to engage in implementation science and to investigate the effectiveness of different partnership models and their impact on scientific innovation and outcomes. There are also opportunities to investigate social media and social networks to better leverage them for scaling-up of environmental health communication efforts. The computational opportunities associated with social media data scraping could provide algorithms for leveraging social media for environmental health issues. Educational gaming and immersive media applications (see Chapter 10) also provide an opportunity to engage individuals about environmental health issues, which could increase their scientific literacy as well as increase their motivation to engage in recommended practices as they deal with environmental risk factors and other challenges. Another area where there is great promise is at the neurocognitive level; specifically, investigating how the brain responds to environmental health messages will provide new insights for message designers. As researchers continue to gather neuro response data, new theories will be built that can inform best practices and approaches to building effective environmental health messages.

## Conclusion

In recent years, there has been a stronger focus on the need for environmental health scientists to translate and communicate their findings for lay audiences. To meet this need, different partnerships with community partners, advocates, and communication scientists have evolved as demonstrated by large programs of research like the BCERP. While community partners and advocates are outstanding partners with trusted relationships with their stakeholder communities, communication scientists also play a crucial role in addressing communication needs. The value of communication science to environmental health risk message design is that it provides a systematic approach that prioritizes theoretically informed messages, audiences, measurable objectives, and rigorous evaluation of message effects.

This chapter emphasizes the EPPM, HSM, TNSB, and DOI as four theoretical frameworks that have particular utility for environmental health messages. The EPPM is a fear appeal theory that employs threat and efficacy as key message components, while the HSM focuses on how people process information. The TNSB emphasizes the role of norms in influencing behavior. All three of these theories provide useful insight by informing environmental health message designers about information and message strategies that might be influential for specifically defined audiences. In other words, these theories provide a roadmap for what to emphasize in environmental health messages. The fourth theory, DOI, also provides insight into how information, behaviors, programs, and products spread or get adopted across social groups. This DOI is helpful in the environmental health context because adoption of a new idea or a risk reduction behavior is a frequent goal of communication efforts, and facilitating the adoption of an innovation is challenging. The theories presented in this chapter are certainly not comprehensive of what is available for environmental health and risk communication efforts, but they do provide a sampling of available theories and some key constructs that are fundamental to message design efforts. Readers interested in a comprehensive review of health and risk communication related theories should refer to The Oxford Encyclopedia of Health and Risk Message Design and Processing series (Parrott 2018).

This chapter also presents the overall process for developing effective messages using examples from the BCERP. Methods that are important for success are formative research efforts to learn about the audience for message development purposes, process evaluation to assess how the messages have been communicated during campaign or community dissemination efforts, and summative evaluation to assess effects and effectiveness of communication efforts. It is important to emphasize that engaging in formative research and evaluation processes are best practices in communication science, but resources are not always available or designated to engage in these processes. Prioritizing these best practices through budgeting for these processes is a gap that needs to be bridged by environmental health programs because higher quality messages are a more likely outcome when these processes are followed. For example, communication scientists in the BCERP have partnered with other scientists and advocates to engage in focus group research, message testing, and collaborative projects across multiple geographic areas to understand audience needs and preferences so that messages are sensitive to culture and literacy levels. These efforts demonstrate the role that a communication science approach can have in contributing to environmental health literacy efforts.

## References


Afifi, W. A., & Weiner, J. L. (2004). Toward a theory of motivated information management. *Communication Theory, 14*(2), 167–190.

Atkin, C. K., Smith, S. W., Ferguson, V., & McFeters, C. (2008). A comprehensive analysis of breast cancer news coverage in leading media outlets. *Journal of Health Communication, 13*, 3–17.

Bandura, A. (1986). *Social foundations of thought and action: A social cognitive theory.* Englewood Cliffs: Prentice Hall.

Boushey, G. (2012). Punctuated equilibrium theory and the diffusion of innovations. *The Policy Studies Journal, 40*(1), 127–146.

Brashers, D. E., Neidig, J. L., Haas, S. M., Dobbs, L. K., Cardillo, L. M., & Russell, J. A. (2000). Communication in the management of uncertainty: The case of persons living with HIV or AIDS. *Communication Monographs, 67*, 63–84.

Cialdini, R. B., Reno, R. R., & Kallgren, C. A. (1990). A focus theory of normative conduct: Recycling the concept of norms to reduce littering in public places. *Journal of Personality and Social Psychology, 58*, 1015–1026.

Clark-Hitt, R., Dean, M., Smith, S., Nazione, S., Atkin, C. K., & Whitten, P. (2010). Relative frequency of breast cancer website information topics: Environmental risk, prevention, detection, treatment, awareness, social support, and survivorship. *Journal of Health and Mass Communication, 2*, 36–55.

Collins, C., Harshbarger, C., Sawyer, R., & Hamdallah, M. (2006). The diffusion of effective behavioral interventions project: Development, implementation, and lessons learned. *AIDS Education and Prevention, 18*(Supplement A), 5–20.

Chaiken, S., Liberman, A., & Eagly, A. (1989). Heuristic and systematic information processing within and beyond the persuasion context. In J. S. Uleman & J. A. Bargh (Eds.), Unintended thought (pp. 212–252). New York: Guilford Press.

Chaiken, S., & Ledgerwood, A. (2011). A theory of heuristic and systematic information processing. In P. A. M. Van Lange, A. W. Kruglanski, & E. T. Higgins (Eds.), Handbook of theories of social psychology: Volume one (pp. 246–266). Thousand Oaks, CA: SAGE

Cho, H., & Boster, F. J. (2005). Development and validation of value-, outcome-, and impression-relevant involvement scales. *Communication Research, 32*, 235–264. https://doi.org/10.1037/t08081-000

Dearing, J. W. (2008). Evolution of diffusion and dissemination theory. *Journal of Public Health Management and Practice, 14*(2), 99–108. https://doi.org/10.1097/01.phh.0000311886.98627.b7

Dearing, J. W. (2009). Applying diffusion of innovation theory to intervention development. *Research on Social Work Practice, 19*(5), 503–518.

Dearing, J. W., Maibach, E. W., & Buller, D. B. (2006). A convergent diffusion and social marketing approach for disseminating proven approaches to physical activity promotion. *American Journal of Preventive Medicine, 31*(4S), S11–S23.

Dearing, J. W., Smith, D. K., Larson, R. S., & Estabrooks, C. A. (2013). Designing for diffusion of a biomedical intervention. *American Journal of Preventive Medicine, 44*(1S2), S70–S76.

Eagly, A. H., & Chaiken, S. (1993). The psychology of attitudes. San Diego, CA: Harcourt Brace

Glynn, C. J., Huge, M. E., & Lunney, C. A. (2009). The influence of perceived social norms on college students' intention to vote. *Political Communication, 26*, 48–64. https://doi.org/10.1080/10584600802622860

Hitt, R., Perrault, E., Smiths, S., Keating, D., Nazione, S., Silk, K., & Russell, J. (2015). Scientific message translation and the heuristic systematic model: Insights for designing educational messages about progesterone and breast cancer risks. *Journal Cancer Education*, 1–8.

Hogan, T. P., & Brashers, D. E. (2009). The theory of communication and uncertainty management: Implication from the wider realm of information behavior. In T. D. Afifi & W. A. Afifi (Eds.), *Uncertainty, information management, and disclosure decisions: Theories and applications*. New York: Routledge.

Johnson, B. T., & Eagly, A. H. (1989). The effects of involvement on persuasion: A meta-analysis. *Psychological Bulletin, 106*, 290–314.

Knapp, M. L., Stohl, C., & Reardon, K. K. (1981). "Memorable" messages. *Journal of Communication, 31*, 27–41.

Lapinski, M. K., Rimal, R. N., DeVries, R., & Lee, E. L. (2007). The role of group orientation and descriptive norms on water conservation attitudes and behaviors. *Health Communication Theory, 15*, 127–147.

Lapinski, M. K., Anderson, J., Shugart, A., & Rodd, E. (2013). Social influence in childcare centers: A test of the theory of normative social behavior. *Health Communication, 29*(3), 219–232.

Lapinski, M. K., Zhuang, J., Koh, H., & Shi, J. (2015). Descriptive norms and involvement in health and environmental behaviors. *Communication Research, 44*(3), 367–387. https://doi.org/10.1177/0093650215605153

Leventhal, H. (1971). Fear appeals and persuasion: the differentiation of a motivational construct. *American Journal of Public Health, 61*(6), 1208–1224. https://doi.org/10.2105/ajph.61.6.1208

Louis, W., Davies, S., Smith, J., & Terry, D. (2007). Pizza and pop and the student identity: The role of referent group norms in healthy and unhealthy eating. *Journal of Social Psychology, 147*, 57–74. https://doi.org/10.3200/socp.147.1.57-74

Maloney, E. K., Lapinski, M. K., & Witte, K. (2011). Fear appeals and persuasion: A review and update of the Extended Parallel Process Model. *Social & Personality Psychology Compass, 5*, 206–219. https://doi.org/10.1111/j.1751-9004.2011.00341.

Nabi, R. L., & Moyer-Gusé, E. (2013). The psychology underlying media-based persuasion. In K. E. Dill (Ed.), *The Oxford handbook of media psychology*. New York: Oxford University Press.

Neuberger, L., Silk, K. J., Yun, D., Bowman, N. D., & Anderson, J. (2011). Concern as motivation for protection: Identifying mothers' concerns about their daughters' breast cancer risk. *Journal of Health Communication, 16*, 1055–1071.

Parrott, R. L. (2018). *The Oxford encyclopedia of health and risk message design and processing* (Vol. 1–4). New York: Oxford University Press.

Rah, J. H., Hasler, C. M., Painter, J. E., & Chapman-Novakofski, K. M. (2004). Applying the theory of planned behavior to women's behavioral attitudes on and consumption of soy prod-

ucts. *Journal of Nutrition Education and Behavior, 36*(5), 238–244. https://doi.org/10.1016/s1499-4046(06)60386-2
Reno, R. R., Cialdini, R. B., & Kallgren, C. A. (1993). The transsituational influence of social norms. *Journal of Personality & Social Psychology, 64*, 104–112.
Rogers, E. M. (1995). *Diffusion of innovations* (5th ed.). New York: Free Press.
Rogers, E. M. (2003). Diffusion of innovations (5th ed.). New York, NY: Free Press.
Rimal, R. N., & Real, K. (2005). How behaviors are influenced by perceived norms: A test of the theory of normative social behavior. Communication Research, 32(3), 389-414. https://doi.org/10.1177/0093650205275385
Rimal, R. N. (2008). Modeling the relationship between descriptive norms and behaviors: A test and extension of the theory of normative social behavior (TNSB). Health Communication, 23, 103–116.https://doi.org/10.1080/10410230801967791
Rimal, R. N., Lapinski, M. K., Turner, M. M., & Smith, K. S. (2011). The attribute-centered approach for understanding health behaviors: How perceived benefits and similarity moderate the impact of descriptive norms on behaviors. *Journal of Health Communication, 10*, 433–450. https://doi.org/10.1080/10810730591009880
Silk, K. J., Bigsby, E., Volkman, J., Kingsley, C., Atkin, C. K., Ferrara, M., & Goins, L. A. (2006). Formative research on adolescent and adult risk perceptions for breast cancer. *Social Science and Medicine, 63*, 3124–3136.
Silk, K. J., Perrault, E. K., Neuberger, L., Rogers, A., Atkin, C., Barlow, J., & Duncan, D. M. (2014). Translating and testing breast cancer risk reduction messages for mothers of adolescent girls. *Journal of Health Communication, 19*, 226–243.
Smith, S. W., & Ellis, J. B. (2001). Memorable messages as guides to self-assessment of behavior: An initial investigation. *Communication Monographs, 68*, 154–168.
Smith, S. W., Atkin, C. K., Munday, S., Skubisz, C., & Stohl, C. (2009a). The impact of personal and/or close relationship experience on memorable messages about breast cancer and the perceived speech acts of the sender. *Journal of Cancer Education, 24*, 129–134.
Smith, S. W., Munday, S., LaPlante, C., Kotowski, M. R., Atkin, C. K., Skubisz, C., & Stohl, C. (2009b). Topics and sources of memorable breast cancer messages: Their impact on prevention and detection behaviors. *Journal of Health Communication, 14*, 293–307.
Smith, S. W., Hamel, L. M., Kotowski, M. R., Nazione, S., LaPlante, C., Atkin, C. K., Stohl, C., & Skubisz, C. (2010). Action tendency emotions associated with memorable breast cancer messages and prevention and detection behaviors. *Health Communication, 25*, 737–746.
Smith, S. W., Clark-Hitt, R., Nazione, S., Russell, J., Silk, K., & Atkin, C. K. (2013). The effects of motivation, ability, and heuristic cues on systematic processing of information about breast cancer environmental factors. *Journal of Health Communication, 7*, 845–865.
Smith, S. W., Hitt, R., Russell, J., Nazione, S., Silk, K., Atkin, C. K., & Keating, D. (2017). Risk belief and attitude formation from translated scientific messages about PFOA, an environmental risk associated with breast Cancer. *Health Communication, 32*(3), 279–287. https://doi.org/10.1080/10410236.2016.1138350.
Studlar, D. T. (1999). Diffusion of tobacco control in North America. *Annals of the American Academy of Political and Social Science, 566*, 68–79.
Torres, R. Y., & Marks, R. (2009). Relationships among health literacy, knowledge about hormone therapy, self-efficacy, and decision-making among postmenopausal health. *Journal of Health Communication, 14*(1), 43–55. https://doi.org/10.1080/10810730802592247.
Totzkay, D., Walling, B., Silk, K. J., Smith, S. W., Jackson, M., Jang, Y., & Park, S. (2017). *Content analysis of breast cancer websites for information on environmental risks.* Poster presented at the 4th Biennial D. C. Health Communication Conference.
Volkman, J., & Silk, K. J. (2008). Adolescent females and their mothers: Examining perceptions of the environment and breast cancer. *Journal of Health Psychology, 13*(8), 1180–1189 (ISI: 3.87).
Wejnert, B. (2002). Integrating models of diffusion of innovations: A conceptual framework. *Annual Review of Sociology, 28*, 297–326.

Whitten, P., Munday, S., LaPlante, C., & Smith, S. W. (2008). Communication assessment of the most frequented breast cancer websites: Evaluation of design and theoretical criteria. *Journal of Computer-Mediated Communication, 13*, 880–911.

Whitten, P., Nazione, S. A., Smith, S., & LaPlante, C. K. (2011). An examination into audience targeting and the use of storytelling or statistical evidence on breast cancer websites. *Patient Education and Counseling, 85*, e59–e64.

Wilder, D. A. (1990). Some determinants of the persuasive power of in-groups and out-groups: Organization of information and attributions of independence. *Journal of Personality and Social Psychology, 59*, 1202–1213.

Witte, K. (1992). Putting the fear back into fear appeals: The extended parallel process model. *Communication Monographs, 59*, 239–349. https://doi.org/10.1080/03637759209376276

Yun, D., Silk, K., Bowman, N., Neuberger, L., & Atkin, C. (2009). Mothers' intentions to teach adolescent daughters about breast cancer risk reduction activities: The influence of self-efficacy, response efficacy, & personal responsibility. *Communication Research Reports, 26*(2), 134–145.

Zuckerman, A., & Chaiken, S. (1998). A heuristic-systematic processing analysis of the effectiveness of product warning labels. *Psychology & Marketing (1986–1998), 15*(7), 621–642. https://doi.org/10.1002/(SICI)1520-6793(199810)15:7 <621::AID-MAR2>3.0.CO;2-H

# Part II
# Raising EHL in the Research Context among Diverse Audiences

# Chapter 4
# Engaging with Ethnically Diverse Community Groups

**Monica Ramirez-Andreotta**

**Abstract** Research has shown that in order to successfully address environmental health disparities, environmental health literacy (EHL) efforts need to be informed by the ethnically diverse communities they aim to serve. Understanding a community's perspectives and informational needs is crucial to conducting and improving environmental health research and EHL initiatives. Socioeconomic variables, linguistic isolation, and measures of political engagement are just some of the factors that effect the spatial distribution of environmental hazards. These variables create additional engagement and educational challenges that need that to be addressed. Efforts to engage ethnically diverse community groups should be rooted in environmental justice, ecosocial theory, self-efficacy building, and empowerment education. Methods that can be used to effectively raise EHL among ethnically diverse community groups include: informal science education, community-based participatory research, public participation in scientific research, traditional ecological knowledge, knowledge mediators, tailored and linguistically appropriate messaging, information design, and cultural models of risk communication. At end of this chapter, research examples are provided highlighting the variety of creative and transformative methods that can be applied to increase EHL in culturally diverse community groups.

**Keywords** Environmental justice · Ecosocial theory · Self-efficacy building · Empowerment education · Informal science education · Community-based participatory research · Citizen science · Traditional ecological knowledge · Knowledge mediators · Culturally-tailored messaging

---

M. Ramirez-Andreotta (✉)
Department of Soil, Water and Environmental Science,
College of Agriculture and Life Sciences, Tucson, AZ, USA
e-mail: mdramire@email.arizona.edu



S. Finn, L. R. O'Fallon (eds.), *Environmental Health Literacy*,
https://doi.org/10.1007/978-3-319-94108-0_4

## Background

To date, scholars have defined literacy in terms of science (National Research Council 1996a), health (Institute of Medicine 2004), critical health (Chinn 2011), public health (Gazmararian et al. 2005), and the environment (North American Association for Environmental Education 2015). Recently, efforts have sought to merge existing explanations and define EHL. Building on the definition proposed by the Society for Public Health Education, Finn and O'Fallon (2017) have defined EHL as "raising scientific literacy, environmental literacy, and numeracy among the general public while increasing awareness of specific exposures and their potential health effects". EHL efforts have spanned the fields of risk communication, education, and health literacy and at its core, is an understanding of the connection between environmental exposures and human health (Finn and O'Fallon 2017). However, health literacy efforts traditionally assess basic literacy skills such as reading, writing, and arithmetic, and are geared toward compliance with recommended clinical care. Though the aforementioned skills are critical, EHL should be viewed as a form of health action (personal, social, environmental) and focus on individual and/or community efforts regarding knowledge integration, informed choices, actions to improve personal and community health, and methods to reduce risk (Ramirez-Andreotta et al., 2016b, c). EHL efforts aim to educate communities and individuals about the connection between environmental exposures and human health. These efforts can range from informational materials explaining a specific contaminant and how it may affect an individual's health, local water and fish advisories, community mapping, and biomonitoring/exposure assessment data report back.

Unfortunately, environmental health disparities exist in our society and there are differences in health outcomes that are closely linked with disproportionate exposure to toxic substances and social, economic and environmental disadvantage (National Institute of Environmental Health Sciences 2015). Environmental factors such as air, water, soil, and food are fundamental determinants of our health and well-being. Just like access to green space and high-quality foods can improve human health, environmental factors can also lead to disease and health disparities when the environments where people live, work, learn and play are toxic, burdened by chemicals, and social inequities. These health inequalities are considered unnecessary, avoidable and unfair/unjust (Commission on Social Determinants of Health, World Health Organization 2008).

Research has shown that to address environmental health disparities, EHL efforts need to be informed by the ethnically diverse communities they aim to serve. This research has confirmed that understanding a community's perspectives and informational needs is crucial to conducting and improving environmental health research and EHL initiatives, especially in the context of contaminated sites. Across America, one in four Americans lives within 3 miles of a hazardous waste site (U.S. General Accounting Office 2013), meaning that one's zip code can be more important than their genetic code in terms of disease risks (Graham 2016).

Socioeconomic variables, along with others such as linguistic isolation and measures of political engagement, are additional factors in the spatial distribution of environmental hazards and these variables further exacerbate community engagement and education efforts.

In this chapter, the emphasis is on the importance of culture in the context of EHL. As effective research to reduce environmental health disparities have demonstrated, EHL efforts need to be informed by the ethnically diverse communities they aim to serve in order to be sustainable. One cannot attempt to promote or increase EHL without a manifold understanding of: (1) the culture, (2) local socioeconomic context and environmental exposures within the community and (3) individual health factors and susceptibility. Raising EHL in culturally diverse groups requires listening, culturally sensitive strategies and inclusion efforts. Promoting EHL in diverse communities is also an opportunity to increase educational efforts in vulnerable communities and build the self-efficacy of community members. Raising EHL in these diverse communities can increase their knowledge, skills, and evidence to address environmental injustices that lead to the health disparities one sees today (Finn and O'Fallon 2017).

## Environmental Justice

Understanding the cultural, socioeconomic and environmental factors underlying environmental health disparities led to the birth of environmental justice (EJ), which is defined by the US EPA as: "the fair treatment and meaningful involvement of all people regardless of race, color, national origin, or income with respect to the development, implementation, and enforcement of environmental laws, regulations, and policies" (EPA 2016). With roots in the civil rights movement, the EJ movement emerged from local communities' struggles with toxic contamination in the U.S. Advocacy and protests in Warren County, North Carolina about local exposures of concern, and in other areas in the 1970s–1980s were pioneering efforts. Shortly after, definitive studies emerged demonstrating the link between minority status, low socioeconomic status, and community proximity to toxic landfills (United Church of Christ 1987). The current literature on EJ comprises a wide range of quantitative studies consistently concluding that environmental risk burdens, known or potential, are distributed inequitably across racial/ethnic minorities and individuals with lower socioeconomic status (Bullard and Johnson 2000; Chakraborty et al. 2014). Lack of privilege, limited political influence, and linguistic isolation also play a role in environmental injustice (Brown 1995). Class differences, health (Grineski 2007) and rural health (Hartley 2004) disparities, as well as information disparities (Emmett and Desai 2010) also need to be considered.

To improve understanding of the convergence of contaminant exposures and sociodemographic information, the United States Environmental Protection Agency (U.S. EPA) developed EJSCREEN, "an environmental justice mapping and screening tool that provides EPA with a nationally consistent dataset and approach for combining environmental and demographic indicators" (USEPA 2016). The EJSCREEN includes: 11 environmental indicators, six demographic indicators, and 12 EJ indexes. This mapping and screening tool helps individuals, communities, public health officials, health care providers, regulators and policymakers to visualize the environmental and demographic data that can inform their efforts. For example, using the EJSCREEN, a U.S. EPA Community Involvement Coordinator and/or a Project Manager can visualize the demographic characteristics in a community (e.g. percent low-income, minority, linguistic isolation) or map the EJ indexes, which combines the environmental and demographic information to determine whether the community in which they work is struggling to achieve EJ. This mapping and screening tool promotes EHL by visualizing the following environmental indicators, along with demographic data:

- National-Scale Air Toxics Assessment (NATA) air toxics cancer risk
- NATA respiratory hazard index
- NATA diesel PM
- Particulate matter
- Ozone
- Traffic proximity and volume
- Lead paint indicator
- Proximity to Risk Management Plan sites (places where there is potential for a chemical accident)
- Proximity to Treatment Storage and Disposal Facilities (TSDFs)
- Proximity to National Priorities List (Superfund, hazardous waste) sites
- Wastewater Dischargers Indicator (Stream Proximity and Toxic Concentration)

As illustrated by the EJ SCREEN and the publications cited above, in the United States, people of color, low-income communities, and tribal populations have been, and continue to be, disproportionately exposed to environmental conditions that can harm their health. EJ SCREEN is one way to visualize sources of pollution and who may be impacted. Recalling sources of pollution and then comprehending what environmental factors influence health are just the first steps in addressing these disproportionate exposures and adverse health outcomes. For, becoming more literate about environmental health issues is an evolutionary process (Finn and O'Fallon 2017). To improve the protection and health of these ethnically diverse communities it is important for these communities to evolve in their understanding and to be able to apply and analyze this knowledge to then synthesize, create, and evaluate new forms of EHL knowledge. In order to evaluate and create new forms of knowledge that would be relevant and useful for specific communities, ethnically diverse community members and practitioners need to work together and incorporate local knowledge and cultural understanding of how the environment affects their health. Given the importance of culturally relevant knowledge creation, the theoretical frameworks listed below are critical to EHL.

## Theoretical Frameworks

These frameworks were selected because they can increase access and inclusion and build human capacity, which are critical to EHL. This discussion begins at the intrapersonal level, meaning an individual's sense of self-efficacy as well as knowledge, attitude, and skills of the individual. Motivation and sense of control are at the core of self-efficacy and these factors affect health and environmental health literacy.

### *A Community's Ecology and Ecosocial Theory*

Understanding a community's ecology is essential and the first step when engaging ethnically diverse community groups in EHL efforts. A community's ecology includes: ethnicity, language, culture, and economics; essentially all factors that influence how health disparities develop and how solutions can be applied (Caron and Serrell 2010; Kreuter et al. 2004).

A community's ecology can be taken further and to include ecosocial theory. As stated by Krieger "Ecosocial theory is an emerging multilevel theory of disease distribution that seeks to integrate social and biologic reasoning, along with a dynamic, historical, and ecological perspective, to address population distributions of disease and social inequalities in health". First proposed by Krieger in 1994, ecosocial theory asks the question: Who and what drives current and changing patterns of social inequalities in health? To adequately reach diverse populations and increase EHL, and ecosocial theory needs to be considered because if one does not have a firm understanding of the demographics, ecology, or nuances/idiosyncrasies of the affected community, EHL efforts will miss the mark. EHL efforts can range from developing a factsheet for a community living near a contaminated site, setting up a public meeting, organizing a series of science cafes or seminars on an environmental health topic, teaching families how to use environmental monitoring tools, to developing an 8th grade classroom curriculum. If these methods for raising EHL are not culturally grounded, researchers might present information in the wrong language, set-up community gatherings at times families work, select a venue for a seminar that community members avoid or is too far away from their home. Conversely, researchers may propose a behavioral change that may reduce exposure to a contaminant, but completely dismiss a prominent cultural practice. This has been the case with some fish advisories. For example, members of the Akwesasne Mohawk community were told to reduce their fish consumption because of high levels of contaminants, which in turn had unintended health and cultural consequences (Hoover 2013), which diminished EHL efforts. Another example relates to EHL efforts with agricultural farmworkers who are told to report potential pesticide poisonings and go to the doctor. This is unreasonable advice when 70% of the agricultural-worker population do not have health insurance

(National Center for Farmworker Health 1999), exposures may be mistaken for flu, and/or because workers may fear losing their jobs if they report work related poisonings.

Utilizing this theoretical framework promotes an understanding of "the combined impact of societal determinants of health, both social and physical, at multiple levels and scales, [which] explicitly recognizes that scientific knowledge and hypotheses are socially situated, and that insights and experiences of affected communities along with the ideas of researchers, past and present, are germane to fruitful scientific inquiry and public health practice" (Krieger 2004). The eco-social context needs to be considered to implement effective EHL with long-term effects, especially in environmental health research because learning and becoming literate in EH, or any topic, requires the educator to understand the social context and constructs in which they target audience lives and operates.

## *Self-Efficacy*

Self-efficacy is defined as the as the confidence and ability to have control over one's own life (Janz et al. 2002) or "as an individual's belief (or confidence) about his or her abilities to mobilize motivation, cognitive resources, and course of action needed to successfully execute a specific task within a given context' (Bandura 1997). EHL efforts to engage ethnically diverse community groups needs to begin with self-efficacy because if an individual does not feel as though they have control over their life or experiences, they will not be able to take part in an evolutionary learning process such as the measurable stages of learning described in Bloom's Taxonomy of Educational Objectives (Finn and O'Fallon 2017; Bloom 1956).

If an individual is confident and believes that they have control over their lives, they are said to have high self-efficacy. An individual who feels that they have no control over their lives and that their life is subject to chance or external forces is said to have low self-efficacy. If one applies the self-efficacy concept to environmental health literacy efforts, one can expect that in general people will be more likely to understand their risk, adopt healthy behaviors and support efforts to mitigate environmental exposures if they are confident in their ability to do so. Meaning, there is a connection between increased EHL and increased self-efficacy. Valuing health, along with self-control over health, is noted as an indicator of preventative health behaviors (Lau et al. 1986). Establishing efficacy is therefore an important step in determining whether the targeted population will value the public health intervention and increase their EHL. In general, ethnically diverse communities are not "privileged" or granted membership in one or more dominant social identity groups. This lack of privilege, along with the associated class ranking (e.g. lower ranking people have fewer resources and opportunities than those of relatively higher rank), creates a low sense of self-efficacy. Because of this, it is suggested that increases in self-efficacy makes an important contribution to environmental health literacy, particularly among ethnically diverse individuals. Later in this chapter, methods in which to increase self-efficacy will be presented, such as peer-education and participatory approaches.

While a sense of control and confidence is important, one must also consider the factors that motivate individuals to engage in acts of environmental public health prevention and interventions. As stated above, valuing health can be a predictor of preventative health behaviors. If there is a general sense of value, then an individual may be more motivated to take on, or incorporate environmental health interventions and prevention strategies. This motivation and level of engagement is relevant to EHL efforts because they indicate recognition, comprehension, and application of EH principles. In general, residents of environmentally compromised communities typically want additional information regarding their environmental health. When working with communities neighboring hazardous waste sites or living in environmentally compromised spaces, individuals are intrinsically motivated to understand their environment and how it may affect their health, hence an increased literacy. For example, these communities want more information about how contaminants are transported through water, soil and air, how to assess the amounts of toxics they are exposed to, and what is the relation between their exposure and their risk of disease (Ramirez-Andreotta et al. 2016a, b).

Concerned community members often learn about health-related topics via informal settings and engage in free choice learning (Dickerson et al. 2004). Importantly, such learning is personal, contextualized, correlated with individual interests and motivation (Falk et al. 2007; NRC 2009) and takes time (Rennie and Johnston 2004). Through this inquiry, they advance their EHL, by (1) seeking out, comprehending, and evaluating environmental health information; (2) using environmental health information to make informed choices and reduce health risks; and (3) improving quality of life and protecting the environment (Finn and O'Fallon 2017). For example, interviews conducted with parents living near an abandoned mining site who had participated in the University of Arizona's Metals Exposure Study in Homes (a children's biomonitoring study) revealed that families used the data to cope with their challenging circumstances, the majority of participants described changing their families' household behaviors, and participants reported specific interventions to reduce family exposures (Ramirez-Andreotta et al. 2016a, b). The strength of this study is that it provided insight into what people learn and gain from such results communication efforts, what participants want to know, and what type of additional information participants need to advance their environmental health literacy. Participants also reported actions and interventions to reduce their child's exposure, suggested news ways in which to communicate results to participants, and asked new research questions – demonstrating their learning is evolving.

## *Popular or Empowerment Education*

Popular education is the empowerment of citizens through direct dialogue, knowledge acquirement, and engaged participation to resolve a problem. This approach captures the spirit of Paulo Freire. Freirian theory and practice recognizes that individuals are the subjects of their own learning and not just "empty vessels" to be filled with knowledge generated by experts. Popular education can serve as a powerful vehicle for increasing EHL efforts because this theoretical framework calls for action beyond knowledge recognition and comprehension and strongly encourages participation in the solution-generating dialogue, which can propel learners to the analysis, synthesis, and evaluation stages of learning. Freire stressed that individuals should not only be involved in the efforts to identify their problems, using a listening–dialogue–action approach, but also be "… engaged in the conscientization to analyze the societal context for these problems" (Minkler and Wallerstein 2008). In order to stimulate change and sustain literacy efforts, the targeted individuals and groups must be part of the problem-solving dialogue. Recently, in a citizen science program coupled with a peer education model, researchers observed significant increases in self-efficacy for doing science and for environmental actions related to environmental resiliency and climate adaptation among minorities (Sandhaus et al. 2018), see below under Methods).

# Methods

Rooted in the aforementioned theoretical frameworks, the methods below can be used to raise EHL among ethnically diverse community groups. This section will highlight an approach/method, followed by a research/outreach example (Fig. 4.1).

## *Free-Choice Learning and Informal Science Education*

Free-choice learning is about deciding what, where and how one wants to learn over the course of a lifetime. It's self-motivated learning that takes place all the time, outside of a classroom setting. Informal science education (ISE) are science learning opportunities that people experience across their lifespan outside of school (Bonney et al. 2009; NRC 2009). Concerned community members often engage in free choice learning and informal education to learn about health-related topics (i.e., Dickerson et al. 2004). These learning approaches are personal, contextualized, correlated with individual interests and motivation (Falk et al. 2007; NRC 2009) and take time (Rennie and Johnston 2004). Free-choice learning and ISE are vital to EHL because, as noted earlier, people have a greater motivation to engage and learn

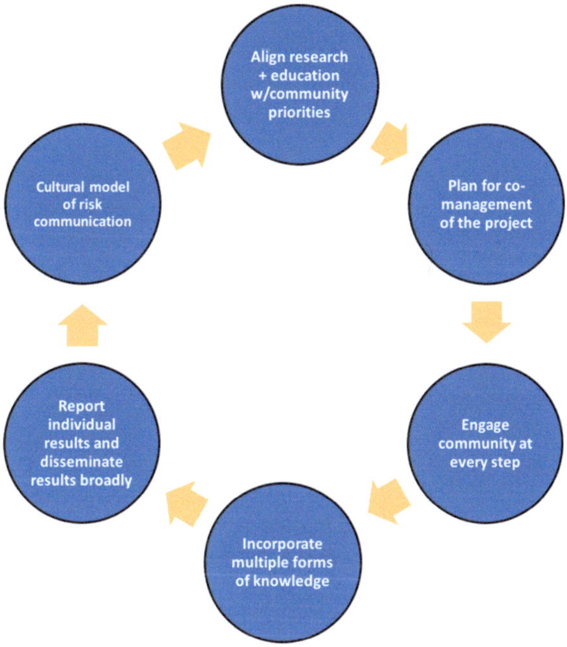


**Fig. 4.1** Engaging ethnically diverse community groups in environmental health literacy efforts (Israel et al. 1998; Pandya 2012; Brody et al. 2014; Ramirez-Andreotta et al. 2015). At the core, is a solid understanding of a community ecology environ-socio-cultural contexts and efforts to build individual's self-efficacy)

when the subject matter is directly related to their lives, and if the learning process is interactive (Falk et al. 2007).

In the context of communities impacted by hazardous waste, environmental health literacy efforts are of great value (1) if they help communities make informed choices to reduce hazardous exposures and (2) if these efforts draw upon local knowledge (Corburn 2005). It has been established that individuals living near contamination are motivated to engage in environmental health research and work with outside partners, such as public health officials and researchers (e.g., Brown 1995). However, what is missing from the traditional community involvement programs are efforts to cultivate ISE learning opportunities via public participation in environmental research and risk assessment projects (Ramirez-Andreotta et al. 2014).

## *Community-Based Participatory Research*

Community-based participatory research (CBPR) principles ensure that knowledge generation is combined with taking action. CBPR projects share many of the core principles and characteristics of participatory action research and popular education, are community driven, and foster co-learning experiences for researchers and community members [Minkler and Wallerstein 2008]. CBPR is an approach to research that shares power with community partners in all aspects of the scientific process to inform public health interventions, improve decision making, and/or stimulate action and policy changes [Minkler and Wallerstein 2008], intentionally crossing research and advocacy boundaries to successfully address complex environmental health challenges [Brown 2013]. Scientific data can be a boundary object; information that may be used in different ways, but is durable enough to maintain its identity even when it intersects various social worlds [Star and Griesemer 1989]. The data itself can be used by a scientist to support a hypothesis and also be used by communities to inform environmental policy reform and intervention [Brown 2013]. In this sense, science can be responsive to community needs, practical, and applied while following the robust principles of the scientific method. Although in practice it might be challenging to include community partners in all stages of the research, it is fundamental that community members are given the opportunity to help shape the research questions and interpret the findings [Green and Mercer 2001]. Ideally, CBPR allows for all partners in the research process to be equally involved, and the unique strengths that each partner brings to the project are recognized and appreciated (e.g. Minkler and Wallerstein 2008; O'Fallon and Dearry 2002; Israel et al. 1998). In recent years, community academic partnerships using a CBPR approach have played an increasingly important role in the area of environmental justice (e.g. Brown et al. 2012; Cohen et al. 2012).

## *Public Participation in Scientific Research*

Public participation in scientific research (PPSR), which includes citizen science, is a form of ISE, and is broadly defined as "…intentional collaborations in which members of the public engage in the process of research to generate new science-based knowledge" (Bonney et al. 2009). PPSR or citizen science projects are primarily being organized as a means to collect field data, to monitor a variety of environmental conditions and to increase a participant's scientific literacy (Bonney et al. 2009), to (Sibertown 2009). Public participation in scientific research lies along a spectrum that ranges from scientist-driven (conventional) research projects with moderate community engagement to 'crowd-sourcing' initiatives that depend on wide networks of CS members to collect data (e.g. GLOBE at Night, BudBurst, The Cornell Lab of Ornithology's NestWatch, FeederWatch, and Backyard Bird Count), to 'co-created' projects that are community-driven and formulated to support capacity that directly addresses their own research needs (e.g. Gardenroots, Public Lab). CS projects can be divided into five categories that differ by level of public participation: contractual, contributory, collaborative, co-created, and collegial.

PPSR and CS efforts provide excellent and exciting opportunities to engage ethnically diverse community groups because these approaches to research democratize the research method. PPSR and CS can broaden and deepen engagement of underrepresented communities in science and increase EHL among these populations because it present opportunities for community members to partner and work together throughout the scientific method and solution generating processes.

Experience and scientific research dedicated to environmental justice and action has shown that the questions asked, partnership type, and research approach used are important elements in research in ethnically diverse community groups. Asking questions that fail to address community concerns and using methods that dismiss public participation can perpetuate the environmental health literacy knowledge gap and separation between professional scientists and members of the public, especially those who are disenfranchised (Soleri et al. 2016; Pandya 2012; Ottinger 2010). For environmental health efforts to be successful in underserved minority populations, they need to address community identified priorities, align with community values, and build trust (i.e. a targeted community needs to have confidence in the institution that is conducting the study). Approaches to environmental health literacy efforts need to move beyond a narrow scientific focus, and adopt a systems-thinking approach (Seiffert and Loch 2005). When a systems-thinking approach is applied, the work can successful integrate social and ethical considerations with EHL efforts (Levine et al. 2009).

## *Traditional Ecological Knowledge*

TEK is an important form of research and knowledge generation and can be defined "as subset of traditional knowledge maintained by Indigenous nations about the relationships between people and the natural environment…preserved though oral tradition and through cultural expressions such as arts, crafts, and ceremonies and the cultivation, collection, and preparation of traditional foods (Finn et al. 2017). Others have further described TEK as resembling adaptive management practices of ecological resources (Berkes et al. 2000). When comparing western scientific approaches to TEK, there are similarities in that they both aim to understand the world around them, evolve as new information is available, and are dynamic processes (Finn et al. 2017). While a western scientific approach utilizes a reductionist approach and seeks to build a collection of knowledge that tends to be discipline-specific, TEK uses an approach that resembles system-thinking (Seiffert and Loch 2005) and embraces the interconnectedness among humans and the environment. When working with indigenous communities one needs to acknowledge, value, and honor TEK and other forms of knowledge generation because one cannot separate, or exclude the additional components of of environmental health, as understood in these communities, such as social dimensions. For, similar to CBPR and PPSR, TEK is an approach to research that can help foster community-professional partnerships and ensure community identified research needs guide research and EHL efforts. Researchers and government agencies have already recognized how TEK can improve resource management and conservation efforts at site-specific locations (Alcorn 1989).

TEK can be viewed as a Tribal form of citizen science, a vehicle for capturing environmental exposure and health, with the exception that the conduct of research is not solely to generate new knowledge and literacy, it is about valuing the data and equally acknowledging local ecological knowledge in environmental health research decision-making (Finn et al. 2017). In this instance, working with Tribal and indigenous communities benefit from a Tribal form of citizen science, respect for TEK, and a CBPR approach to research (see this section above under "Methods").

## *Knowledge Mediators*

Knowledgeable mediators are individuals who serve as the broker, interpreter and/or translator of information between individuals in the targeted population and the persons and organizations outside of the targeted community (May et al. 2003). Individuals serving in the role of a knowledge mediator not only utilize knowledge provided by the community and translate it with service providers; they utilize the knowledge of the service providers and translate it with community residents. These individuals have dual competency, which includes a capacity for understanding as well as communicating in locally appropriate ways (May et al. 2002)! They share similar social backgrounds or life experiences and are socially accepted by the targeted community. Because they are indigenous to the targeted community, they can effectively reach, teach and/or sharing information with their peers. These individuals can be considered gatekeepers for hard to reach populations and are critical for EHL efforts because the community will listen to them!

Knowledge mediators go by different terms. To name a few, they include: *promotoras*, community health workers, and health educators. In Hispanic communities, the term promotora is used and traditionally refers to a female, Hispanic community member who has leadership qualities that allow her to effectively promote a particular issue in her indigenous community. May et al. (2003) identified five general domains of practice of *promotoras*: information and referral, education, emotional support, community and capacity building, and advocacy.

The community health worker model has been implemented for decades to address public health disparities in communities (Hunter et al. 2004) and has been leveraged to address pollution prevention efforts. In the Body and Soul project, Boston Safe Shop Model and the Arizona-Sonora Border Pollution Prevention program (see Section "Knowledge Mediators – Demographic: Urban, Mexican/Mexican American" below), culturally and linguistically appropriate technical assistance informed hard-to-reach small businesses to improve their practices to reduce occupational and community exposures (Moreno Ramirez et al. 2015; Roelofs et al. 2010). In general, knowledge mediators bridge communities in two ways: horizontally, by facilitating social networks within the community, and vertically, by connecting targeted residents with researchers or critical services from outside the community (May et al. 2003). These "bridging acts" are critical for EHL because they connect ethnically diverse community groups with environmental health information to then make informed choices and reduce health risks, improve quality of life, and take steps to protect the environment.

## Tailored and Linguistically Appropriate Messaging

Effective communication is vital to EHL efforts. In order to achieve effective communication, it is critical to assess community needs and tailor messages to correspond with a specific audience in an appropriate context (Tushman and Scanlan 1981). Tailored messaging means the communicator recognizes and understands their target audience and then selects the appropriate information transfer mechanism (NRC 2008). Linguistically appropriate messaging is knowing the predominant language spoken, the unique dialects spoken in the targeted community and what type of information is desired. For example, a regulatory agency might appreciate an executive summary of the scientific evidence. Conversely, a member of the affected community may prefer information specific to the chemical of concern, the risks they pose, and possible methods to reduce exposure. To select which information-transfer mechanism to use for a targeted community, it is crucial to learn the community's ecology and the social context in which the environmental contamination and human health risks are embedded (Green et al. 2002; Caron and Serrell 2010). To accomplish this, information transfer from the targeted community to the practitioner (e.g. academic, government agency, community-based organization) should also be occurring in order for the values, knowledge, and opinion of the targeted community to be heard and considered. As described above in section 2.1, by understanding the community's ecology, one is able to establish and ensure a two-way dialogue with affected communities. Altman et al. (2008) observed that personal and collective environmental history influenced how one interpreted and responded to personal exposure results. This highlights the fact that personal responses to exposure information will vary depending on social and eco-historical experiences of the population.

## Data Visualization and Information Design

Data visualization is any effort to help people understand the significance of data by placing it in a visual context. Information design is about presenting information in a way that fosters efficient and effective understanding of it. Regardless of audience type, it is important that the information communication mechanism use visual aids and provides sound, unbiased scientific information to inform and build human capacity. People can interpret visuals much faster than words (Covello 2015). This makes data visualizations and information design essential for EHL. Information visuals need to be clear, impactful, meaningful, and demonstrative to various end-users and decision-makers. Design can also affect an individual's interpretation, uptake of information, and how they see their choices affecting their environment and future (Nicholson-Cole 2005). Since imagery can elicit different responses from people depending on background, previous knowledge, motivations, and personal beliefs, it is critical to engage and learn about one's targeted audience. It is

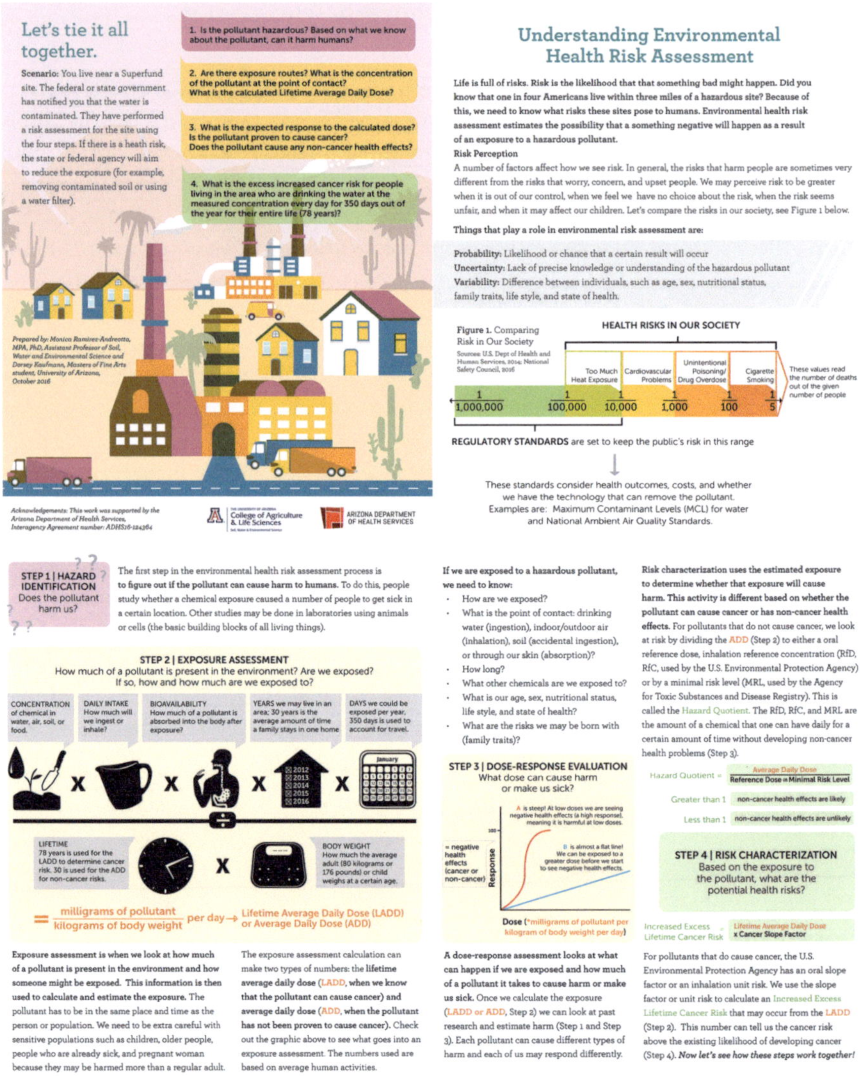


**Fig. 4.2** An example of factsheet that balances text and graphics prepared by Ramirez-Andreotta and Kaufmann (2016). Available at: https://superfund.arizona.edu/ University of Arizona Superfund Research Program. (2018)

also imperative for a science communicator/information designer to ask questions about what kind of images portray the appropriate level of scientific information, how to create an appropriate visual representation of the information, and audience perception (Nicholson-Cole 2005) (Fig. 4.2).

For example, an infographic, or *information graphic*, is a self-contained, pictorial representation of information, knowledge, or data.[33] Formally defined as "a visualization of data or ideas that tries to convey complex information to an audience

**Fig. 4.3** An example of a fish advisory sign developed by the Washington State Department of Health (n.d). The text is written in nine different languages. Available at: https://www.doh.wa.gov/Portals/1/Documents/Pubs/334-330.pdf

in a manner that can be quickly consumed and easily understood," an effective infographic combines data with design to improve comprehension via visual learning (Cox and Pezzullo 2016). Infographics have been especially useful in public health promotion efforts related to topics as distinct as Ebola awareness and fish advisories. Infographics frequently employ color to convey a sense of risk with red indicating highest risk and green indicating safer levels of exposure (see Fig. 4.3).

## Technical vs. Cultural Models of Risk Communication

Traditionally, a technical model of risk communication has been employed. This traditional method is solely the translation of technical data (Andrews 2006). It aims to inform, change behavior, and assure populations that the scientifically determined risk is acceptable through a one-way communication (expert →laypeople) (Cox and Pezzullo 2016). Though the goal is to educate a target audience, this one-way communication process has led to clashes between residents and public officials at contaminated sites. Such models assume that any failure of technical risk communication is due to public fear, irrationality, and other emotional factors (Cox 2016), completely dismissing the fact that the targeted community did not get to inform or participant in the risk assessment process.

Given the limitations of the technical, uni-directional risk communication approach, a cultural model of risk communication should be adopted. A cultural model of risk communication can be defined as a two-way conversation where scientists involve the affected public in assessing their risk (Cox and Pezzullo 2016). Citizen participation in risk communication puts risk into context, makes comparisons with other risks, encourages a dialogue, and involves the affected group in the judgment of acceptable and unacceptable risks (NRC 1996a, b).

When working with diverse community groups who are living in environmentally compromised areas or near hazardous waste sites (i.e. National Priorities Listed/ Superfund sites), a cultural model of communication is critical to advance environmental health literacy efforts. For example, community members should be given the opportunity to contribute to the risk analysis, be part of the risk assessment process by learning the meaning and source of each variable, modifying the exposure assessment to fit their site-specific parameters (e.g. exposure duration, exposure frequency, intake rate), and learning how their behavior or a remediation strategy can change their exposure (i.e. "If I drink less water from this source or if the concentration of the contaminant decreases then my daily dose decreases") (Ramirez-Andreotta et al. 2015, 2016c). When done correctly, a cultural model of risk communication effort can dramatically improve EHL efforts because the community will have applied the environmental health information to understand how their behavior changes or a clean-up strategy reduces their exposure and associated health risks.

To implement a cultural model of risk communication collaboration among citizens, experts, and agencies is necessary. Data to inform the risk assessment model is more complete when it stems from and is informed by the targeted communities' experiences. This approach recognizes the relevant experience of the local community and makes use of this local knowledge in risk studies (Cox and Pezzullo 2016; Corburn 2005; Dewey 1954).

## Research Examples

Below are examples demonstrating the application of the methods presented above. Each example in this section describes activities designed to raise EHL in a specific culturally diverse community group. These examples showcase a myriad of creative methods and activities that can be applied to increase EHL in culturally diverse community groups. Each example provides practical details on how the environmental health education goal was attained with a targeted culturally diverse community and the researcher's success in raising EHL.

## Risk Communication and Breast Cancer – Demographic: African American Women

This case study highlights the importance of a cultural model of risk communication that provides information that is culturally relevant, appropriate, and accurate. African American women are disproportionately affected by breast cancer and suffer a higher burden of basal-like breast cancers, an aggressive subtype that has no targeted therapy (Allicock et al. 2013). Through focus groups, Allicock et al. (2013) investigated women's knowledge, beliefs, and attitudes about breast cancer to learn about risk perceptions. It was observed that "race was a strong thread that influenced how risk was perceived; knowledge, beliefs, and attitudes about breast cancer were shaped by women's core knowledge about the disease, age perceptions, understanding of cancer causes, and connotations of the disease" (Allicock et al. 2013). Participants reported that breast cancer is typically depicted as a white woman disease and that perhaps the lack of visual representation of African Americans in breast cancer information transfer translated to a lowered perceived relevance of the disease, and subsequently risk. Researchers recognized that there may not be one best method to effectively communicate breast cancer risk to African Americans, rather it was critical that the information was targeted to them and came from culturally relevant sources, consistent across sources, and convenient for women to access (Allicock et al. 2013). The data collected from this focus group highlights how risk perception can impact one's understanding of disease and distribution of burden.

## *Knowledge Mediators – Demographic: Urban, Mexican/ Mexican American*

This case study highlights the integral role of *promotoras* in pollution prevention efforts with small and home-based businesses in a predominately Hispanic area of Tucson, Arizona. Using handheld GPS units, *promotoras* walk neighborhoods to identify high-risk areas based on environmental hazards. The maps indicated a large number of small and home-based businesses in and around the predominately Hispanic neighborhoods. If pollution prevention practices are not applied at these small and home-based businesses, occupational and community exposures to volatile organic compounds (VOCs) can occur. To increase the targeted communities access to pollution prevention (P2) practices, a partnership between *promotoras* from a non-profit organization, Sonora Environmental Research Institute, Inc. (SERI) and researchers from the University of Arizona (UA) worked together to reach these small and home based businesses. From 2008 to 2011, a *promotora*-led P2 program reached a total of 640 small and home-based businesses. P2 Activities included: technical trainings for *promotoras* and businesses, generation of culturally and linguistically appropriate educational materials, and face-to-face peer education via multiple on-site visits. Pre- and post- surveys were administered to measure P2 best practices implemented by businesses and perceptions towards pollution prevention. Two hundred and eight-eight (45%) of the total businesses reported modifying their behavior ranging from recycling to substitution of products; these behavior modifications varied from business to business (Moreno Ramirez et al. 2015). For example, a total of 84 of the 124 nail salons reported they would try acetone-free nail polish remover, which is equivalent to 16,308 kilograms of VOC emissions reduced per year when acetone is replaced with non-acetone nail polish remover. For automotive repair and paint and body shops, 203 businesses reported they began covering solvent degreaser containers when not in use. Covering a cold solvent degreaser and associated drainage facility reduces emissions between 13% and 38%. Using the conservative estimate of 13%, it was calculated that emissions were reduced by 45 kg/year. This gives a reduction in VOC emissions of 763 kg/year. One of the 17 shops reported that they switched entirely to an aqueous degreaser reducing a total of 10,886 kg of VOC emissions per year at this business (Moreno Ramirez et al. 2015). The outcomes of this project demonstrate the vital role the knowledge mediator played in engaging traditionally hard to reach businesses and how culturally and linguistically appropriate technical assistance reduced occupational and community exposures.

## Knowledge Mediators – Demographic: Urban, Mixed Minority Groups

Similar to the Tucson Promotora case study described above, the Boston Public Health Commission Safe Shops Project also worked with knowledge mediators to incentivize hard-to-reach small businesses to improve their environmental practices and to become good neighbors in neighborhoods that include a mix of residences and businesses in close proximity. The Safe Shop program also used knowledge mediators who provided culturally and linguistically appropriate technical assistance to raise the level of EHL among small businesses owners. The Safe Shops field staff conducted walk-through assessments, provide training business-specific training sessions that included best practices and technology for worker protection and pollution prevention, and held one-on-one meetings with business owners to develop improvement plans (Roelofs et al. 2010). In addition, they also assessed worker and owner needs for health care, business loan assistance, and child care resources. After a pre- and post-survey with all participating businesses, results revealed that compliance with lab waste area containers increased from 45% of shops to 85%; and that properly labeling waste oil, antifreeze, paints, and solvents increased from 30% to 85% (Roelofs et al. 2010). This project also worked with nail salons technicians; Roelofs et al. (Roelofs et al. 2010) describes noteworthy outcomes while working with these nail technicians, which can be challenging. They examined the products being used in Boston nail salons and then distributed material safety datasheets (N = 45), trained 30 community health care providers how to recognize work-related health issues for nail technicians, and made 5 salon referrals to a community partner who could provide child care and business development resources (Roelofs et al. 2010). During these efforts, it was observed that the nail technicians believed infection control masks were suitable and effective in reducing their exposure to the chemicals and dust produced during treatments, specifically acrylic nails. The Safe Shops Project educated nail technicians that they should instead be using National Institute of Occupational Safety and Health–approved N95 dust masks with impregnated carbon for low-level solvent control. Nail technicians expressed that they would not use the mask based on appearance. Based on this feedback, the Safe Shops Project identified a seamstress to prepare attractive covers for the N95 masks. The outcomes of this project demonstrate the vital role the knowledge mediator played in improving occupational health by reducing exposures to solvents, facilitating social networks within the community, and by connecting targeted residents with critical services from outside the community like child care and business development resources.

## Body and Soul Project (Knowledge Mediators) – Demographic: Urban, African American

This case study highlights the importance of knowledge mediators who attend or work at local churches. Body & Soul is a church-based fruit and vegetable (FV) intake program implemented at churches with predominantly African American membership. Based upon the community's ecology, the Body & Soul program has four critical components: pastoral involvement, educational activities, church environmental changes, and peer counseling (Allicock et al. 2012). Body & Soul is a collaborative effort among the National Cancer Institute, the American Cancer Society (ACS), and two university research groups. In recent studies, the intervention was delivered by volunteer members of African-American churches with less training, monitoring, and support (i.e. real-world conditions) to test whether these local knowledge mediators could be successful in reaching the targeted group as shown in previous studies ran by a trained professional staff (Resnicow et al. 2004). Results of this case-control study revealed that the intervention group consumed significantly more FV at follow-up than the control group, suggesting that the volunteer members of African-American churches can better reach their constituents (Resnicow et al. 2004). The results suggest that research-based interventions, delivered collaboratively by community volunteers and a health-related voluntary agency, can be effectively implemented under real-world conditions. Though this case study is specific to nutritional health, it provides an evidenced-based framework that can be implemented to raise literacy and address environmental health challenges in predominately African-American communities.

**Image 4.1** The abandoned Humboldt smelter nestled in the community of Dewey-Humboldt, Arizona

## Citizen Science – Demographic: Rural, Predominately White[1]

This case study highlights how to implement a co-created citizen science project with a community neighboring a contaminated site. Dewey-Humboldt, Arizona, is a rural community adjacent to the Iron King Mine and Humboldt Smelter Superfund (IKMHSS) site. In August 2008, community members expressed concerns about the site's impact on their soil at a public meeting. Specifically, they wanted to know if it was safe to consume vegetables from their home gardens, and if so, how much they could eat. In response, a community-academic partnership was formed and together they investigated the uptake of arsenic in commonly-grown vegetables, and evaluated arsenic exposure and potential risk to local vegetable gardeners Ramirez-Andreotta et al. 2013). Because the research question came from community members who then worked alongside the academic investigator throughout most stages of the study, Gardenroots was a co-created citizen science project (Image 4.1).

Recruitment for the Gardenroots project was primarily achieved through an announcement in the Dewey-Humboldt town newsletter, through personal interactions at local events, and by community "champions" who initially posed the research question. More than 40 community members signed up to attend a 1.5-hour training session on how to properly collect soil, water, and vegetable samples from their home gardens for laboratory analysis. Participants received a home instructional manual and collection toolkit, and were asked to deliver their samples to the local UA Cooperative Extension office. A series of informal science education experiences were organized throughout the program to maintain participant motivation.

The presentation and delivery of study findings at a final report-back gathering was instrumental to ensuring that participants understood the data and could use the information to make decisions. Each participant received a personalized booklet containing their household sampling results, including a chart that illustrated how many cups of vegetables per week an individual could safely eat from their garden

[1](part of text below pulled from https://www.niehs.nih.gov/research/supported/translational/community/gardenroots/index.cfm (Kerry Mandernach from MDB, Inc. Contractor to NIEHS took lead, Ramirez-Andreotta help write/edit).

based on different levels of estimated risk. Ramírez-Andreotta led presentations on how to interpret the results, and provided handouts outlining recommended safe gardening practices to help reduce an individual's arsenic exposure. Outcomes for such citizen science efforts can be analyzed at the programmatic, individual and community level. Below are some noteworthy outcomes:

- Programmatic – The Gardenroots project has improved understanding of (1) the uptake of arsenic in common homegrown vegetables grown in soils near a mining site; (2) the amount of arsenic introduced to an individual through ingestion of homegrown vegetables, incidental ingestion of soil, and water, and the potential risks posed by those exposure routes; and (3) how to implement a co-created citizen science project with a community neighboring a hazardous waste site.
- Individual – Surveys indicated that most of the participants understood the scientific findings and would continue to eat homegrown vegetables while modifying their gardening practices to reduce their arsenic exposure from water and soil. During the report-back gathering, some participants recalculated their potential exposure and risk by changing the exposure assessment variables, such as the amount or number of days eating a vegetable, to better suit their behavior and way of living. Participants also demonstrated an increased understanding of soil contamination, food quality, and the scientific process by posing additional research questions. For example, they inquired about the quality of chicken eggs, and whether raised-bed building materials like cinder blocks could contribute arsenic to their garden soils.
- Community-level – Gardenroots participants increased their community networking in natural resource-related issues such as soil and water quality, and subsequently participated in other natural resource-related projects. They also leveraged the study results to encourage government officials to take action about the contaminated soil and water. The Gardenroots project revealed that water from the local public water system exceeded the U.S. Environmental Protection Agency (EPA) arsenic drinking water standard, prompting participants to work together to notify other households that were connected to the public water supply. They also reported their test results to federal and state environmental agencies, and as a result the municipal water supplier received a notice of violation for exceeding the EPA arsenic drinking water standard.

This example highlights how to implement a citizen science project in an underserved community neighboring contaminated sites as well as the various and diverse outcomes that grew out of a participatory approach to research.

## *Promotora Science (Citizen Science and Knowledge Mediators) – Demographic: Urban, Mexican/Mexican American*

This example highlights how to combine frameworks and methods, specifically community ecology, knowledge mediators, and citizen science efforts, to increase self-efficacy for environmental action and conducting science in a Mexican/Mexican

American community. South Tucson metropolitan area is considered an environmental justice community. When compared to the U.S., over 90% of the target area is low income, minority, linguistically isolated, and disproportionately located in proximity to hazardous waste sites, chemical treatment, storage facilities, and high traffic areas. This area also lacks tree canopy, has high surface temperatures, and high heat vulnerability (Pope 2013). Further, the Arizona Department of Health Services has designated the area as medically underserved, nearly 60% of the residents are members of sensitive populations, including both children and the elderly, and according to the U.S. Census Bureau (2010), over 40% of adults lack a high school education, compared to just 15.8% for the entire Tucson area. A partnership between *promotoras* from the non-profit organization, SERI and researchers from the University of Arizona was formed and in March 2016, a Climate Change Adaptation and Resiliency training was held. This program encompassed all aspects of sustainability and climate change and was informed by the targeted population's ecological and social context. *Promotoras* signed up for this program and received a comprehensive training in environmental quality, climate change, and sample collection that was 40 h long (5 h a day for 2 weeks; Monday to Friday). In addition to the topics covered, major highlights of the training include: (1) a five-hour session on how to build a passive rainwater harvesting system and (2) two five-hour sessions on environmental quality and field sampling and laboratory methodologies. During the training, *promotoras* collected harvested rainwater and soil samples from residents living in southern metropolitan Tucson. Once the samples were collected, the *promotoras* promptly arrived at the University of Arizona's Integrated Environmental Science and Health Risk Laboratory to prepare the samples for analysis. In the lab, the *promotoras* entered their sample identifications (chain of custody) and field notes into a spreadsheet, placed the water samples in the refrigerator, and sieved, weighed out, and dried soil samples. After the training and sample collection, *promotoras* went out in pairs and conducted home visits (N = 32) to share sustainability and climate change information with families in the targeted area (Image 4.2).

Through a mixed-method evaluation approach, data indicated that the designed environmental sustainability and climate change adaptation training was successful in increasing the EHL of the participating *promotoras*, and that the *promotoras* themselves were able to go into their communities and effectively educate others on these environmental sustainability practices (Sandhaus et al. 2018). Researchers furthered observed that self-efficacy for environmental action increased across both the *promotoras* and the home visit participants, while the *promotoras* gained efficacy in doing science, indicating that a citizen science approach can be an effective tool in science literacy (Sandhaus et al. 2018). This example highlights how the aforementioned frameworks can be combined to create a program that increases EHL across various types of community members (i.e. knowledge mediator and families). Designing a training that was tailored and had linguistically appropriate messaging and combining knowledge mediators with citizen science led to significant increases in self-efficacy for doing science and for environmental actions related to environmental monitoring and climate adaptation among minorities (Sandhaus et al. 2018). As stated in section 2.2, increasing self-efficacy is critical

**Image 4.2** *Promotoras* prepping soil samples for analysis, March 2016

for EHL because a sense of self-control is noted as an indicator of preventative health behaviors (Lau et al. 1986) (Image 4.3).

## PPSR/Tribal Citizen Science – Demographic: Rural, Mohawk Community

This example highlights how researchers and members of the Mohawk community worked together using a CBPR and PPSR approach to address polychlorinated biphenyls (PCBs) exposures via local seafood consumption. Akwesasne is a Mohawk community of about 15,000 people. In 1954, the St. Lawrence Seaway was built, which brought development and industry to the area. Years later in 1981, two dormant sludge pits filled with polychlorinated biphenyls (PCBs) were discovered behind a GM plant, adjacent to the Akwesasne community. This site was then placed

**Image 4.3** *Promotoras* using weighing out soil subsamples, March 2016

on the US EPA's National Priorities List in 1984 (US EPA n.d.). In parallel, a local Mohawk midwife, Katsi Cook, invited a Wildlife Epidemiologist to test fish and wildlife in the area; these studies revealed elevated levels of PCBs in duck, sturgeon, and turtle (Hoover 2016). Based on the data and her role in her community as a midwife, Katsi Cook began to set the stage for scientific studies to determine whether the PCB contamination found in their food source was impacting the health of mothers and their infants. Cook ultimately proved to be one of the "champions" who emerged to initiate this participatory approach to research, which she termed "barefoot epidemiology" (Hoover 2016). Using CBPR and PPSR approach, Cook partnered with researchers, under the condition that the researchers would hire and train local residents for the project. Cook ensured that the mothers were co-investigators in the study and she started the "First Environment Research Project" (FERP) as a means of organizing Mohawk women fieldworkers and coordinating the data for the health studies. FERP employees collected blood and breast milk samples, and for some studies conducted cognitive assessments, body measurements, and nutritional surveys (Hoover 2016). Through this community-academic partnership, it was revealed that there as was a connection between the levels of PCBs in participants' breast milk and blood to fish consumption (Fitzgerald et al. 1998). Though there were reported challenges, the community-academic partnership used CBPR principles to ensure transparency and the sharing of responsibility, decision-making, and data sharing. The CBPR approach and, in particular, the training for local residents increased EHL among the Akwesasne and set the stage for further collaboration to reduce health disparities in the region.

## Future Directions for Engaging Ethnically Diverse Community Groups

In conclusion, for EHL efforts to be successful and sustainable, projects need to consider the community's ecology, build self-efficacy among and within the culturally diverse community groups, and embrace popular education practices. As this chapter describes, communities, especially those living in close proximity to pollution sources, are intrinsically motivated to learn about and address environmental health issues. Methodologically, this chapter presented a series of approaches to reach and increase EHL among culturally diverse community groups. They include the:

- Effectiveness of free-choice learning
- Collaborative approaches and shared decision making inherent to projects based on CBPR and PPSR principles
- Cultural appropriateness and sensitivity that knowledge mediators bring to projects
- Importance of tailored messaging for specific subpopulations
- Use of cultural models of risk communication that more readily convey risk in culturally meaningful terms
- Advantages that data visualizations and information design entail particularly for low literacy or non-English speaking subpopulations.

All of these approaches call for inclusiveness and the democratization of science and all have proven effective for engaging with communities and raising EHL.

The field of EHL is emerging and due to the current state of health disparities, reaching culturally diverse community groups is timely and fertile ground. It is therefore appropriate to mention in this chapter that racial minorities comprise about 38% of the US population and this group makes up only 16% of staff and board members of major US environmental NGOs, foundations, and government agencies (Taylor 2014). Those involved in EHL and decision-making ideally need to reflect the diversity of the general and/or affected population; until this occurs, however, the use of knowledge mediators and cultural models of risk communication should be preferred methodologies for addressing health disparities in diverse community groups. In addition, given the disproportionate level of exposures and resulting health disparities among ethnic and racial subpopulations, it is imperative that scientists raise EHL across all populations by working alongside culturally diverse community groups. As illustrated in the examples highlighted above, there are many opportunities for EHL in environmental health public health research and programs. EHL efforts can create substantial change and if done well, can be transformative. Additionally, literacy efforts in environmental public health can potentially help to achieve global health justice. EHL programs are shifting the way scientists and the public think about reducing health disparities. It is time to change the paradigm and the way scientists and culturally diverse communities look at environmental health solutions.

## References


Alcorn, J. B. (1989). Process as resource: The traditional agricultural ideology of Bora and Huastec resource management and its implications for research. *Advances in Economic Botany, 7*, 63–77.

Allicock, M., Campbell, M. K., Valle, C. G., Carr, C., Resnicow, K., & Gizlice, Z. (2012). Evaluating the dissemination of Body & Soul, an evidence-based fruit and vegetable intake intervention: challenges for dissemination and implementation research. *Journal of Nutrition Education and Behavior, 44*(6), 530–538. https://doi.org/10.1016/j.jneb.2011.09.002.

Allicock, M., Graves, N., Gray, K., & Troester, M. A. (2013). African American Women's perspectives on breast Cancer: Implications for communicating risk of basal-like breast Cancer. *Journal of Health Care for the Poor and Underserved, 24*(2), 753–767. https://doi.org/10.1353/hpu.2013.0082.

Altman, R. G., Morello-Frosch, R., Brody, J. G., Rudel, R., Brown, P., & Averick, M. (2008). Pollution comes home and gets personal: Women's experience of household chemical exposure. *Journal of Health and Social Behavior, 49*(4), 417–435.

Andrews, R. N. L. (2006). *Managing the environment, managing ourselves: a history of American environmental policy*. New Haven: Yale University Press.

Bandura, A. (1997). *Self-efficacy: The exercise of control*. New York: Freeman.

Berkes, F., Colding, J., & Folke, C. (2000). Rediscovery of traditional ecological knowledge as adaptive management. *Ecological Applications, 10*(5), 1251–1262. https://doi.org/10.2307/2641280.

Bloom, B. S. (1956). *Taxonomy of educational objectives: The classification of educational goals* (1st ed.). New York: Longmans, Green.

Bonney, R., Ballard, H., Jordan, R., McCallie, E., Phillips, T., Shirk, J., et al. (2009). *Public participation in scientific research: Defining the field and assessing its potential for informal science education*. Washington, DC: Center for Advancement of Informal Science Education.

Brody, J. G., Dunagan, S. C., Morello-Frosch, R., Brown, P., Patton, S., & Rudel, R. A. (2014). Reporting individual results for biomonitoring and environmental exposures: lessons learned from environmental communication case studies. *Environmental Health, 13*(1).

Brown, P. (1995). Race, class, and environmental health: A review and systematization of the literature. *Environmental Research, 69*, 15–30.

Brown, P. (2013). Integrating medical and environmental sociology with environmental health: Crossing boundaries and building connections through advocacy. *Journal of Health and Social Behavior, 54*, 144–163.

Brown, P., Brody, J. G., Morello-Frosch, R., Tovar, J., Zota, A. R., & Rudel, R. A. (2012). Measuring the success of community science: The Northern California household exposure study. *Environmental Health Perspectives, 120*, 326–331.

Bullard, R., & Johnson, S. (2000). Environmental justice: Grassroots activism and its impact on public policy decision making. *Journal of Social Issues, 56*, 555–578.

Caron, R. M., & Serrell, N. (2010). Community ecology and capacity: Keys to progressing the environmental communication of wicked problems. *Applied Environmental Education and Communication, 8*, 195–203.

Chakraborty, J., Collins, T. W., Grineski, S. E., Montgomery, M. C., & Hernandez, M. (2014). Comparing disproportionate exposure to acute and chronic pollution risks: A case study in Houston, Texas. *Risk Analysis, 34*, 2005–2020.

Chinn, D. (2011). Critical health literacy: A review and critical analysis. *Social Science & Medicine, 73*(1), 60–67.

Cohen, A., Lopez, A., Malloy, N., & Morello-Frosch, R. (2012). Our environment, our health: A community-based participatory environmental health survey in Richmond, California. *Health Education & Behavior, 39*, 198–209.

Commission on Social Determinants of Health, World Health Organization. (2008). Available at: http://www.who.int/social_determinants/thecommission/finalreport/en/. Accessed on 13 Oct 2017.

Corburn, J. (2005). *Street science: Community Knowledge and environmental health justice*. Cambridge: MIT Press.

Covello, V. T. (2015, August). *Summary of neuroscience research: 7 implications for effective risk, crisis, and organizational change communications*. US Environmental Protection Agency Community Involvement Conference, Atlanta, GA.

Cox, R. (2016). *Environmental communication and the public sphere*. California: Sage Publications.

Cox, R., & Pezzullo, P. C. (2016). *Environmental communication and the public sphere*. Thousand Oaks: SAGE.

Dewey, J. (1954). *The public and its problems*. Chicago: Gateway Books.

Dickerson, S., Reinhart, A. M., Feeley, T. H., Bidani, R., Rich, E., Garg, V. K., & Hershey, C. O. (2004). Patient internet use for health information at three urban primary care clinics. *Journal of the American Medical Informatics Association, 11*(6), 499–504.

Emmett, E. A., & Desai, C. (2010). Community first communication: Reversing information disparities to achieve environmental justice. *Environmental Justice, 3*(3), 79–84.

Falk, J. H., Storksdieck, M., & Dierking, L. D. (2007). Investigating public science interest and understanding: Evidence for the importance of free-choice learning. *Public Understanding of Science, 16*, 455–469.

Finn, S., & O'Fallon, L. (2017). The emergence of environmental health literacy—From its roots to its future potential. *Environmental Health Perspectives, 125*(4), 495–501. https://doi.org/10.1289/ehp.1409337.

Finn, S., Herne, M., & Castille, D. (2017). The value of traditional ecological knowledge for the environmental health sciences and biomedical research. *Environmental Health Perspectives, 125*(8). https://doi.org/10.1289/EHP858.

Fish Consumption Advisories. Washington State Department of Health. Available at: https://www.doh.wa.gov/CommunityandEnvironment/Food/Fish/Advisories. Accessed on 17 Oct 2017.

Fitzgerald, E. F., Hwang, S.-A., Bush, B., Cook, K., & Worswick, P. (1998). Fish consumption and breast milk PCB concentrations among Mohawk women at Akwesasne. *American Journal of Epidemiology, 148*(2), 164–172.

Gazmararian, J. A., Curran, J. W., Parker, R. M., Bernhardt, J. M., & DeBuono, B. A. (2005). Public health literacy in America. *American Journal of Preventive Medicine, 28*(3), 317–322.

Graham, G. N. (2016). Why your ZIP code matters more than your genetic code: Promoting healthy outcomes from mother to child. *Breastfeeding Medicine, 11*, 396–397. https://doi.org/10.1089/bfm.2016.0113.

Green, L. W., & Mercer, S. L. (2001). Can public health researchers and agencies reconcile the push from funding bodies and the pull from communities? *American Journal of Public Health, 91*, 1926–1929.

Green, L., Fullilove, M., Evans, D., & Shepard, P. (2002). "Hey, mom, thanks!": Use of focus groups in the development of place-specific materials for a community environmental action campaign. *Environmental Health Perspectives, 110*(s2), 265–269.

Grineski, S. E. (2007). Incorporating health outcomes into environmental justice research: The case of children's asthma and air pollution in Phoenix, Arizona. *Environmental Hazards, 7*, 360–371.

Hartley, D. (2004). Rural health disparities, population health, and rural culture. *American Journal of Public Health, 94*, 1675–1678.

Hoover, E. (2013). Cultural and health implications of fish advisories in a Native American community. *Ecological Processes, 2*(1), 4.

Hoover, E. (2016). "We're not going to be Guinea pigs;" Citizen science and environmental health in a native American community. *Journal of Science Communication, 15*(01), A05.

Hunter, J. B., de Zapien, J. G., Papenfuss, M., Fernandez, M. L., Meister, J., & Giuliano, A. R. (2004). The impact of a Promotora on increasing routine chronic disease prevention among women aged 40 and older at the U.S.-Mexico border. *Health Education & Behavior, 31*, 18S–28S.

Institute of Medicine. (2004). *Health literacy: A prescription to end confusion*. Washington, DC: National Academies Press.

Israel, B. A., Schulz, A. J., Parker, E. A., & Becker, A. B. (1998). Review of community-based research: Assessing partnership approaches to improve public health. *Annual Review of Public Health, 19*, 173–202.

Janz, N. K., Champoin, V. L., & Strecher, V. J. (2002). The health belief model. In K. Glanz, B. K. Rimer, & F. M. Lewis (Eds.), *Health behavior and health education: Theory, research, and practice* (3rd ed.). San Francisco: Jossey-Bass.

Kreuter, M. W., de Rosa, C., Howze, E. H., & Baldwin, G. T. (2004). Understanding wicked problems: A key to advancing environmental health promotion. *Health Education & Behavior, 31*, 441–454.

Krieger, N. (1994). Epidemiology and the web of causation: Has anyone seen the spider? *Social Science and Medicine, 39*, 887–903.

Krieger, N. (2004). Ecosocial theory. In N. B. Anderson (Ed.), *Encyclopedia of health and behavior* (pp. 293–294). Thousand Oaks: SAGE Publications. https://doi.org/10.4135/9781412952576.n87.

Lau, R. R., Hartman, K. A., & Ware, J. E., Jr. (1986). Health as a value: Methodological and theoretical considerations. *Health Psychology: Official Journal of the Division of Health Psychology, American Psychological Association, 5*(1), 25–43. https://doi.org/10.1037/0278-6133.5.1.25.

Levine, R., González, R., & Martínez-Sussmann, C. (2009). *Learner diversity in earth system science*. Paper prepared for the committee for the review of the NOAA education program. Washington, DC: National Research Council. www7.nationalacademies.org/bose/NOAA%20 Diversity.pdf. Viewed 12 Dec 2011.

May, M. L., Contreras, R., Callejas, L., Ledezma, E., & Porter, C. (2002). *Final report: Mujer y Corazon: Community health workers and their organizations on the U.S.-Mexico border—An exploratory study*. College Station: Texas A&M University School of Rural Public Health.

May, M. L., Bowman, G. J., Ramos, K. S., et al. (2003). Embracing the local: Enriching scientific research, education, and outreach on the Texas-Mexico border through a participatory action research partnership. *Environmental Health Perspectives., 111*(13), 1571–1576.

Minkler, M., & Wallerstein, N. (2008). *Community-based participatory research for health from process to outcomes* (2nd ed.). San Francisco: Jossey-Bass.

Moreno Ramirez, D., Ramirez-Andreotta, M. D., Estrella-Sánchez, R., Wolf, A. M., Kilungo, A., Spitz, A. H., & Betterton, E. A. (2015). Pollution prevention through peer education: A community health worker and small and home-based business initiative on the Arizona-Sonora border. *International Journal of Environmental Research and Public Health, 12*, 11209–11226. https://doi.org/10.3390/ijerph120911209.

National Center for Farmworker Health. (1999). Available at: www.ncfh.org

National Institute of Environmental Health Sciences. (2015). *Advancing environmental justice*. https://www.niehs.nih.gov/research/supported/assets/docs/a_c/advancing_environmental_justice_508.pdf. Accessed on 10 Apr 2018.

National Research Council. (1996a). *National science education standards*. Washington, DC: National Academy Press.

National Research Council. (1996b). *Understanding risk: Informed decisions in a democratic society*. Washington, DC: National Academy Press.

National Research Council. (2008). *Public participation in environmental assessment and decision making*. Washington, DC: National Academy Press.

National Research Council. (2009). Learning science in informal environments: People, places, and pursuits. Committee on Learning Science in Informal Environments. In P. Bell, B. Lewenstein, A. W. Shouse, & M. A. Feder (Eds.), *Board on Science Education, Center for Education. Division of Behavioral and Social Sciences and Education*. Washington, DC: The National Academies Press.

Nicholson-Cole, S. A. (2005). Representing climate change futures: a critique on the use of images for visual communication. *Computers, Environment and Urban Systems, 29*(3), 255–273.

O'Fallon, L. R., & Dearry, A. (2002). Community-based participatory research as a tool to advance environmental health sciences. *Environmental Health Perspectives, 110*(Suppl. 2), 155–159.

Ottinger, G. (2010). Buckets of resistance: Standards and the effectiveness of citizen science. *Science, Technology & Human Values, 35*, 244–270, https://doi.org/10.1177/0162243909337121.

Pandya, R. E. (2012). A framework for engaging diverse communities in citizen science in the US. *Frontiers in Ecology and the Environment, 10*, 314–317. https://doi.org/10.1890/120007.

Pope, J. (2013). *Pima Association of Governments. Green infrastructure prioritization tool.* Retrieved from http://www.pagnet.org/documents/committees/epac/2013/EPAC-2013-11-01-Presentation-Green-Infrastructure-Tool.pdf

Ramirez-Andreotta MD and Kaufmann D. 2016. Understanding Environmental Health Risk Assessment. Available at: https://superfund.arizona.edu/sites/superfund.arizona.edu/files/8.5x14v9final.pdf

Ramirez-Andreotta, M. D., Brusseau, M. L., Artiola, J. F., & Maier, R. M. (2013). A greenhouse and field-based study to determine the accumulation of arsenic in common homegrown vegetables. *Science of the Total Environment, 443*, 299–306.

Ramirez-Andreotta, M. D., Brusseau, M. L., Artiola, J. F., Maier, R. M., & Gandolfi, A. J. (2014). Environmental research translation: Enhancing interactions with communities at contaminated sites. *Science of the Total Environment, 497*, 651–664.

Ramirez-Andreotta, M. D., Brusseau, M. L., Artiola, J. F., Maier, R. M., & Gandolfi, A. J. (2015). Building a co-created citizen science program with gardeners neighboring a Superfund site: The Gardenroots case study. *International Public Health Journal, 7*(1), 139–153.

Ramirez-Andreotta, M. D., Brody, J., Lothrop, N., Loh, M., Beamer, P., & Brown, P. (2016a). Reporting back environmental exposure data and free-choice learning. *Environmental Health, 2*(15). PMID: 26748908.

Ramirez-Andreotta, M. D., Brody, J. G., Lothrop, N., Loh, M., Beamer, P. I., & Brown, P. (2016b, July 8). Improving environmental health literacy and justice through environmental exposure results communication. *International Journal of Environmental Research and Public Health, 13*(7), 690–717. PMID: 27399755.

Ramirez-Andreotta MD, Lothrop N, Wilkinson ST, Root R, Artiola JF, Klimecki W, Loh MM. (2016c). Analyzing patterns of community interest at a legacy mining waste site to assess and inform environmental health literacy efforts. *Journal of Environmental Studies and Sciences, 6*(3), 543–55.

Rennie, L., & Johnston, D. J. (2004). The nature of learning and its implications for research on learning from museums. *Science Education, 88*(S1), S4–S16. https://doi.org/10.1002/sce.20017.

Resnicow, K., et al. (2004). Body and soul a dietary intervention conducted through African-American churches. *American Journal of Preventive Medicine, 27*(2), 97–105.

Roelofs, C., Shoemaker, P., Skogstrom, T., Acevedo, P., Kendrick, J., & Nguyen, N. (2010). The Boston safe shops model: An integrated approach to community environmental and occupational health. *American Journal of Public Health, 11*, 52–55.

Sandhaus S, Ramirez-Andreotta MD, Kilungo A, Wolf AM, Sandoval F, Henriquez P. (2018). Combating climate injustices: An informal science and popular education approach to addressing environmental health disparities. pedagogy in health promotion: The Scholarship of Teaching and Learning, https://doi.org/10.1177/2373379917751476.

Seiffert, M. E. B., & Loch, C. (2005). Systemic thinking in environmental management: support for sustainable development. *Journal of Cleaner Production, 13*(12), 1197–1202.

Sibertown, J. (2009). A new dawn for citizen science. *Trends in Ecology & Evolution, 24*, 467–471.

Soleri, D., Long, J. W., Ramirez-Andreotta, M. D., Eitemiller, R., & Pandya, R. (2016). Finding pathways to more equitable and productive public-scientist partnerships. *Citizen Science: Theory and Practice, 1*(1):9, 1–11, DOI: https://doi.org/10.5334/cstp.46

Star, S., & Griesemer, J. (1989). Institutional ecology, translations and boundary objects: Amateurs and professionals in Berkeley's Museum of Vertebrate Zoology 1907–39. *Social Studies of Science, 19*, 387–420.

Taylor, D. E. (2014). *The state of diversity in environmental organizations*. Ann Arbor: Green 2.0.

Tushman, M. L., & Scanlan, T. J. (1981). Boundary spanning individuals: Their role in information transfer and their antecedents. *Academy of Management Journal, 24*(2), 289–305.

U.S. Census Bureau. (2010). *Current population survey*.

U.S. EPA. (2016). *Environmental justice*. Available at: https://www.epa.gov/environmentaljustice. Accessed on 30 Sept 2017.

U.S. EPA. (n.d.). *Superfund site: General motors (Central foundry division)*. Massena, NY. Available at: https://cumulis.epa.gov/supercpad/SiteProfiles/index.cfm?fuseaction=second.Cleanup&id=0201644#bkground. Accessed on 18 Oct 2017.

U.S. General Accounting Office. (1983). *Siting of hazardous waste landfills and their correlation with racial and economic status of surrounding communities*. Washington, DC: United States General Accounting Office.

United Church of Christ. (1987). *Toxic wastes and race in the United States: A national report on the racial and socio-economic characteristics of communities surrounding hazardous waste sites*. New York: United Church of Christ, Commission for Racial Justice.

University of Arizona Superfund Research Program. (2018). Understanding environmental health risk assessment. Available at: https://superfund.arizona.edu/sites/superfund.arizona.edu/files/8.5x14v9final.pdf

Washington State Department of Health. (n.d.). *Lower Duwamish river advisory*. Available at: https://www.doh.wa.gov/Portals/1/Documents/Pubs/334-330.pdf

# Chapter 5
# Advancing Environmental Health Literacy Through Community-Engaged Research and Popular Education

**Catalina Garzón-Galvis, Michelle Wong, Daniel Madrigal, Luis Olmedo, Melissa Brown, and Paul English**

**Abstract** This chapter describes the process of increasing and sustaining environmental health literacy (EHL) within communities impacted by environmental hazards and associated health conditions through the comprehensive engagement of community members in environmental health research and education projects. The chapter discusses the use of popular education approaches to facilitate more effective collaboration and optimize mutual co-learning among community members and their project partners. It also explores how, by using this approach, community members can contribute their own knowledge and awareness of environmental and health conditions to advance the research and education process, thereby increasing the EHL of their academically-credentialed partners.

Examples from several community-engaged research and education projects are shared to illustrate the versatility of these approaches as a means of raising EHL in a variety of contexts, including researcher-initiated scientific studies, community-initiated trainings, and co-initiated research projects. The case studies demonstrate how popular education and community-engaged research approaches can contribute to building and sustaining more equitable partnerships between impacted communities and their partners (who may be based in varied institutional settings such as academia, government agencies, and nonprofit organizations), with an emphasis on mutual co-learning and knowledge co-creation among community and academically-credentialed partners as key drivers of increased EHL.

C. Garzón-Galvis (✉) · M. Wong · D. Madrigal · P. English
California Environmental Health Tracking Program, Richmond, CA, USA
e-mail: catalina.garzon@phi.org

L. Olmedo
Comite Civico del Valle, Brawley, CA, USA

M. Brown
University of California, Berkeley, CA, USA



S. Finn, L. R. O'Fallon (eds.), *Environmental Health Literacy*,
https://doi.org/10.1007/978-3-319-94108-0_5

**Keywords** Community-based participatory research · Community-engaged research · Community engagement · Community participation · Knowledge co-creation · Mutual co-learning · Popular education

## Introduction

Environmental Health Literacy has been described as a process for expanding an understanding of the connection between environmental exposures and health outcomes (Finn and O'Fallon 2017). The process of raising EHL can occur at both an individual and community level, resulting in a greater capability to take personal and collective action to reduce exposures and to "address environmental injustices that lead to health disparities" (Finn and O'Fallon 2017: 496). Individuals and communities may be drawn towards this process as a way of critically reflecting on their own socioeconomic circumstances, rather than simply incorporating specialized knowledge about environmental factors derived from external sources (Nutbeam 2000; Nutbeam 2008; Finn and O'Fallon 2017).

Community-engaged research is a promising, empowerment-oriented approach for raising EHL in which members of communities directly impacted by environmental health disparities are involved in the collection, analysis, and dissemination of data related to these disparities (English et al. 2017). By actively participating in environmental health research, both individuals and communities can develop an increased understanding of the social and environmental determinants of health, as well as the range of actions that can be taken to address them. Moreover, the process of community-engaged research can better position participants to more effectively implement interventions and other actions that can improve local environmental and health conditions.

This chapter focuses on how community-engaged research can combine knowledge derived from academic or professional experience with knowledge about local environmental and health conditions that is gained through lived experiences, thereby raising the EHL of community and academically-credentialed partners alike. The term popular education is used in this chapter to refer to a group-based approach to learning that acknowledges and incorporates lived experiences as a valuable source of knowledge (Wallerstein and Duran 2008). By recognizing that all individuals are experts about their own lives and communities of identity, a popular education approach compels one to challenge prevailing power dynamics and hierarchies that value the expertise of academically-credentialed researchers over that of communities directly impacted by environmental health disparities.

This chapter is based on the concept that health literacy is an asset that impacted communities also bring, rather than a deficit to be filled or addressed by outsiders (Nutbeam 2008). In this asset-based definition, health literacy is a way to build the confidence, knowledge, and ability of individuals and communities to gain greater control over their health as well as the environmental and social factors that impact it. Increased health literacy resulting from community-engaged research and popular education processes can support a longer-term outcome of greater empowerment in decision-making to improve health and quality of life at a personal and community level (Nutbeam 2008; Wiggins 2011).

## Background

### *EHL and Communities Impacted by Environmental Health Disparities*

In an environmental health context, an impacted community can be defined based on geographic proximity to a particular environmental hazard or pathway of exposure. Impacted communities can also be defined based on shared identity, such as a neighborhood where health disparities are concentrated or a group of individuals grappling with similar health outcomes. Low-income communities, communities of color, and other historically disadvantaged communities are more likely to be impacted by environmental hazards and adverse health outcomes (Brown 1995). Institutionalized racism and other structural inequalities contribute to disproportionate environmental exposures and elevated rates of associated health conditions in these communities (Paradies et al. 2015).

The prevailing framework for understanding and addressing health conditions primarily emphasizes the role of genetics and individual behaviors as factors contributing to adverse health outcomes. This can contribute to a lack of awareness at an individual and institutional level about the impacts that social and environmental determinants, such as poverty or proximity to pollution sources, have on health outcomes. As a result, residents of impacted communities can have negative experiences with government agencies or perceive that these agencies – such as environmental regulatory agencies, health service providers, or medical institutions – are dismissive about their concerns or even complicit in perpetuating the conditions that they are facing (Chavez et al. 2008).

Communities overburdened by environmental hazards often also face other pressing social and economic challenges like high rates of poverty, unemployment, crime, and housing and food insecurity. Given this context, some communities may have limited awareness of or interest in the environmental health issues impacting them. Lack of access to adequate health care and educational opportunities in these communities can also impede access to information about environmental factors that affect health and about interventions that can help reduce environmental exposures. Language barriers and limited access to information and communications technologies such as computers, mobile phones, or the internet can further affect the amount and quality of environmental health information that residents of these communities obtain (Finn and O'Fallon 2017).

At the same time, residents of these communities may be more aware of how structural or systemic inequities impact them than the academically-credentialed agency staff, health professionals, and researchers they interact with, who often come from more privileged socio-economic backgrounds and communities (Muhammad et al. 2017). These disparate levels of awareness and concern about the impacts of structural inequities on community environmental and health conditions can undermine trust and relationship-building between community and academically-credentialed partners. Explicitly integrating an analysis of how structural inequities shape community environmental and health conditions into community-engaged projects can create an opportunity for establishing authentic dialogue and building trust among all partners (Minkler 2004).

### *Lived Experiences with Environmental Hazards and Exposures as a Source of Knowledge*

The lived experiences of residents with environmental exposures and related health conditions are a key starting point to increasing EHL in communities contending with environmental health disparities. Personal experiences with managing chronic health conditions like asthma or caring for relatives who are sick can contribute to a heightened awareness of and interest in better understanding how environmental conditions can impact health. Lived experiences with environmental hazards and exposures can also contribute to a greater motivation to learn more about the connections between health and environmental conditions. Those directly impacted by these conditions are also more likely to apply their improved understanding of the relationship between health and environment to engage in behavior change, community activism, or otherwise take personal and collective action to address these conditions (Brown et al. 2011).

The lived experiences of residents of impacted communities are also a key source of knowledge that community-engaged research often aims to document or leverage. Community-engaged research affirms that community partners have valuable knowledge to contribute to the research process and that academically-credentialed partners can benefit from learning from their community partners. It also affirms that academically-credentialed partners may have lived experiences besides their professional or educational experiences that they can share and that community partners can benefit from. Mutual co-learning occurs when community and academically-credentialed partners value, share, learn from, and incorporate each other's lived experiences and knowledge into the research process (Israel et al. 2008).

Using approaches that engage community members in the processes of knowledge co-creation and dissemination can increase the EHL of all partners, as well as make the focus and outcomes of research more relevant to community priorities (Wright et al. 2011). Knowledge co-creation, or knowledge mobilization, is the process of engaging diverse stakeholders in generating knowledge together based on mutual learning through meaningful dialogue and collaboration (Abma et al. 2017).

In contrast to the more conventional concepts of "knowledge transfer" or "knowledge translation," knowledge co-creation challenges the premise that knowledge is generated by and transmitted via a one-way flow of information from academically-credentialed experts to community members, practitioners, and others.

## Theoretical Frameworks

### *Political Economy of Health*

Traditional clinical and public health frameworks tend to focus on individual responsibility in changing behaviors and attitudes in order to improve health. In contrast, a political economy of health framework emphasizes that an individual's ability to change behaviors and improve health is shaped by the local environment, community setting, and historical context. This historic view reveals how resource allocation and shifts in government policies and power dynamics have shaped health in communities. Demographic factors such as gender, class, and race impact life expectancy in the political economy of health framework.

To design and implement effective community-engaged research projects and other programs intended to improve health outcomes, it is essential to have a clear understanding of barriers that undermine health. A political economy of health framework can be useful for describing how social, economic, and environmental factors impact the health of individuals. This framework emphasizes the importance of exploring the ways that structural inequities influence health and examines how history has contributed to health conditions (Minkler et al. 1994).

The political economy framework is useful for health care providers, health educators, health promoters, local government officials, researchers, and others working with communities facing health disparities. This framework can support authentic dialogue about how structural inequities impact local environmental and health conditions. This framework can also help to create more effective community-focused solutions that address these inequities.

### *Community-Engaged Research Principles and Practices*

In an environmental health context, community-engaged research can be broadly defined as an approach to research that involves those directly impacted by environmental and health disparities in determining what is researched and how the research is carried out (Israel et al. 2008). This approach to research engages members of the impacted community as partners in making key decisions throughout the research process, from defining the research question and devising research methods to

**Fig. 5.1** Community-engaged research process. (*Adapted from Smith* 2011)

interpreting results and disseminating findings. Community-engaged research practitioners contend that the research process is improved and more impactful when communities play a significant role in its design, implementation, and use. The rigor, relevance, validity, and reach of research results can be improved by involving those directly impacted by the research topic in decisions about the research process (Balazs and Morello-Frosch 2013). Figure 5.1 shows the stages in a community-engaged research process.

According to Arnstein (1969), levels of community participation in decision-making can vary from nonparticipation to tokenism to citizen power, the latter of which includes partnership, delegated power, and citizen control (see Fig. 5.2). In theory, community-engaged research aims to involve community partners at the highest rungs of this ladder of participation by giving them the opportunity to be involved in decisions at each stage of the research process. In reality, community partners are not always involved in making key decisions at each stage of the process or may only be involved at certain stages of the process, as determined by or as decided in discussions with research partners (Banks et al. 2013). English et al. 2017 describe three levels of community participation in the research process, ranging from crowdsourcing or involvement in data collection to extreme citizen science that also engages community members in research design as well as interpretation and dissemination of findings.

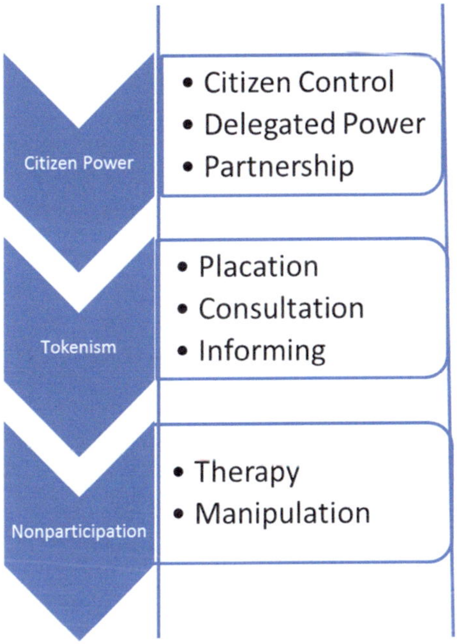


**Fig. 5.2** A Ladder of Community Participation. (*Source: Arnstein* 1969)

The "Decide, Announce, Defend" approach to decision-making historically used by government agencies and research institutions has often limited participation of community members to, at best, tokenistic levels of participation in this process (informing, placation, and/or consultation as described by Arnstein). Environmental justice activists and other residents of impacted communities have been instrumental in developing more meaningful structures for community participation in governmental and research decision-making. For example, the strategy of filing administrative complaints with environmental regulatory agencies based on civil rights legislation has resulted in more transparent and inclusive decision-making processes in the creation, implementation, and enforcement of environmental laws and policies (Cole and Foster 2001). Similarly, human subjects' protocols and informed consent procedures at universities and other research institutions have been expanded and refined based on grievances advanced by individuals and communities whose rights were exploited or violated during the research process.

Given this historical context, the expansion of community-engaged and participatory research approaches in the environmental and public health fields has called for developing and applying tools to optimize meaningful community participation at various stages of the research process. These participation tools have ranged from integrating community organizing tactics and group facilitation techniques into research processes, to the use of social media and cell phone applications to support broader community participation in data collection efforts. Technological innovations, such as the development of more accessible and affordable environmental sensors, have also enabled more widespread initiation and participation of community members in collecting and analyzing data (English et al. 2017). One outcome of this participation has been the increase in the EHL of those involved in the research.

## *Popular Education Principles and Practices*

Popular education is a key approach to optimizing the meaningful participation of impacted community members that has been applied within community-engaged research and education processes. A popular education approach affirms that each person is an expert in their own lives and communities of identity, and it is consistent with a more expansive definition of the environment advanced by the environmental justice movement as places where we live, work, play, pray, and go to school. As an approach to increasing EHL among all partners, popular education involves challenging a conventional linear or "banking" approach to health education in which the teacher or academically-credentialed "expert" deposits information or knowledge into the mind of the student or community member (Wallerstein and Duran 2008). Due to its emphasis on visual, experiential, and kinetic or activity-based forms of learning, popular education is an especially effective approach in working with children and youth, those with limited formal education, those with learning disabilities and other non-traditional learners, those with limited literacy or fluency in a predominant language, and other historically disadvantaged populations (see Chaps. 11 and 12 – Sullivan and Vandiver).

Championed by Brazilian educator Paulo Freire, popular education is grounded in a more cyclical and iterative understanding of how learning takes place, in which taking collective action based on a shared understanding of patterns across the lived experiences of community members is the ultimate goal (Freire 1970). As shown in Fig. 5.3, Freire's approach to empowerment education is based in fostering dialogue, questioning the root causes of one's lived experiences (i.e., developing a critical consciousness), and taking collective action to address these root causes and improve one's lived reality (Wallerstein and Bernstein 1988). Consistent with a political economy of health framework, this approach also involves developing a shared analysis of power relations and dynamics between the community and institutions like government agencies and corporations that can play a role in changing local conditions.

A prevailing use of popular education with impacted communities in the environmental health field is in peer-based health education or *promotores de salud*

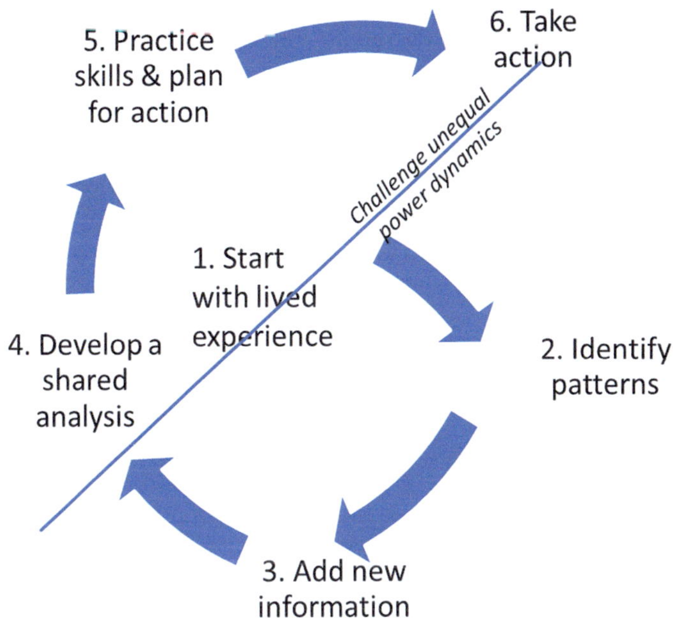


**Fig. 5.3** A popular education spiral of learning. (*Source: Freire* 1970)

models that engage individuals impacted by a specific environmental hazard or health condition in developing and disseminating health education materials. *Promotores* are typically members of a community who speak the same language and share similar lived experiences as those they are reaching out to and actively draw on these experiences in sharing information with other community members. These models often employ a train-the-trainers approach to build the capacity of lay health educators from impacted communities who in turn educate and engage other community members in interventions to protect and promote health. For example, Liebman et al. 2007 discuss how the Migrant Clinicians Network used a popular education approach to train *promotores* in rural New Mexico to share ways to reduce pesticide exposures at home and at work with other farmworkers and their families, with an emphasis on protecting children's health. Results of pre- and post-assessments before and after the *promotores*' educational interventions indicate that these activities led to increased knowledge among participants about actions that they could take to reduce exposure and improved behaviors to minimize incidences of accidental pesticide poisonings in the home.

## *Applications and Examples of Popular Education Within Community-Engaged Research*

Popular education can be used to build the capacity of residents to engage in research as well as to document community knowledge and integrate community-generated data into the research process. For example, Minkler et al. (2010) discuss how the Environmental Health Coalition involved *promotores* in southwest San Diego, CA, in mapping industrial land uses, conducting a community survey to identify health concerns, and identifying land use changes based on this research that could be integrated into local planning efforts. The *promotores* first completed a popular education training that built their research skills and their understanding of the connections between land use, air quality and health. They then applied their increased EHL and skills to engage a broader range of community residents by carrying out research activities, who in turn educate decision-makers about community concerns and advocate for land use changes to protect health as identified by residents.

A popular education approach can also create opportunities to explicitly discuss and address unequal power relations in research, as well as to determine how to apply research results to actions that address the environmental and health conditions being studied. Sadd et al. (2014) describe how the Los Angeles Environmental Health and Justice Collaborative analyzed data on the locations of air emissions sources and sensitive land uses collected by government and regulatory agencies. Academically-credentialed partners first determined that "analyzing the government's own data to assess racial and other disparities would be a powerful way to draw regulatory attention to environmental justice issues" (Sadd et al. 2014: 2015). Community members then undertook a ground-truthing process to validate government data, which deepened their understanding of agency data collection procedures and strengthened their ability to critically engage with regulatory agencies to articulate the limitations of this data. The results were used to support a policy campaign to get the City of Los Angeles to designate impacted neighborhoods as "green zones" offering incentives for greening the operations of local businesses and better enforcing regulations to address the cumulative impacts of environmental hazards.

## *How Community-Engaged Research Using Popular Education Can Increase EHL*

As applied within a community-engaged research process, a popular education approach can be used to engage members of impacted communities throughout the research process from study design to dissemination of results to action. When used in community-engaged research, a popular education approach affirms the lived experiences of impacted community members as a valued source of knowledge, which the research process can help to document. A popular education approach to community-engaged research also demystifies the decisions that researchers make throughout the research process, which is necessary for community members to be meaningfully involved in these decisions. Within a community-engaged research process, this approach can help to maximize the integration of knowledge based on lived experiences with academically-derived knowledge. It is through this integration that mutual co-learning and knowledge co-creation between community and academically-credentialed partners occurs, with increased EHL among all partners as a key outcome.

Using popular education as a tool to engage communities in the research process can deepen a shared understanding of the social and environmental determinants of health. This can also support the involvement of residents in identifying and taking public health actions to address environmental health disparities, which can further contribute to increasing EHL in impacted communities. Within a community-engaged research process, popular education provides greater opportunities to increase EHL by exploring, considering, learning, and sharing both community-based and academically-derived knowledge about how the environment and social factors impact health. Popular education creates avenues for different levels of engagement, with opportunities and support of community and academically-credentialed partners to move toward deeper engagement as EHL increases through mutual co-learning. Community-engaged research provides a structure for community and academically-credentialed partners to apply their increased EHL in a tangible and practical manner by enabling the sharing and co-creation of knowledge.

## Methods

Community-engaged research and popular education deploy a variety of tools and techniques to demystify technical information and more meaningfully engage impacted residents in the process of learning and documenting shared knowledge. Popular education techniques include question-posing, group dialogue, power mapping, role-playing, action planning, and collaborative problem-solving (see DDD and PI 2010). Critically analyzing and practicing taking action to change power dynamics is integral to many of these techniques (see Freire 1970). Wallerstein and Bernstein 1988 describe how posing a series of questions during facilitated dialogues can be applied in health education and promotion settings through the use of the SHOWeD framework. Each question in this framework, shown in Fig. 5.4, corresponds to a step in the popular education spiral of learning and culminates in planning to take action. This series of questions has been widely adapted and integrated into community-engaged research methods such as *photovoice*, in which residents take pictures to document their lived experiences with an issue and then use these pictures to discuss and share community solutions to address the issue (Hergenrather et al. 2009). *Question-posing* within facilitated group dialogue can be particularly useful in eliciting, sharing, and documenting community knowledge on an issue as well as in identifying key community concerns or issues to focus research on.

| | |
|---|---|
| Naming the Problem | "What do we See here?" |
| Defining the Problem | "What's really Happening here?" |
| Sharing Experiences | "How does this relate to Our lives?" |
| Questioning Root Causes | "Why is this happening?" |
| Exploring Solutions | "How can we become Empowered with our new understanding? |
| Taking Action | "What can we Do about this?" |

**Fig. 5.4** The SHOWeD Framework. (*Source: Wallerstein and Bernstein* 1988)

*Role-playing* as a popular education technique is often used to provide opportunities for participants to practice using new skills or knowledge as well as to apply and articulate their existing knowledge in new contexts. For example, participatory theater and Theater of the Oppressed approaches (see Chap. 9) engage participants in acting out scenarios based on their lived experiences in which they confront those in positions of power to change the outcome of a disempowering or traumatic situation (Sullivan and Lloyd 2006). *Power mapping* involves participants in identifying those individuals or entities with the power or authority to make decisions on a particular issue, and in developing strategies to persuade those in power to change their decisions or actions. *Action planning* often includes elements of power mapping and engages participants in identifying what they themselves can do to advance change on a particular issue. Lastly, *collaborative problem-solving* often includes elements of both role-playing and power mapping by having participants practice carrying out their actions or strategies to advance change in a scenario based on a real-world context in which they will engage those in power, such as a meeting with decision-makers (see Gonzalez 2014).

Co-created knowledge is more likely to result in social impact such as policy and systems changes that can improve health and quality of life, since it is often generated in a context in which it will be directly used or applied (Abma et al. 2017).

To set the stage for mutual co-learning and to achieve the goal of increasing EHL among all partners, community knowledge and expertise must be valued on

**Table 5.1** Outcomes of popular education projects. (*Source: Wiggins 2011*)

| Health-related outcomes | Empowerment-related outcomes |
|---|---|
| Changes in health-promoting behavior | Increases in self-esteem and self-confidence |
| Gains in health knowledge | Increases in multiple components of psychological empowerment |
| Increased health literacy | |
| Improvements in physical markers of health risk factors (e.g., exposure reduction) | Development of critical consciousness |
| | Increases in sense of community |
| Actual physical changes in health | Increased participation |
| Improved food security | Increases in personal and collective control |
| | Actions of solidarity to improve the community |
| | Increased motivation to bring about change through advocacy |

par with knowledge and expertise derived from formal educational systems (Minkler 2004). The goal of increasing the EHL of all partners must be shared among all involved, including academically-credentialed partners, and requires that all partners approach their collaboration with cultural humility and a willingness to learn from their experiences in working together throughout the research process. Cultural humility is an underlying principle of community-engaged research that recognizes that no one else can master or speak for another's culture or community of identity and, as such, challenges prevailing notions of expertise (Israel et al. 2008) Table 5.1.

## *Case Study: Imperial County Community Air Monitoring Project*

This case study outlines the Imperial County Community Air Monitoring Project (Imperial Air Project), a multi-year, federally-funded, community-engaged project conducted in Imperial County, California (English et al. 2017). Located at the U.S.-Mexico border, this mainly desert and agricultural county is impacted by a range of environmental health concerns, particularly air pollution and asthma. In this community-initiated project, residents played a key role in establishing a network of 40 particulate matter (PM) air monitors to better understand community-level air quality. The case study describes how community members were engaged in this project, highlighting how popular education methods were interwoven into the project to facilitate engagement. It also demonstrates how the integration of scientific and community knowledge improved the project and led to increased EHL among all partners.

## Background

Outdoor air pollution is a serious public health issue, leading to the premature deaths of over 3 million people worldwide annually (WHO 2014). One type, PM pollution, consists of materials such as dust, dirt, soot, smoke, acids, and metals. Major sources of PM include motor vehicles, industrial operations, fires (such as from fireplaces, wildfires, brush or trash burning), and dust (such as from construction, agriculture, and landfills or windblown from open lands).

Exposure to PM is known to be harmful to human health. It is related to increased risk of mortality and excess hospitalization, even at levels below regulatory limits (Vaduganathan et al. 2016; Di et al. 2017). PM exposure is also associated with increased respiratory disease, decreased lung function, and asthma attacks in susceptible individuals (Anderson et al. 2012).

Information about local PM levels is essential for community members who wish to reduce personal exposures or advocate for emissions reductions. It is also helpful for conducting analyses to examine geographic and temporal patterns and to model the spatial distribution or dispersion of pollutants. In the U.S., government agencies use stationary air monitors to measure ambient PM levels and ensure they remain within regulatory limits. While this information is made available to community residents with varying degrees of accessibility, it is often incomplete. PM levels can vary within small geographic areas, yet regulatory monitors are too costly to install at a density that captures this variation.

With the advent of portable, low-cost, next-generation air monitoring technology, it is now feasible to deploy a network of monitors to collect air quality data at a finer geographic resolution. To maximize the utility of this data for residents, community input must be integrated into project activities, including decisions about where the monitors are installed, how the data are analyzed, and how information is presented and disseminated.

## Population Affected

Because air is ubiquitous and essential to human life, air pollution impacts everyone. However, children, the elderly, individuals with chronic diseases, and outdoor workers are particularly susceptible to PM exposures and related health impacts. In Imperial County – home to a primarily Latino population (82%) and some of the highest rates of unemployment and poverty in the nation (U.S. Census 2010) – residents endure a disproportionate burden of PM-related diseases. For example, the county surpasses all other California counties as having the highest rates of emergency visits and hospitalizations for asthma among school-age children (CEHTP 2017). PM pollution itself is a well-known environmental issue in the county, which has exceeded the California standard for $PM_{10}$ (particles 10 μm or smaller) for several decades, with periods lasting over six months (CARB 2012).

The Imperial Air Project, which utilized a tiered community engagement process (described below), involved residents from cities and towns throughout Imperial County. These individuals included local activists, parents, high school students, senior citizens, school staff, and teachers. While not a condition for involvement in the study, all participants were concerned about air pollution and either had asthma themselves or knew someone with asthma.

## *Origin of Project and Project Goals*

Although PM pollution is a serious problem in Imperial County, there are only 5 regulatory PM monitors for an area greater than 4000 square miles. Many towns are located miles from the nearest air monitor, prompting concerns that regulatory monitoring data are not representative of local conditions and therefore are not useful in real-time for deciding when to take action to reduce exposure (such as limiting outdoor physical activity or staying indoors).

At a meeting about air pollution convened by a local community-based environmental justice organization called Comite Civico del Valle (CCV), residents cited the need for monitors in their own communities. This community input eventually became the inspiration for the Imperial Air Project, which was collaboratively initiated and designed by the project's three main partners: CCV, the California Environmental Health Tracking Program (CEHTP),[1] and University of Washington (UW). The goal of the project was to develop a community air monitoring network that would provide air quality data that (1) could be used in real-time so that residents could be better informed about when to take action to reduce exposure, and (2) would support scientific analyses to identify air pollution trends, patterns, and hotspots to inform policy and other decision-making. Funding for the project was provided by a "Research to Action" grant from the National Institute of Environmental Health Sciences (NIEHS), which required projects to include community engagement and use research results to support actions to improve health.

## History of Collaboration/Partner Roles

The ability to pursue a funding opportunity through NIEHS to implement this community vision was enabled by a positive relationship between CCV and CEHTP, which was developed while working together on previous projects, as well as CCV's membership on CEHTP's advisory group. The addition of UW as a partner came through CEHTP's knowledge of their air monitoring expertise and previous community-based work.

[1] CEHTP is a collaboration of the Public Health Institute, a non-profit organization, and the California Department of Public Health

A key aspect of the Imperial Air Project is the strong collaborative partnership among the main project partner organizations. Each partner encompassed a unique, complementary role in the project. CEHTP, as the main recipient of the project funding, provided overall project management and coordination, as well as expertise in community-based participatory research, health education, data visualization and other communications, and environmental health sciences. CCV provided local knowledge about environmental health concerns, strong relationships and networks within and outside of the county, and expertise in outreach, organizing, advocacy, and policymaking. UW provided expertise in air pollution measurement, analysis and spatial modeling, and low-cost sensor technology. Air monitoring and public health researchers from the University of California Los Angeles (UCLA) and George Washington University (GWU) also provided consultation to the project.

## Structures for Community Engagement

The project's tiered community participation structure was designed to ensure that activities were conducted with meaningful community engagement. CCV's co-investigator role and active participation in the initiation, design, implementation, and evaluation of project activities ensured strong community engagement throughout. Next, the project's Community Steering Committee (CSC) – consisting of about 20 community members that included local advocates and concerned residents – contributed additional guidance, participation, and decision-making at key stages. Finally, a broader group of community participants representing towns throughout the county had more limited engagement, participating in several activities, including a process to decide where to put the monitors.

## Application of a Popular Education Approach in Project Activities

When examining the overall project with an EHL lens, the fundamental knowledge desired by both the community and the academically-credentialed partners was a deeper understanding of community-level air quality in Imperial County. The project aimed to enhance literacy in that area through the collection, analysis, interpretation, and dissemination of this data. A critical part of this was to ensure community input to the network design; if the monitors were placed in locations that were not meaningful to residents, the resulting data would have limited utility and impact for decision-making. Key aspects of this community engagement process, enhanced by using a popular education approach, are summarized below. These steps follow the popular education spiral of learning as displayed in Fig. 5.3 (Freire 1970).

*Step 1: Sharing Lived Experiences and Step 2: Identifying Patterns* To select locations for the monitors, the CSC and the broader group of community participants shared their *lived experiences* with each other and the project team, which enabled them to *identify patterns* related to air quality issues. Building upon this, they also *used their knowledge* of who in their community were most vulnerable to air pollution and where they spent time in the community to help identify, assess, and select priority locations for air monitors. This process was conducted over a two-day period and also used a popular education approach. After sharing their knowledge and identifying commonalities, participants learned new information about air monitoring concepts, visited candidate monitoring sites to collect data on site characteristics, assessed the results, and strategized together to make recommendations to the project team on where to place monitors.

*Step 3: Add New Information* To establish the monitoring network, monitors were then deployed at community-selected sites, *adding new information* by collecting air quality data. As part of this process, CCV staff used their community knowledge and existing relationships to obtain permission from owners of the sites to install the monitors, which greatly facilitated the approval process for monitor siting. CCV staff was trained on how to assemble, install, operate, and maintain the monitors, and were able to assume responsibility for all of these tasks by the end of the process.

*Step 4: Develop a Shared Analysis* The project team set up ongoing calculations of the new monitoring data in order to produce metrics that would inform the community about real-time air quality levels. These metrics were selected based on input by the CSC about what would be most useful and understandable to the community. A spatial *analysis* of initial data also helped *identify patterns* of air pollution in the county, which informed the placement of a second round of air monitors. Finally, the monitoring data were used to conduct more complex *analyses* to assess the quality of the monitors and characterize community-level air quality throughout the county. CCV and the CSC provided feedback on the analyses and contributed to the interpretation of the results, providing insights on potential factors contributing to those patterns and discussing implications and uses of the findings.

*Step 5: Practice Skills, Strategize* CCV and the CSC engaged in several *strategizing* sessions with the rest of the project team to identify *plans* for displaying and disseminating the data and conducting other community *actions*. As part of this, the CSC participated in focus groups and user testing, which informed the data display plan by identifying data visualization methods that would be understandable and appropriate for the community. To disseminate the data, CCV recommended that the air monitoring data be made available on the community-based environmental reporting website that CCV operated (ivanair.org), thus both leveraging and enhancing an existing community resource. Finally, as part of a community *action strategy*, CCV and the CSC identified strategies for using project data and results to inform environmental and health-related behaviors, programs, and policies and to generate support for sustaining the monitoring network.

*Step 6: Take Action* A range of *actions was taken* by CCV, the CSC, and other community members in response to the new information generated by the monitoring network and other knowledge gained through participation in the project around air quality, health, and air monitoring. CCV conducted targeted outreach about the monitoring network at schools, resulting in increased awareness and use of the data. For example, the monitoring data were used for the "flag programs" at the 14 schools where a monitor was installed. For this program, the school displays a color-coded flag corresponding to the current air quality level–thereby informing the broader community of the air quality status–and keeps students indoors when pollution levels are high. CSC members also engaged in individual actions. For example, one member established a flag program at his senior center, while another founded an environmental leadership program at the high school she was attending. Motivated by her participation in the project and her personal experience as a parent of an asthmatic child, another CSC member applied for and received a postdoctoral fellowship to examine the relationship between air quality (using project data and regulatory monitoring data), asthma outcomes, and agricultural burning. As a mechanical engineer, this represented a new field of study for her. Finally, *using what was learned* in the project, CCV and the rest of the project team have initiated and supported air monitoring efforts in other communities in California and have provided technical assistance to regulatory agencies tasked with implementing community-level air monitoring. To this point, the project's community-engaged approach and successful results were a key influence on the introduction and passage of California Assembly Bill 617, which requires local air districts to deploy community air monitoring systems in their jurisdictions (AB 617, Chapter 136, Statutes of 2017).

The Imperial Air Project is an example of a larger scale, multi-component project where–due to the tiered participation structure and a complex research design–there were often multiple levels of community engagement occurring at the same time. Although this broad project overview excluded a number of project activities for the sake of clarity, in some cases, the activities occurred in parallel and corresponded to different steps of the popular education spiral, while some activities (such as the site selection process) required their own popular education approach (a spiral within a spiral). This case study highlights the versatility of popular education techniques and the ways in which these, in combination with community engagement strategies, can be integrated into a project to raise the EHL of participants.

## Impacts on Environmental Health Literacy and Community Empowerment

Community engagement in the project resulted in new knowledge about air quality in Imperial, increasing the EHL of the community and academically-credentialed project partners, the broader group of community participants, and other members of the community. This information was co-created through the collaborative establishment of the community air monitoring network.

Community engagement in this project also resulted in empowerment of community partners and members, as evidenced by evolving roles in the project. For example, several community participants joined the CSC after participating in the site selection process, one CSC member became CCV staff, and CCV staff began taking on roles originally assigned to the academically-credentialed partners (such as monitor deployment and maintenance). Beyond project activities, empowerment of CCV was demonstrated by their increased leadership role in government advisory boards, the increased number of requests for CCV staff to attend and present project findings at regional, statewide, and national scientific meetings (including NIEHS and US EPA), and their new role as mentors for other community-based organizations and regulatory agencies interested in conducting or supporting community air monitoring. Evaluation results also indicate that community partners and members felt that their EHL related to environmental health and air monitoring had increased due to their participation and that they felt more empowered to take action.

## Challenges and Limitations

The project team encountered multiple obstacles that challenged the team's ability to effectively establish mechanisms of engagement or threatened the community's interest or ability to participate in the project. These challenges included the length of the project, the long-distance partnerships (aside from CCV the other main project partners are more than 500 miles away from Imperial County), the complex and numerous project activities requiring planning around who, how, when, and what kind of input or engagement should occur for each activity, and the limited staffing among all partners. CCV's strong engagement in the project helped to mitigate some of these challenges. As a project partner, they had meaningful participation at all team meetings and decision-making processes. Thus, CCV served both as the primary community contributor in the project and as a bridge between the broader community and the project. Furthermore, to enable CCV to sustain such a high level of participation and contribution to the project, it was imperative that they receive equitable financial compensation alongside their academically-credentialed partners, and this understanding was reflected early on in the budget for the grant proposal.

Despite the challenges noted above, the level of community engagement and empowerment achieved in this project went well beyond what any of the partners had expected or planned for at the beginning of the project. Therefore, some unexpected challenges arose when the academically-credentialed partners were not expecting or immediately prepared to meet CCV's requests for additional knowledge sharing (related to monitor assembly and software installation, database configuration, data validation and conversion calculations, QA/QC scripts, and other technical activities).

Facilitated by the effective communication methods and trust developed amongst the partners, this challenge was resolved by co-developing and implementing specific plans to address these requests. Through this process, CCV was able to gain the additional knowledge and capacities they had identified, aside from some highly technical capacities that could not be fully developed during the project due to the time required to gain them. This example demonstrates the success of the community engagement process, resulting in increased EHL and empowerment within CCV and open communication among project partners that enabled them to reassess and challenge the traditional roles and hierarchies in the research process.

## Case Study: HERMOSA Study

This case study discusses an opportunity to investigate an emerging environmental health topic by a previously established community-engaged research partnership with youth from a farmworker community. In the Health and Environmental Research on Make-up of Salinas Adolescents (HERMOSA) study, youth researchers from Salinas, CA, a majority Mexican-American farmworker town, and academically-credentialed researchers from UC Berkeley and Clinica de Salud del Valle de Salinas investigated exposure to endocrine disrupting chemicals in adolescent girls (Harley et al. 2016). This case study describes the approach to a project that was initiated by academically-credentialed partners but was then conducted in partnership with youth researchers.

## Background

Endocrine disrupting chemicals (EDCs) in personal care products have been a growing source of concern because of their ubiquity and their potential health risks. Studies in animals and humans provide evidence that these chemicals may interfere with the human endocrine system. Of particular concern is how they disrupt estrogen and thyroid hormones that regulate development processes. National surveys have found detectable levels of EDCs such as phthalates, parabens, BP-3 and triclosan in 96%, 90%, 97%, and 75% of the population, respectively (CDC 2015). Public awareness of EDCs and concern about their effects is also growing, which is demonstrated by the expanding selection of personal care products that explicitly exclude these as ingredients. Despite these concerns, there are few protections and regulations for personal care products in the United States.

To further explore this issue, a new project was proposed with a previously formed community-based participatory research program for youth. The proposed project was a collaboration between youth research partners and academically-credentialed researchers to learn more about how adolescent girls are exposed to

EDCs in their make-up and personal care products and to evaluate ways to reduce exposure to EDCs.

## Population Affected

Although all people are vulnerable to the effects of endocrine disrupting chemicals in personal care products, there is special concern for windows of susceptibility that may have an outsized influence on disease development. Puberty is one of the windows, and there is concern for the impact that endocrine disruptors may have on the development of breast cancer. Since the community partner, the CHAMACOS Youth Community Council, is located in a majority Latino community, this project focused on Latino adolescents. The youth researchers included males and females; the participants in the research study were female.

## History of Collaboration/Partner Roles

The CHAMACOS Youth Community Council was an element of the community-engagement program of a longitudinal birth cohort study in an agricultural region of California (the CHAMACOS Study). The CHAMACOS Study enrolled 600 pregnant women from 1999–2000 and will continue to follow them and their children until 2024 to better understand the connection between environmental exposure and health outcomes with a focus on neurodevelopmental outcomes of the children (Eskenazi et al. 2003). Study participants were recruited from prenatal care clinics serving the farmworker population and the resulting cohort consists mostly of Mexican-American mothers from low-income farmworker families. More than 150 publications have come from the CHAMACOS study, finding associations of pesticides and other environmental exposures with birth outcomes, neurodevelopment, behavior, asthma and respiratory outcomes, timing of puberty, and obesity. From the beginning, the CHAMACOS study included activities to engage community members in the research ranging from presentations to more interactive participatory research projects (O'Fallon and Dearry 2001).

In 2010, investigators involved in the CHAMACOS Study launched the Youth Community Council (YCC) to engage high school adolescents from the community in environmental health, justice, and literacy issues. The first activity of the YCC was a PhotoVoice project in which students took pictures of local environmental risk factors and assets and described their lived experiences to identify what impacts health in their community. The PhotoVoice methodology allowed youth to identify important issues from their perspective, and used the SHOWeD framework described in the Methods section of this chapter to dig into the root causes of why those issues exist and persist. From this activity, youth identified a wide array of issues, many of which focused on the built environment of their city (Madrigal et al. 2014). Based on this assessment, they participated in a community effort to provide more opportunities for healthy activities in safe spaces. Following up on this activity, youth researchers worked with the local health department to conduct an assessment on walkability to further investigate issues that impacted health. Through their involvement in this program, the youth researchers were able to work on issues of common concern to members of their community thereby raising their EHL while learning new methods of gathering and producing information. During this time, the relationship between youth researchers and the academically-credentialed researchers also developed as methods of research practice and priorities were shared and discussed.

## Origin of Project and Project Goals

Two years after the establishment of the YCC, a grant announcement was issued by the University of California's Breast Cancer Research Program for projects related to breast cancer that used participatory research methods. The academically-credentialed researchers – aware of the EDCs and knowledgeable about the abilities of the youth researchers – proposed a research project to the entire group: Conduct an intervention study with high school students with the goal of lowering EDC levels through use of personal care products with lower-EDC ingredients. One critical component of this project was the funding source, which required both a research principal investigator and a community principal investigator. The community principal investigator was a staff member of Clinica de Salud del Valle de Salinas who had a strong leadership role in the CHAMACOS study and was a member of the community. This joint leadership codified the collaborative approach taken by the project for all involved.

### *Structures for Community Engagement*

The project proposal was originally conceived by the academically-credentialed researchers, and then shared with the youth partners to gauge their interest. Once the project was shared, the youth researchers expressed enthusiasm and the final grant proposal was written by academically-credentialed researchers as well as community partner staff. After the project was selected for funding, the methodology was jointly developed between academically-credentialed researchers and youth researchers at bi-monthly meetings held during the school year. Youth researchers helped academically-credentialed researchers to better understand purchasing patterns of personal care products and the relevant behaviors of their adolescent peers. Academically-credentialed researchers proposed a methodology to conduct the study, described the purpose of each element of the study, and discussed with the youth the various options for carrying out the study. Through this process youth researchers became better acquainted with the culture and customs of the academic research process.

## *Applications of a Popular Education Approach in Project Activities*

With emerging environmental concerns that are largely hidden from view, such as endocrine disrupting chemicals, an adapted popular education approach is necessary to incorporate the knowledge and lived experiences of community members. Awareness of endocrine disrupting chemicals in consumer products is relatively new and the impact of EDCs on human health is still being clarified, although public awareness is increasing. The idea for this project was proposed to the youth researchers in a meeting with academically-credentialed researchers; for that reason, when considering the popular education spiral, this project started with "*Step 3: adding new information.*" From there the youth participants were able to explore their lived experiences as these related to personal care products and identify patterns before entering the shared analysis, practice, and action steps.

*Step 3: Add New Information* The concept of endocrine disrupting chemicals was presented to youth researchers with activities that helped youth learn about common EDCs present in products they use. Using mobile applications that scan bar codes of personal care products used by youth in their homes, they were able to see how many of these products contain phthalates, parabens, BP-3, and triclosan. After the activity, the academically-credentialed researchers described the outline of the project. Adolescent girls would complete surveys about their personal care product use, use low-EDC products for a three-day intervention period, and provide urine samples before and after the intervention to see if levels of EDCs decreased. The youth researchers helped with the design and implementation of the research project. While the topic of EDCs was not familiar, the youth recognized that use of make-up and personal care products is an important issue for adolescents and that exposure during this time period of rapid reproductive development could have long-term health consequences. This recognition represents an early stage of EHL. The study provided the opportunity to work on a health research project that would provide new knowledge and new experiences, and members of the YCC were eager to participate in the venture.

*Step 1: Share Lived Experiences* In the first session with youth researchers, youth were asked to write down all personal care products they use on a weekly basis.

*Step 2: Identify Patterns* Once lists of often-used products had been established, academically-credentialed researchers facilitated a discussion with youth regarding the options available to their families, who made decisions on which products to purchase, and what factors influenced these decisions.

*Step 3: Add New Information (On-going)* After the initial project proposal, academically-credentialed researchers continued to provide experiences and activities that would bring new information to youth researchers to help them better understand the context and relevance of EDCs in cosmetics at bi-monthly meetings. Speakers from the Campaign for Safe Cosmetics and from the California Department of Public Health spoke to the YCC about EDCs. From these talks, youth researchers learned about the relatively weak safety regulations imposed on personal care products and strategized to minimize use of personal care products with high levels of EDCs.

*Step 4: Develop a Shared Analysis* After obtaining this new information, examining the lived experiences of youth researchers, and identifying patterns in their personal care product use, the group (youth and academically-credentialed researchers) developed protocols for the intervention study. The basic outlines of the study and the necessary scientific methodologies were provided by the academically-credentialed researchers, but the details and implementation of the project were conducted in deep collaboration with youth partners. Through this process, academically-credentialed researchers learned about the youth's lived experiences and knowledge, and youth researchers learned about research methodology and practices (Harley et al. 2016). The following list of activities provides selected examples of how this collaboration looked in practice.

**Selecting a Name for the Project**
In one of the first sessions, youth and researchers spent 2 h in a brainstorming and discussion process to decide on a name for the program. Priority characteristics identified were to select a name that resonated with the local community, that it be descriptive of the project, and that it be memorable. After hours of debating between several acronyms the group settled on the HERMOSA Study as an acronym (HERMOSA is Spanish for "beautiful"). Allowing the youth to determine the name and design the logo increased their feelings of ownership over the project.

**Recruiting Participants**
Youth researchers recruited their peers into the program because they were both well-informed about the project and had trust as members of the community. If potential research participants had questions, they were able to discuss them with the youth researchers.

**Developing the Intervention Protocol**
As the protocol was developed, academically-credentialed researchers presented major elements of the study and youth researchers gave feedback. Discussions between the youth and academic researchers helped the group identify priorities from academic researchers on rigor and sound methodology, and the youth were able to assess which elements of the proposed protocol would be acceptable to their peers. With this information, a study design was developed that both gathered

necessary data and provided useful information for the study participants. The protocol was pilot tested with girls from the YCC who were effectively the first study participants. Testing allowed youth researchers to experience the entire study from the point of view of participants and to provide feedback to assure that the goals of the study were being met. Having academically-credentialed researchers explain the purpose of each element of the study and having youth researchers absorb the information, ask questions, and make recommendations ensured a process that optimized priorities for both the youth and academically-credentialed researchers involved.

**Carrying out Intervention Protocol**
During the summer of 2013, the YCC carried out the entire study protocol with 100 adolescent girls from the Salinas area. This included a home visit and two office visits for each student. Due to careful consideration of the methodology, there was a 100% retention rate of participants throughout the three phases of the study. Working on the study gave youth researchers the opportunity to present information to their peers many times throughout the summer, allowing them to gain a deeper understanding of the relationship that study participants had to their personal care products.

*Step 5: Practice Skills, Strategize* After the intervention component of the study was carried out, the youth and academic researchers pivoted to education and action. In addition to conducting the study to learn more about a possible way to lower levels of EDCs in adolescent girls, a critical element of the project was to raise the EHL of the local community and take action to lower EDC levels at a larger scale. To reach these goals, youth researchers developed a series of presentations, trainings, health education materials, and social media communications to raise awareness about EDCs in the community. These health education materials were enhanced with data that came from an analysis included in the study and were reviewed by academically-credentialed researchers.

*Step 6: Take Action* The action goal at the state and national level was another critical element of the campaign. The youth researchers chose to create a petition for local major retailers to encourage them to display low-EDC products more prominently in their stores. They also developed educational materials, including a wallet-sized shopping guide for choosing products with lower levels of chemicals, and participated in health fairs and other events to demonstrate do-it-yourself (DIY) natural beauty product recipes that could be made at home. To share this information the youth spoke with radio, print and TV journalists, an experience that increased the visibility of the study with community members and family members. Finally, youth researchers presented the study findings to policymakers at the California Safer Consumer Products Commission in Sacramento who are responsible for regulations on chemical ingredients in personal care products throughout the state.

## Impacts on Environmental Health Literacy and Community Empowerment

Through their participation in the design and execution of the study, youth researchers gained a deeper understanding of environmental health and about the topic of endocrine disrupting chemicals in personal care products (Madrigal et al. 2016). As a result, EHL was raised for these youth through the process of their full community engagement in the study. During this process, youth participants used their own experiences and knowledge to help create a successful research investigation. Youth researchers became educators for their community to share information on a topic that is not well understood, namely, exposure to endocrine disrupting chemicals in personal care products. They also became advocates by contacting local companies to provide a greater number of lower-EDC product alternatives, thereby joining larger efforts to prevent exposure to EDCs. The leadership of the youth brought many more community members into the conversation on the topic.

The academically-credentialed researchers also gained a deeper understanding of exposure to endocrine disrupting products from the youth researchers by having constant dialogue about this issue throughout the project. By hearing from the youth researchers about their family's personal care product purchasing habits and perceptions about related health issues, they were able to better understand the context in which the study was taking place. In addition to resulting in an improved study design, having community members actively engaged in the research made the knowledge generated by the study more likely to be distributed and acted upon.

## Challenges and Limitations

Similar to many emerging environmental health concerns, EDCs were not identified as an urgent priority by youth researchers at the beginning of the project. The project was seen as an opportunity to learn to explore a health topic that was of interest, but was not initially a priority concern for community partners. Balancing the benefits of working together on the project and other concerns for youth in the community was always a consideration. Throughout the project, time was taken during project meetings to discuss other issues salient to the lives of the youth partners. Coordinating the project with academically-credentialed partners was also challenging because this required a more deliberate process to explain the purpose of each element of the study methodology. This additional time can be difficult to schedule when tending to many competing research priorities.

## Cast Study: Ditching Dirty Diesel Collaborative

This case study on the work of the Ditching Dirty Diesel Collaborative (DDD) discusses how a robust popular education approach can be used to strengthen community-engaged research and collaborative problem-solving among community groups, agency staff, and academically-credentialed researchers. DDD's work to reduce the impacts of freight transport illustrates how popular education can support efforts to improve compliance with existing regulations and to engage impacted communities in advancing actions to improve health. This case study demonstrates how popular education activities can be used to disseminate and deepen an understanding of environmental health information within and across community organizations and public health agencies. Lastly, DDD's work also shows how popular education can be used by community leaders and advocates as a tool to educate decision-makers and others with the power to institute policy and systems change about community concerns and solutions.

## Background

Freight transport, or goods movement, is the movement of raw materials and products from where they are made to where they are sold. Freight trucks, trains, ships and cargo handling equipment are largely powered by diesel fuel, whose combustion releases toxic air emissions such as diesel particulate matter (DPM 2.5 and 10). Over 70% of airborne cancer risk in California is attributable to exposure to diesel emissions (ARB 1998). Other health impacts of exposure to diesel exhaust include reduced lung function in children, cardiovascular disease, and respiratory illness such as asthma (ARB 2006). Community impacts of living near ports, freeways, railyards, distribution centers, and other freight transport infrastructure include noise and vibrations, disrupted sleep and concentration, and pedestrian safety hazards (DDD and PI 2006).

## Populations Affected

In California, low-income people, people of color, children, the elderly, and those with preexisting medical conditions are disproportionately affected by the health impacts of freight transport (DDD and PI 2006). In the San Francisco Bay Area, freight traffic is concentrated along major rail lines, freeways, and warehousing areas such as the I-880 corridor servicing the Port of Oakland, the nation's fifth busiest containerized port (DDD and PI 2011). In Oakland, freight trucks are prohibited on the I-580 freeway, concentrating freight pollution and impacts in flatlands neighborhoods like West and East Oakland which are predominantly low-income and people of color communities. Other freight-impacted communities in the region include North Richmond, San Leandro, and Bayview-Hunters Point in San Francisco.

## History of Collaboration and Partner Roles

Founded in 2004, the Ditching Dirty Diesel Collaborative (DDD) is a regional coalition of community-based organizations, environmental health advocates, and public health departments working to reduce exposure to diesel pollution in the San Francisco Bay Area. The strategic direction of DDD has historically been set by a Steering Committee comprised of individuals and organizations from freight-impacted communities. DDD members also include environmental health nonprofit organizations, local public health departments, and environmental regulatory agencies like US EPA Region IX's Air Division. DDD's core activities are carried out by issue-based committees, which include community leaders, agency staff, and advocates so as to foster collaborative problem-solving to address freight impacts.

## Origin of Project and Project Goals

At DDD's founding gathering in October 2004, over 100 people participated in a participatory mapping exercise to identify freight traffic "hot spots" and issues of concern throughout the San Francisco Bay Area. This mapping exercise helped identify patterns in the impacts that different communities were experiencing, which then led to identifying common solutions to address shared concerns (DDD 2017). Building on the outcomes of this gathering, DDD established two issue-based committees to lead its core campaigns and activities, an Anti-Idling Committee and a Freight Transport Committee. DDD's core activities include community and decision-maker education, capacity-building, community-engaged research, and policy advocacy to reduce diesel truck idling and other impacts of freight transport operations in nearby communities. Grounded in its community-driven coalition structure, DDD has applied a popular education and leadership development approach to engage freight-impacted communities in regional planning and decision-making on air quality, land use and transportation issues.

### *Structures for Community Engagement*

Members of DDD's Anti-Idling Committee developed an educational campaign with and for truck drivers to provide them with information about the health impacts of idling. This campaign also aimed to increase awareness of and compliance with state anti-idling regulations that imposed limits on idling near schools and residential areas (ARB 2005). To develop educational materials on the health impacts of idling, DDD members held meetings and focus groups with truck drivers and businesses that identified saving money on fuel and maintenance costs as a key message to emphasize along with protecting health (DDD 2017) Fig. 5.5.

During Anti-Idling Days of Action in 2005 and 2008, DDD members distributed bumper stickers and placed door hangers on truck cabs in freight-impacted communities around the region to educate truck drivers about the state's anti-idling laws and how idling could affect health. DDD also developed an educational flyer on California's anti-idling law for the California DMV's Commercial Driver's Handbook in 2008 which was widely distributed in English and Spanish to commercial drivers throughout the state. This information seemed new and of interest to many of the truck drivers that DDD members spoke to, particularly that diesel fumes get more concentrated in truck cabins while they are driving thereby exposing them and their passengers including children and pets (see EBASE and PI 2009). DDD also developed an anti-idling toolkit for schools to educate and mobilize students, teachers, parents, school bus and delivery vehicle drivers, and diesel vehicle drivers to reduce idling at and around schools (DDD 2014). DDD members then distributed the toolkit to schools as well as environmental and public health organizations working with youth and their families.

**Fig. 5.5** Educational materials in DDD's anti-idling education campaign. *Source: http:www.ditchingdirtydiesel.org*

Participation in the coalition's anti-idling education activities built the capacity of DDD members to carry out related community-engaged research and organizing efforts. Communities for a Better Environment (CBE) led a diesel truck count study and particulate matter monitoring project with academically-credentialed researchers that supported an organizing campaign to reduce idling and reroute trucks away from schools and other sensitive land uses in East Oakland (CBE 2010a, b). Building on the prior success of a truck route campaign led by the West Oakland Environmental Indicators Project (WOEIP) (Gonzalez et al. 2011), CBE's campaign involved local residents, truckers, regulatory agencies, elected officials and Port of Oakland staff in changing the truck route ordinance in East Oakland to reduce diesel exposure and other freight impacts. DDD members also issued a research report estimating the costs to health of freight transport in California that was used to support statewide advocacy efforts to inform California's Goods Movement Action Plan process (DDD and PI 2006).

## Application of a Popular Education Approach Within the Project

Members of DDD's Freight Transport Committee developed and piloted popular education activities on the health impacts of freight transport in a curriculum guide available in English and Spanish (DDD and PI 2010). The guide includes six chapters that draw on the lived experiences of participants with freight traffic and decision-making. Each of these chapters, described in Fig. 5.6, corresponds to a step in the popular education spiral of learning.

1. *What Is Freight Transport?* Start with our lived experiences and knowledge about freight traffic in our community; learn more about and locate our community within the global freight transport system
2. *How Does Freight Transport Affect Us?* Identify patterns in lived experiences with freight traffic by mapping "hot spots" and "magnets" for diesel trucks and trains; connect these patterns to health impacts
3. *Why Are Trucks and Trains in my Neighborhood?* Add new information about how freight-related land uses attract freight traffic into communities; link this to patterns in what we know about freight impacts
4. *Who Are the Decision-Makers?* Develop a shared analysis of what's causing and contributing to these patterns, based on learning more about how land use and transportation planning decisions are made
5. *What Solutions Can Work for Our Community?* Practice skills and strategize using power mapping and action planning, based on a shared analysis of land use and transportation decisions as a root cause
6. *How Can We Get Decision-Makers to Support Our Solutions?* Practice taking action based on what we've learned, using role-playing and group problem-solving to build skills for advocating for our solutions

**Fig. 5.6 DDD's "Gearing Up for Action" Curriculum.** (Source: http://www.ditchingdirtydiesel.org/publications)

*Step 1: Share Lived Experiences* The activities in the curriculum guide were initially carried out by DDD members as part of several community-engaged research and planning projects to document diesel truck traffic, related emissions, and proposed land use changes in West Oakland and North Richmond (see WOEIP and PI 2003; NHNR and PI 2005). DDD members developed these activities to deepen residents' understanding of freight transport decision-making and to demystify the technical language of transportation engineers and planners. These activities aim to affirm what community members already know about this issue and to hone their skills in sharing their *lived experiences* and concerns about the impacts of freight traffic with planners and decision-makers.

*Step 2: Identify Patterns* After the guide was released, DDD members developed and piloted an accompanying training-of-trainers to build the capacity of community organizations and health educators to apply and use the activities in their work. In June 2013, DDD members organized and led a training-of-trainers in Oakland CA attended by over two dozen community organizers, advocates, health educators, and outreach workers (De Leña 2013). The training-of-trainers helped DDD members to *identify patterns* in the organizing work and resource needs of local community organizations to help inform the coalition's capacity-building efforts. DDD members worked with interested training participants to develop and deliver tailored community workshops to incorporate freight-related health impacts into their ongoing work.

*Step 3: Add New Information* Planning the trainings-of-trainers and community workshops based on the guide provided an opportunity for mutual co-learning among DDD members by *adding new information* about varied public health actions that can reduce exposure to diesel exhaust. In 2014 DDD members partnered with CBE and the Alameda County Healthy Homes Department (ACHHD) to develop a Spanish-language workshop for recipients of ACHHD's services in East Oakland. The workshop – attended by over 20 Latino residents, most of them mothers of asthmatic children – incorporated DDD's materials on outdoor air pollutants and community-level actions to reduce exposures, ACHHD's materials on indoor asthma triggers and personal actions to reduce exposures, and CBE's materials on community organizing efforts underway to improve air quality in East Oakland (Garzón 2014).

*Step 4: Develop a Shared Analysis* DDD members also distributed the guide to state and national coalitions focusing on freight transport issues, thereby supporting mutual co-learning in broader coalition-building efforts. At a 2010 gathering of the Moving Forward Network (MFN), DDD members facilitated a training-of-trainers based on the guide attended by over 30 participants from organizations across the country (Garzón 2010). The training helped coalition members to make connections across freight transport issues faced by diverse communities around the country, while building their capacity to carry out these activities in their own communities (see for example CACWNY 2012).

The activities in the guide were also integrated into the leadership development work of DDD member organizations, including several youth academies and projects led by WOEIP and the Rose Foundation's New Voices Are Rising Program. The activities helped West Oakland youth participating in WOEIP's summer academy to *develop a shared analysis* that connected freight transport with local business operations, air quality, and public health. The New Voices program also adapted activities from the guide for its Summer Advocacy Training Institute with East Bay high school students to build their capacity to engage in local and regional climate, air quality, transportation, and land use planning and decision-making.

*Step 5: Practice Skills, Strategize* DDD also adapted activities in the guide to support community and stakeholder engagement in developing a Health Impact Assessment (HIA) of the Alameda County Goods Movement Plan (the Plan). DDD members partnered with the Alameda County Public Health Department to conduct the HIA, which assessed the potential impacts to health of proposed changes in the county's freight transport system (e.g., strategies) to be included in the Plan. In 2014 and 2015, DDD held a series of workshops using modified activities from the guide to engage youth, community members, and other stakeholders in informing key decisions in the development of this HIA and to share preliminary results to inform upcoming research activities (DDD 2016). Workshop participants *practiced skills* by identifying potential health impacts of selected strategies to be included in the Plan, learned more about the decision-making process that would be used to develop and approve the Plan, and then *strategized* to develop recommendations to address these impacts that could be included in the HIA.

*Step 6: Take Action* To build community capacity to engage with decision-makers, whenever possible DDD held its HIA workshops prior to the workshops that consultants for the Alameda County Transportation Commission (ACTC) conducted as part of the agency's formal stakeholder engagement process for the county's Plan. DDD members also met with ACTC staff and consultants and attended ACTC meetings to provide them with feedback on their proposed stakeholder engagement process and research activities for the Plan. Lastly, DDD members submitted written public comments to ACTC on the draft Plan based on the HIA results and recommendations informed by the HIA workshops.

## Impacts on Environmental Health Literacy and Community Empowerment

Through the work of its issue-based committees, DDD leveraged a popular education approach to build community capacity to engage with decision-makers and to strengthen meaningful community engagement in freight transport planning. These activities have ranged from leading educational and organizing campaigns to improve truck routes and compliance with existing anti-idling regulations near schools and residential areas, to integrating popular education into community-engaged research and planning efforts to impact local and regional freight transport decision-making (see Garzón et al. 2014). Enabled by its collaborative structure, DDD's engagement of community-based organizations, environmental health advocates, and local health departments in developing and delivering these activities in a variety of contexts has created opportunities to deepen and sustain Environmental Health Literacy among local agency staff, academically-credentialed researchers, and community leaders.

The mutual co-learning and knowledge co-creation that has occurred among DDD members is evidenced by the integration of both lived experience and technical knowledge in the research and educational materials that DDD has developed. This work has also contributed to an increased capacity of community-based DDD member organizations and training participants to deliver and build on this content through their own community workshops, youth leadership development programs, community-engaged research, and organizing campaigns. DDD has also started providing technical assistance to other communities that are early on in their formation and organizing campaigns on freight transport issues. Finally, this work has supported coalition-building at multiple scales (local, regional, state and national), strengthened relationships with agency staff and decision-makers, and contributed to positive changes in the local and regional freight transport system that reduce its impacts on community health and quality of life.

## Challenges and Limitations

Challenges and limitations have included sustaining the engagement of participants in DDD's popular education activities in its ongoing capacity-building and policy advocacy work. Freight transport is still often perceived as a technical and specialized issue in impacted communities, and some of the solutions that DDD has emphasized, such as changes in land use and transportation planning decisions, can seem far-off in terms of addressing immediate priorities and concerns. DDD has also experienced some attrition in its membership in recent years due to cuts in foundation funding for freight-related work and changes in members' organizational priorities. Garnering sufficient resources to support the leadership of community-based organizations on DDD's Steering Committee has been a continued challenge given the integral yet under-funded role they play in guiding, developing, and carrying out the coalition's activities. To address this asymmetry in resources, agency and intermediary members of DDD have continued to play a key role by providing in-kind support in advising and carrying out coalition activities so that coalition funding can support the role of DDD members from community organizations.

## Conclusions and Future Directions

The case studies in this chapter demonstrate how popular education techniques can be applied within a diversity of community-engaged research and education projects on environmental health issues in both rural and urban settings to optimize community engagement as well as health- and empowerment-related outcomes. The case studies illustrate how diverse projects – ranging from researcher-initiated academic studies to community-initiated trainings and co-initiated research projects – have leveraged both popular education and community-engaged research approaches to increase the EHL of all partners involved and support varied actions to improve health. These public health actions have included a youth-led campaign to get more local stores to carry lower-EDC personal care products, a school flag campaign to reduce children's exposure during days with high air pollution levels, and a coalition campaign to address the community health impacts of proposed changes in the freight transport system as part of a countywide planning process.

In sum, popular education and community-engaged research are promising empowerment-based approaches to increasing and sustaining the EHL of impacted communities and their academically-credentialed partners. Along with increased EHL, the outcomes of projects applying these approaches include increased individual, community and partnership capacity to take action to improve environmental health conditions, as well as changes in individual behavior, policies, and institutional practices that shape health. While popular education and community-engaged research approaches are effective when used in a comprehensive and community-initiated way, these approaches are still worthwhile to integrate whenever possible in public health studies and other projects or initiatives to optimize meaningful community participation. Below is a summary of recommendations for future directions in the applications of community-engaged research and popular education as approaches to increasing and sustaining EHL among all partners:

- Community members/partners should continue to play a stronger role in not just community-engaged research, but community-driven research.
- Leveraging existing community capacity helps facilitate increased EHL, but also supports long-term sustainability by increasing the capacity of academically-credentialed partners to make research more relevant and actionable in impacted communities.
- EHL within a community should be recognized as originating more from within that community, based in lived experiences with environmental hazards and health conditions, rather than with outside researchers.
- The EHL of academically-credentialed researchers/partners should be assessed before and after working with community partners on a project, based on an understanding that mutual co-learning is a key outcome of community-engaged research and education work.
- Mutual co-learning between academically-credentialed and community partners should be integrated into scales and tools used to measure EHL.

- As these approaches to engaging impacted communities become more widely applied, evaluation will be key to learn which elements and approaches of a project provide the greatest gains in understanding and agency as these relate to environmental health.
- Funders can play a key role by supporting and facilitating sharing best practices and model projects from the growing network of community-engaged research and popular education practitioners that has as one of its aims to increase environmental health literacy in both impacted communities and in academically-credentialed partners.

**Acknowledgements** We would like to thank our colleagues from project partner organizations involved in the case studies highlighted in this chapter for reviewing and contributing their ideas to help shape earlier drafts: Kimberly Harley, Katherine Kogut, and James Nolan of the UC Berkeley Center for Environmental Research and Children's Health (CERCH); and Adenike Adeyeye, Brian Beveridge, Paul Cummings, Joel Ervice, Frank Gallo, Ms. Margaret Gordon, Michael Kent, Anna Lee, and Jill Ratner of the Ditching Dirty Diesel Collaborative (DDD). We'd also like to thank the Community Steering Committee and the entire project team of the Imperial Air Project at CEHTP, CCV, UW, UCLA, and GWU. Liam O'Fallon and Symma Finn at National Institutes of Environmental Health Sciences and Rick Kreutzer at California Department of Public Health provided insightful comments on previous chapter drafts. Lastly, we extend our heartfelt appreciation to all the community members and project participants who shared the experiences, knowledge, and leadership that made the content of this chapter possible, though the views expressed in this chapter and any errors are of course our own.

## References

(ARB), California Air Resources Board. (1998). *Proposed identification of diesel exhaust as a toxic air contaminant - initial statement of reasons*. Sacramento: California Environmental Protection Agency Retrieved from https://www.arb.ca.gov/toxics/dieseltac/staffrpt.pdf.

(ARB), California Air Resources Board. (2005). *ARB begins enforcement of its new idling rule*. Sacramento, CA: California Environmental Protection Agency Retrieved from https://www.arb.ca.gov/newsrel/nr013105.htm.

(ARB), California Air Resources Board. (2006). *Quantification of the health impacts and economic valuation of air pollution from ports and goods movement in California*. Sacramento, CA: California Environmental Protection Agency Retrieved from https://www.arb.ca.gov/planning/gmerp/plan/appendix_a.pdf.

(ARB), California Air Resources Board. (2012). Air quality trend summaries. Retrieved from http://www.arb.ca.gov/adam/trends/trends1.php

(CACWNY), Clean Air Coalition of Western New York. (2012). Gearing up for action training. Retrieved from https://www.cacwny.org/2012/12/gearing-up-for-action-training/

(CBE), Communities for a Better Environment. (2010a). *East Oakland diesel truck survey report*. Retrieved from Oakland, CA.: http://www.cbecal.org/wp-content/uploads/2013/01/Diesel-truck-study-FINAL-092710.pdf

(CBE), Communities for a Better Environment. (2010b). East Oakland particulate matter 2.5 community air monitoring research report. Retrieved from Oakland, CA: http://www.cbecal.org/wp-content/uploads/2013/01/East-Oakland-PM-Monitoring-Report-FINAL-2010.pdf

(CEHTP), California Environmental Health Tracking Program. (2017). California environmental health tracking program. Retrieved from http://www.cehtp.org

(DDD), Ditching Dirty Diesel Collaborative. (2014). *Anti-idling toolkit for schools*. Oakland: DDD.
(DDD), Ditching Dirty Diesel Collaborative. (2016). *Making a good move for health: A health impact assessment of selected strategies in the Alameda County goods movement plan*. Retrieved from Oakland, CA.
(DDD), Ditching Dirty Diesel Collaborative. (2017). A history of success – Anti-idling. Retrieved from http://www.ditchingdirtydiesel.org/about/success-anti-idling
(DDD), Ditching Dirty Diesel Collaborative, & (PI), Pacific Institute. (2006). *Paying with our health: The real cost of freight transport in California*. Retrieved from Oakland, CA
(DDD), Ditching Dirty Diesel Collaborative, & (PI), Pacific Institute. (2010). *Gearing up for action: A curriculum guide for freight transport justice*. Retrieved from Oakland, CA
(DDD), Ditching Dirty Diesel Collaborative, & (PI), Pacific Institute. (2011). *At a crossroads in our region's health: Freight transport and the future of community health in the San Francisco Bay Area*. Retrieved from Oakland
(EBASE), East Bay Alliance for a Sustainable Economy, & (PI), Pacific Institute. (2009). *Taking a toll: The high cost of health, environment, and worker impacts of the Oakland port trucking system* (1893790193). Retrieved from Oakland, CA: http://pacinst.org/wp-content/uploads/2013/02/taking_a_toll3.pdf
(WOEIP), West Oakland Environmental Indicators Project, & (PI), Pacific Institute. (2003). *Clearing the air: Reducing diesel pollution in West Oakland*. Retrieved from Oakland, CA.
Abma, T. A., Cook, T., Rämgård, M., Kleba, E., Harris, J., & Wallerstein, N. (2017). Social impact of participatory health research: Collaborative non-linear processes of knowledge mobilization. *Educational Action Research, 25*(4), 489–505.
Anderson, J. O., Thundiyil, J. G., & Stolbach, A. (2012). Clearing the air: A review of the effects of particulate matter air pollution on human health. *Journal of Medical Toxicology, 8*(2), 166–175. https://doi.org/10.1007/s13181-011-0203-1.
Arnstein, S. R. (1969). A ladder of citizen participation. *Journal of the American Institute of Planners, 35*(4), 216–224.
Balazs, C. L., & Morello-Frosch, R. (2013). The three R's: How community based participatory research strengthens the rigor, relevance and reach of science. *Environmental Justice, 6*(1), 9–16. https://doi.org/10.1089/env.2012.0017.
Banks, S., Armstrong, A., Carter, K., Graham, H., Hayward, P., Henry, A., et al. (2013). Everyday ethics in community-based participatory research. *Contemporary Social Science, 8*(3), 263–277.
Brown, P. (1995). Race, class, and environmental health: A review and systematization of the literature. *Environmental Research, 69*(1), 15–30.
Brown, P., Mayer, B., Zavestoski, S., Luebke, T., Mandelbaum, J., McCormick, S., & Lyson, M. (2011). The health politics of asthma: environmental justice and collective illness experience. In *Contested illnesses: Citizens, science, and health social movements*. University of California Press.
Centers for Disease Control and Prevention. (2015). *Fourth national report on human exposure to environmental chemicals, updated tables,* February 2015. *Atlanta: US Department of Health and Human Services*. Retrieved from Atlanta, GA: https://www.cdc.gov/biomonitoring/pdf/FourthReport_UpdatedTables_Feb2015.pdf.
Chavez, V., Duran, B., Baker, Q. E., Avila, M. M., & Wallerstein, N. (2008). *The dance of race and privilege in CBPR. Community-based participatory research for health: From process to outcomes* (pp. 91–105).
Cole, L. W., & Foster, S. R. (2001). *From the ground up: Environmental racism and the rise of the environmental justice movement*. New York: NYU Press.
De Leña, A. (2013). *What's Freight Got To Do With It?* Retrieved from Oakland, CA: http://pacinst.org/whats-freight-got-to-do-with-it/

Di, Qian, Wang, Yan, Zanobetti, A., Wang, Yun, Koutrakis, P., Choirat, C.,. .. Schwartz, J. D. (2017). Air pollution and mortality in the Medicare population. New England Journal of Medicine, 376(26), 2513–2522. https://doi.org/10.1056/NEJMoa1702747.

English, P. B., Olmedo, L., Bejarano, E., Lugo, H., Murillo, E., Seto, E., et al. (2017). The Imperial County community air monitoring network: A model for community-based environmental monitoring for public health action. *Environmental Health Perspectives, 125*(7), 074501. https://doi.org/10.1289/ehp1772.

Eskenazi, B., Bradman, A., Gladstone, E. A., Jaramillo, S., Birch, K., & Holland, N. (2003). CHAMACOS, a longitudinal birth cohort study: Lessons from the fields. *Journal of Children's Health, 1*(1), 3–27.

Finn, S., & O'Fallon, L. (2017). The emergence of environmental health literacy—From its roots to its future potential. *Environmental Health Perspectives, 125*(4), 495.

Freire, P. (1970). *Pedagogy of the oppressed*, (pp. 65–80, Myra Bergman Ramos, Trans.). New York: Continuum.

Garzón, C. (2010). Training the trainers on using popular education to demystify freight transport. Retrieved from Oakland, CA: http://pacinst.org/media-center/online-updates/november-2010-online-update/

Garzón, C. (2014). Community workshop on taking action to improve indoor and outdoor air quality. Retrieved from http://pacinst.org/about-us/programs/community-strategies-for-sustainability-and-justice/cssj-program-monthly-updates/

Garzón, C., Dominie, W., & Gordon, M. (2014). Pollution-free housing for all: Coalition-based research, education, and advocacy for healthier housing in transportation and land use planning in the San Francisco Bay Area. *Environmental Justice, 7*(6), 186–190.

Gonzalez, R. (2014). Popular education practices for community organizing. Retrieved from http://www.facilitatingpower.com/popular_education_for_community_organizing

Gonzalez, P. A., Minkler, M., Garcia, A. P., Gordon, M., Garzon, C., Palaniappan, M., et al. (2011). Community-based participatory research and policy advocacy to reduce diesel exposure in West Oakland, California. *American Journal of Public Health, 101*(Suppl 1), S166–S175. https://doi.org/10.2105/ajph.2010.196204.

Harley, K. G., Kogut, K., Madrigal, D. S., Cardenas, M., Vera, I. A., Meza-Alfaro, G., et al. (2016). Reducing phthalate, paraben, and phenol exposure from personal care products in adolescent girls: Findings from the HERMOSA intervention study. *Environmental Health Perspectives, 124*(10), 1600–1607. https://doi.org/10.1289/ehp.1510514.

Hergenrather, K. C., Rhodes, S. D., Cowan, C. A., Bardhoshi, G., & Pula, S. (2009). Photovoice as community-based participatory research: A qualitative review. *American Journal of Health Behavior, 33*(6), 686–698.

Israel, B. A., Schulz, A. J., Parker, E. A., Becker, A. B., Allen, A., GUZ-MAN, J. R., et al. (2008). *Critical issues in developing and following CBPR principles* (pp. 47–66).

Liebman, A. K., Juarez, P. M., Leyva, C., & Corona, A. (2007). A pilot program using promotoras de salud to educate farmworker families about the risk from pesticide exposure. *Journal of Agromedicine, 12*(2), 33–43. https://doi.org/10.1300/J096v12n02_04.

Madrigal, D., Salvatore, A., Casillas, G., Casillas, C., Vera, I., Eskenazi, B., & Minkler, M. (2014). Health in my community: Conducting and evaluating PhotoVoice as a tool to promote environmental health and leadership among Latino/a youth. *Progress in Community Health Partnerships: Research, Education, and Action, 8*(3), 317–329. https://doi.org/10.1353/cpr.2014.0034.

Madrigal, D. S., Minkler, M., Parra, K. L., Mundo, C., Gonzalez, J. E., Jimenez, R., et al. (2016). Improving Latino Youths' environmental health literacy and leadership skills through participatory research on chemical exposures in cosmetics: The HERMOSA study. *International Quarterly of Community Health Education, 36*(4), 231–240. https://doi.org/10.1177/0272684x16657734.

Minkler, M. (2004). Ethical challenges for the "outside" researcher in community-based participatory research. *Health Education & Behavior, 31*(6), 684–697.

Minkler, M., Wallace, S. P., & McDonald, M. (1994). The political economy of health: A useful theoretical tool for health education practice. *International Quarterly of Community Health Education, 15*(2), 111–126. https://doi.org/10.2190/t1y0-8aru-rl96-lpdu.

Minkler, M., Garcia, A. P., Williams, J., LoPresti, T., & Lilly, J. (2010). Si se puede: Using participatory research to promote environmental justice in a Latino community in San Diego, California. *Journal of Urban Health, 87*(5), 796–812. https://doi.org/10.1007/s11524-010-9490-0.

Muhammad, M., Garzón, C., Perez, A., & West Oakland Environmental Indicators Project (2017). Understanding contemporary racism, power and privilege and their impacts on CBPR. In *Community-Based Participatory Research for Health: Advancing Social and Health Equity* (3rd ed.). San Francisco: Jossey-Bass.

Neighborhood House of North Richmond (NHNR) and Pacific Institute (PI). (2005). *Deluged by Diesel: Healthy Solutions for West County*. Oakland, CA.

Nutbeam, D. (2000). Health literacy as a public health goal: A challenge for contemporary health education and communication strategies into the 21st century. *Health Promotion International, 15*(3), 259–267.

Nutbeam, D. (2008). The evolving concept of health literacy. *Social Science & Medicine, 67*(12), 2072–2078.

O'Fallon, L. R., & Dearry, A. (2001). Commitment of the National Institute of Environmental Health Sciences to community-based participatory research for rural health. *Environmental Health Perspectives, 109*(Suppl 3), 469–473.

Paradies, Y., Ben, J., Denson, N., Elias, A., Priest, N., Pieterse, A.,. .. Gee, G. (2015). Racism as a determinant of health: A systematic review and meta-analysis. PLoS One, 10(9), e0138511.

Sadd, J., Morello-Frosch, R., Pastor, M., Matsuoka, M., Prichard, M., & Carter, V. (2014). The truth, the whole truth, and nothing but the ground-truth: Methods to advance environmental justice and researcher–community partnerships. *Health Education & Behavior, 41*(3), 281–290.

Smith, H. (2011). *Building a research and outreach partnership to improve Latino community health in Charlotte-Mecklenburg: Geographers, doctors, and community members*. Charlotte: UNC Charlotte Urban Institute.

Sullivan, J., & Lloyd, R. S. (2006). The forum theatre of Augusto Boal: A dramatic model for dialogue and community-based environmental science. *Local Environment, 11*(6), 627–646.

United States Census Bureau. (2010). US Census Bureau Quick Facts: Imperial County. Retrieved from http://www.census.gov/quickfacts/table/PST045215/06025,0636294,00

Vaduganathan, M., De Palma, G., Manerba, A., Goldoni, M., Triggiani, M., Apostoli, P., et al. (2016). Risk of cardiovascular hospitalizations from exposure to coarse particulate matter (PM10) below the European Union safety threshold. *American Journal of Cardiology, 117*(8), 1231–1235. https://doi.org/10.1016/j.amjcard.2016.01.041.

Wallerstein, N., & Bernstein, E. (1988). Empowerment education: Freire's ideas adapted to health education. *Health Education Quarterly, 15*(4), 379–394.

Wallerstein, N., & Duran, B. (2008). The theoretical, historical, and practice roots of CBPR. *Community-based participatory research for health: From process to outcomes,* 2nd ed., 25–46.

Wiggins, N. (2011). Popular education for health promotion and community empowerment: A review of the literature. *Health Promotion International, 27*(3), 356–371. https://doi.org/10.1093/heapro/dar046

World Health Organization (WHO). (2014). *Burden of disease from ambient air pollution for 2012*. World Health Organization.

Wright, K. N., Williams, P., Wright, S., Lieber, E., Carrasco, S. R., & Gedjeyan, H. (2011). Ties that bind: Creating and sustaining community-academic partnerships. *Gateways: International Journal of Community Research and Engagement, 4*, 83–99.

# Chapter 6
# Returning Chemical Exposure Results to Individuals and Communities

**Julia Green Brody, Phil Brown, and Rachel A. Morello-Frosch**

**Abstract** Biomonitoring and personal exposure assessments – measurements in people's blood, urine, breast milk, and other tissues, or in household dust, air, or other personal space – are vital tools for scientists to discover and monitor the links between environmental chemicals and health. For participants who contribute their samples, these methods generate curiosity: What did you find, was it safe, and what should I do? These motivating questions make the process of reporting back to study participants and communities a powerful opportunity to improve environmental health literacy. Decisions about what to report can be challenging, though, because scientists may not yet fully understand how the chemicals affect health, where they come from, or how to reduce exposures. This chapter explores the ethical foundations, evidence-based methods, and outcomes in environmental health studies that reported results for emerging contaminants.

Decisions about what and how to report back are guided by human research ethics and knowledge from social science and communications research. Report-back balances the Belmont Report values to avoid harm, maximize benefit, and support the autonomy of study participants, and it goes farther, to support the principles of community based participatory research, including co-ownership of data, community problem-solving, and shared knowledge as an avenue to shared power and democratizing science. Research about environmental disasters, embodied health movements, and risk communication can guide development of ethical and effective report-back models.

J. G. Brody (✉)
Silent Spring Institute, Newton, MA, USA
e-mail: brody@silentspring.org

P. Brown
Northeastern University, Social Science Environmental Health Research Institute, Boston, MA, USA

R. A. Morello-Frosch
University of California, Berkeley School of Public Health, Berkeley, CA, USA



S. Finn, L. R. O'Fallon (eds.), *Environmental Health Literacy*,
https://doi.org/10.1007/978-3-319-94108-0_6

**Keywords** Biomonitoring · Human research ethics · Return of results · Health literacy · Environmental health · Community based participatory research · Digital health communications

Now that more environmental health studies have reported personal as well as study-wide results for emerging contaminants, researchers have empirical data on the outcomes. Interviews with participants who received their results show that report-back is appreciated and builds trust and understanding of science. The "exposure experience" of learning about unseen chemicals in one's own body or home often changes people's mental model of pollution from something "out there" to understanding that health-related contaminants often come from everyday environments and consumer products used at home, work, and play. People interpret their results in the context of what they already know about the environmental history of their community, and they add new conceptualizations. They quickly begin brainstorming about sources and strategies to reduce exposures. Researchers benefit, too, by focusing on their data in a new way as they prepare reports and interact with participants about their exposures.

To help participants understand and use their results, reports must be designed thoughtfully in collaboration with community members and study participants who can help tailor the plans to the particular context. Including both graphs and text is helpful. Reports must address what is known and not known and, when possible, provide strategies for reducing exposure at the individual level and through community or national action. Digital methods to generate reports for print, computer, or smartphone can make high-quality personalized reports efficient in studies of any size. By continuing to experiment with report-back methods and improve them, researchers can develop novel ways to advance EHL about personal exposures to toxic chemicals.

## Background

When women on Cape Cod, Massachusetts, first learned that breast cancer incidence was significantly higher there than the statewide rate, their first ideas about "why" turned to pollution from the military base or pesticides like DDT that were used in the 1950s and 1960s on cranberry bogs and sprayed on forests to fight gypsy moths. These factors may, indeed, play a role (Brody et al. 2004; McKelvey et al. 2004), but the emerging science of endocrine disruptors, chemicals that affect the body's natural hormone systems, suggested additional hypotheses about myriad products used at home and leaching into drinking water (Brody et al. 2005). Laboratory studies showed that chemicals in plastics, home pesticides, cleaners, furniture, personal care products, and countless other everyday items could mimic estrogen (a known breast cancer risk factor), block androgens, disrupt thyroid hormones, or otherwise alter normal growth and signaling systems. To learn more about how people are exposed and the levels of exposure, the Cape Cod Household Exposure Study (HES), which began in the late 1990s, tested for 89 endocrine disrupting compounds (EDCs) in 120 homes (Rudel et al. 2003). Trained researchers, including some recruited from affected communities, visited the homes to collect air and dust, and urine samples from women who lived there.

Within a few days, Cheryl Osimo, the study coordinator and a Cape Cod resident, was fielding phone calls and queries at the grocery store from study participants who wanted to know when they would receive their own results. At the time, researchers weren't sure how to respond. Studies that returned results reported back aggregate findings, with the exception of a few well-studied chemicals, like lead and mercury, for which standards were in place for reporting individual results above established clinical health guidelines. But only a few of the chemicals in the Cape Cod household exposure study had been studied enough for scientists to develop health guidelines defining harmful levels of exposure. For 30 of the chemicals analyzed in the Cape Cod study, these household measurements were the first in the U.S., so the scientists didn't even know what exposure levels were typical across the country, and there were no standards for returning individual results for emerging contaminants nor experience with how such results would be understood. Might telling study participants their personal results cause excess worry and do them more harm than good? On the other hand, don't people have a right to know their own results, so they can make decisions about how to protect their families and communities from potential harm? Would people want to know about the possible links to illness even if they couldn't change them?

This chapter explores the experiences of researchers, participants and other stakeholders involved in studies that have grappled with these questions over the past two decades and chosen to report back personal chemical exposures to participants as well as sharing aggregate findings with their communities. Over the last decade, laboratory techniques have advanced tremendously, enabling researchers to measure low levels of a larger number of chemicals in people – in blood, urine, fingernails, and other tissues – and in personal spaces like homes, through air and dust measurements. Measurements of chemicals in body tissues are referred to as exposure biomonitoring. These measurements have become a cornerstone of environmental health science, as studies use them to test the links between chemicals in the body and health outcomes, including cancer, neurodevelopment, fertility, obesity, pregnancy outcomes and many other conditions influenced by environmental factors. They are also vital for public health tracking to see trends in population exposures to environmental chemicals. Since 2003, the U.S. Centers for Disease Control and Prevention (CDC) has been tracking levels of hundreds of chemicals in blood and urine samples from participants in the National Health and Nutrition Examination Survey (NHANES), a representative sample of the U.S. population.

At the same time, environmental health advocates began pushing for return of individual results. They wanted to "put a face on pollution" to make the NHANES statistics more compelling and they believed that people have a right to know their own results, so they could take action. In 2006, California became the first state to establish its own exposure biomonitoring program, and advocates won provisions in the law requiring the program to make biomonitoring results available to any participants who want them (State of California Senate 2006). Canada (Day et al. 2007) and Europe (Becker et al. 2014) also have population biomonitoring programs that allow participants to receive their own results. NHANES currently reports only limited results, such as high levels of lead or mercury (HHS (Department of Health and Human Services) 2017a).

Interviews and observations in studies that do report back exposure results show that nearly all participants want to receive their personal reports, even if the health implications are uncertain (Brody et al. 2014). They learn a great deal from their reports about personal and community exposures and about environmental health science more broadly, and they are motivated to take action to reduce their own exposures. As a result, report-back for personal chemical exposures creates a powerful opportunity to improve environmental health literacy (EHL) (Finn and O'Fallon 2017), increasing awareness and understanding of exposures and empowering people to act on this knowledge to protect their health and that of their families and communities. Advocates have used both personal and study-wide exposure biomonitoring results to persuade manufacturers and retailers to change product formulations and avoid some chemicals entirely (O'Rourke 2005). Additionally, such data has figured into important city- and state-level bans or restrictions on certain chemicals and has been central to the profound policy changes in flame retardant requirements that started with California's reform (Cordner et al. 2013) and continues in the 2017 Consumer Products Safety Commission recommendations to reduce use in children's products (Hawthorne 2017). In addition to influencing environmental health policies and marketplace practices, the growing experience with report-back, along with ethnographic and interview data on how it works, has contributed to a deeper understanding on the part of researchers of the interplay between returning exposure results, EHL and community-engaged research, with a body of literature attesting to report-back benefits to all parties.

## Ethics, Theory, and Empirical Traditions to Inform Effective Personal Exposure Reports

Decisions about how and what to report back from biomonitoring and personal exposure studies begin with principles of human research ethics, based on the 1978 Belmont Report (Department of Health Education and Welfare 1979), and social science theory and empirical research can inform these ethics decisions as well. Many different fields of research can inform the best ways to return results, including research on environmental disasters, embodied health movements, community-based research, risk communication, and basic psychology and cognitive science about how people understand language and graphs and respond with emotions. Theories can then be tested empirically. For example, whether or not people become very worried by receiving personal exposure results, returned to them in specific ways, is an empirical question, and later in this chapter, we discuss research about that concern.

Ethics perspectives have shifted dramatically over the past few decades to favor transparent communications in medicine and biomedical research. The U.S. Department of Health and Human Services, through its *Secretary's Advisory Committee on Human Research Protections (SACHRP)*, currently "recommends a rebuttable presumption in favor of returning individual results. SACHRP intends that this recommendation apply not only to clinical research, but also to other types of research such as genomics research and Social Behavioral research" (SACHRP (Secretary's Advisory Committee on Human Research Protections) 2016). Environmental health studies have been slow compared with medicine and genetics to adopt routine report-back of personal results when their health implications are uncertain, but many consensus statements now recommend report-back (Brody et al. 2014), and it is increasingly becoming the norm. The multi-stakeholder Boston Consensus Conference on Human Biomonitoring concluded, "…the group asserts that study participants should be able to decide whether or not they want to receive their personal results, and that an important element of this report be inclusion of action steps for reducing exposure, when these are available" (Nelson et al. 2009).

## Potential Harms and Benefits Within Traditional Ethics Frameworks

For the past 50 years, human research ethics have been monitored by Institutional Review Boards, relying on the Belmont Report together with local values and accumulated experience, primarily in clinical trials and genetics research. The Belmont Report directs biomedical researchers to consider these dimensions: to avoid harm to participants and maximize beneficence, autonomy (the rights of people to make their own decisions), and justice (fair distribution of benefits and risks of studies) Confronted by the unfamiliar issues posed by environmental measurements and biomonitoring, IRBs have often focused on potential harms and de-emphasized other dimensions of Belmont-based ethics (Ohayon et al. 2017). Specifically, they have expressed concern that participants might be harmed by receiving results, because learning about potentially toxic exposures with uncertain health implications could cause emotional distress or lead people to take actions that are costly, unnecessary or harmful (Ohayon et al. 2017). For example, a tribal collaboration with university researchers and a nongovernmental organization sought to conduct a community-based participatory breast milk biomonitoring study among Alaska Native communities. However, the study was thwarted by the Alaska IRB, despite clear support from tribal leadership, on the grounds that the study might discourage breastfeeding. For nearly 5 years, tribal and university researchers worked to alleviate these ethical concerns by demonstrating that their biomonitoring study protocol would report results to participants while actively encouraging the continuation of breastfeeding (Saxton et al. 2015). Moreover, the Alaska IRB's decision to disallow the breast milk study was not supported by a previous study of nursing mothers who were biomonitored in Boston with no documented effect on their breastfeeding behavior (Wu et al. 2009). While the Alaska case involved extreme speculation on the part of the IRB, other instances may require realistic assessment of possible harms. In rare situations, learning personal results could create legal responsibilities to share results with others to warn them about potential health risks (Goho 2016). When household measurements reveal high levels of lead, a well-known neurotoxicant, this can trigger public health reporting requirements related to property sales; however, for chemicals with uncertain health effects, legal consequences from personal biomonitoring or household measurements would be unusual. Reporting individual results might also be harmful to participants and their communities by creating collective stigma, because entire communities are subsequently perceived as "contaminated" (Quigley 2012). The names of communities like Love Canal can become synonymous with environmental disaster (Gibbs 2011), ruining property values and stigmatizing residents. However, this example also illustrates that reporting on contamination is a necessary step to empowering communities to acquire resources for remediation.

Researchers who have experience reporting personal exposure results to their participants are more likely than IRB members to focus on the potential benefits (Ohayon et al. 2017). Giving people the choice to receive their results supports their autonomy and allows them to take actions to reduce their exposures, consistent with their own values and preferences about health risk.

Earlier traditions in clinical medicine restricted report-back to results that exceeded a clinical health guideline, such as lead levels above an action level. This model for communicating results is expert-driven, since in this method health professionals make decisions on behalf of study participants without consulting them. Yet this approach is becoming outdated as clinical norms for doctor-patient relationships and biomedical research have evolved to be more open; for example, The Open Notes project has experimented with ways to invite patients to review doctor visit notes online with the goal of improving patients' understanding of diverse indicators of their health status, fostering more productive communication, correcting errors, and encouraging shared decision-making (Delbanco et al. 2010). The traditional model of clinical ethics with restricted information-flow is particularly problematic for environmental health, because our knowledge about how chemicals affect human health is evolving quickly, and levels of a chemical that once seemed "safe" later are shown to be harmful. Blood lead levels are important examples – just a few decades ago 50 ug/dL was considered safe, the 10ug/dL standard only took effect in 1991, and the current 5 ug/dL level was adopted in 2012. If participants are not informed about their exposure results, then they can't act on scientific knowledge as it evolves to reduce exposures (Brody et al. 2007).

## Community-Based Participatory Research Ethics and Activist "Data Judo"

The first studies to pioneer report-back for large numbers of chemicals with uncertain health effects were inspired by the ethics of community based participatory research (CBPR) and other community-engaged research traditions. CBPR conceptualizes research as a joint enterprise of researchers and community members (Minkler and Wallerstein 2008). Research is designed not only to advance science but also to address community concerns and problems, and build community competence. Data is co-owned by researchers and the individuals and communities that contribute the data, so report-back is simply returning to people something that belonged to them all along (Quandt et al. 2004). Similarly, CBPR values co-learning and respects the knowledge of lay people, so report-back is jointly designed with the idea that community members will contribute local information that is important to the interpretation of scientific data, for example by identifying the likely sources of chemicals that are detected. CBPR views knowledge as power, so sharing research results helps to equalize the power-imbalance between researchers and communities and empower communities to improve health. These ethical values embedded in CBPR strongly favor a "research right-to-know." However, CBPR ultimately places decisions about whether and how to report in the hands of collaborative research teams, so that potential stigma or other unintended consequences can be weighed and mitigated.

The Silent Spring Institute Household Exposure Study that introduces this chapter bridges CBPR ethics and additional traditions of citizen science, since the study was initiated by local activists and conducted by a multidisciplinary team of scientists governed by activist leaders (Brody et al. 2005). In the Cape Cod study and others, access to biomonitoring and exposure assessment has emerged as a vital tool for communities facing disproportionate rates of disease or environmental injustice. Biomonitoring reveals "toxic trespass," the invisible and involuntary exposure of people to risky or untested chemicals (Morello-Frosch et al. 2009).

In other biomonitoring studies conducted by environmental advocacy groups, (such as "Is It In Us" (Is it in us? 2017) and "Mind Disrupted" (Mind Disrupted 2017)) participants have blood or urine samples tested specifically for the purpose of publicly communicating their results in ways that humanize the implications of toxic trespass. Although participants in these studies are not required to disclose their personal results, and disclosure decisions are made after results are returned, these projects still fundamentally challenge IRB conceptions of research privacy, since the open and often online sharing of personal chemical exposure results is considered central to their effectiveness. "Putting a face on pollution" generates compelling personal stories and news media coverage that can effectively engage elected officials and other decision-makers to change chemicals policies. This strategy is described as data judo, because it uses elite, expert methods to achieve grass-roots organizing and advocacy goals. Personalized information about chemical body burden is designed to expand public support for toxics use reduction and generate activism and personal behavior change, such as through consumer campaigns (Morello-Frosch et al. 2009).

## Social Science and Communications Literature

Decisions about how best to communicate results can draw on large and diverse theoretical and empirical literature, including some of the topics discussed in this volume. Drawing on this body of evidence can guide report-back to build on people's cognitive and emotional strengths, social support networks, and coping and resilience strategies, so that learning about toxics in their bodies and homes supports EHL and constructive action rather than generating misplaced worry or alarm.

As a starting point, the unseen and involuntary nature of the toxic contamination revealed by personal exposure sampling evokes new translations of research on environmental and technological disasters (Brown and Mikkelsen 1997), and contested illness and embodied health movements (Brown 2007). Results reports engage strange chemical names and scientific knowledge and yet identify them in intimate body fluids and private spaces. Experiences from contaminated communities can be informative about how people learn about toxics, moderate stress, and make decisions, particularly in a context of shared contamination in a place-based study or illness-based social groups.

Risk communication literature is also relevant. Public health efforts to moderate public responses to risk or to ramp-up awareness and prompt health-protective action have led to extensive research on how people perceive and evaluate risks, and these ideas have been incorporated into the field of health risk communications. Contrasts between expert and lay risk perceptions are sometimes misinterpreted as implying that lay people are "bad" at understanding uncertainty. However, more sophisticated interpretations show how high concern can result from poor communications by experts and/or represent legitimate expressions of public values (National Research Council 1989). Risk communication literature is a valuable tool for ensuring that report-back meshes with cognitive abilities and personal values in ways that help people calibrate their response. Similarly, research on stress and coping shows how strengthening efficacy can reduce alarm (Lazarus and Folkman 1984). We have found the mental models approach to risk communication particularly helpful both in designing reports and for evaluating how well they are working (Morgan et al. 2002). This method involves eliciting how people understand "the way things work," so that communications can extend existing mental models or replace misconceptions. Figure 6.1 shows an example of a biomonitoring report designed to shift mental models that equate "BPA-free" with safe. Mental models approaches have been used to shape communications about environmental hazards including radon, nuclear wastes, and dioxin (Morgan et al. 2002; Zikmund-Fisher et al. 2013).

In addition, research on data visualization can supplement risk communication science and help to guide specific choices about graphs, tables, illustrations, and

Chemicals We Found

**Your sample had a higher level of a bisphenol than 95% of Americans.**

**WHAT YOU CAN DO**
Choose fresh or frozen instead of canned food or drinks. Be aware that plastics and food cans labeled "BPA-free" may contain BPS, BPF, or other chemical substitutes.

**Fig. 6.1** Communications that connect with people's existing mental models can avoid misconceptions and improve uptake of new information. This excerpt from a results report for the Detox Me Action Kit (Silent Spring Institute 2017) explains how to reduce exposure and connects the recommendation to personal results

other components of reports. Well-designed graphs can draw on innate abilities to judge visual characteristics, such as above versus below or size comparisons, so understanding depends less on literacy and numeracy (Few 2004; Kosslyn 2006).

## Learning and Action Outcomes

Now that more than a dozen environmental health studies have reported back personal as well as study-wide results for EDCs and other chemicals, we have empirical data on the outcomes of this approach. Much of this data comes from the Personal Exposure Report-back Ethics (PERE) Study, which has interviewed participants in biomonitoring and household exposure studies in a wide range of settings. The PERE Study includes household exposure studies that have measured endocrine disruptors and carcinogens from consumer products, agriculture, and industry in rural, suburban, and urban settings. Other research covered by the PERE Study includes measurements in blood, urine, drinking water, and soil in communities affected by contaminants from Superfund sites or industrial pollution, and the PERE Study includes cohort studies of children, pregnant women, and older adults from diverse racial, ethnic, and economic backgrounds. Hour-long semi-structured interviews with adult participants and parents of child participants provide rich information about people's experiences before and after they receive their reports.

In the early studies, researchers were relying on experience from the fields of health risk communication and environmental communication, as described earlier, incorporating community knowledge, and then learning by observing outcomes and continuously improving future report-back methods. CBPR values supported the ethical impulse to fully report results, including those with less certain health effects, and encourage the feedback that contributes to further innovation and improvement. One consistent finding is that, when asked, 90% or more of participants want to receive their personal results ([Anonymous] 2011; Adams et al. 2011; Altman et al. 2008; Barlow and Kushi 2011).

## The Exposure Experience

Exposure science quantifies an objective reality – the chemical body burden in a person or place – that is usually not perceptible through the senses. When these results are reported back, they become a part of embodied and subjective knowledge in what we call the "exposure experience" (Altman et al. 2008). The exposure experience adds meaning and interpretation to scientific data and integrates thoughts and feelings.

The core of the "exposure experience" in biomonitoring and household studies results when previously unknown chemical exposures become known. People process the new information in the context of what they knew before and make decisions about how to act on it. As one participant described, "We were joining a small group of people that actually knew what their chemical body burden was and that this was info that we would not be able to unlearn…It would be a new kind of responsibility" (Dunagan et al. 2013). Another described the experience this way:

> *It just shows you how many chemicals you are around. I mean, this is only a partial list… And I think that something happens on, you know, on a cellular level between the body and environment, and things that we see in our environment are not as dangerous as the things we don't see. …we've changed our environment so much since we've learned how to make things and we don't know how all this affects the body… (Dunagan et al.* 2013*).*

First reactions often included surprise at the number and range of chemicals detected in samples. During interviews, participants grappled with this information, brainstorming about possible sources in their personal histories and environments. They expressed frustration at not knowing the sources and at government and industry decisions that allowed potentially harmful chemicals to be used and to enter their bodies. Despite the unwelcome news, however, participants have not reported persistent, extreme worry or other distress. People focused more on the future and opportunities to reduce exposures.

Participants generally understood the scientific uncertainties and the caveat that biomonitoring results could not be used to explain personal illness, though some reflected with sadness about the possible, though uncertain, connections between chemical exposures and family illnesses. And many were frustrated by the scientific knowledge gaps. Noting that chemicals were detected in every home in the Cape Cod HES, one participant vented that "every home also has corn flakes in the kitchen, but that doesn't tell me anything" (Altman et al. 2008). A woman who joined the HES to learn about breast cancer risk said, "I basically feel I got nothing," since results from her own biomonitoring and the parent Cape Cod Breast Cancer and Environment Study were not yet able to explain elevated rates in the region (Altman et al. 2008).

## Changing Mental Models

Interviews after people receive their exposure results reveal a tremendous amount of learning and major conceptual shifts. People are generally accurate in reporting how their results compare with others – whether their chemical levels were higher or lower than typical. More importantly for EHL, seeing personal results stimulates extensive reflection about sources and pathways of chemical exposure, and, for many participants, this process leads to an expanded definition of pollution and activates efforts to respond.

Typical mental models before people receive their results have two key elements that are transformed by learning personal data. First, "before" images of how people come into contact with harmful chemicals focus on the pollution outdoors. Toxics are identified as coming from the military, waste dumping, industrial emissions, and dirty cities. Residents living away from these sources feel "clean." It's easy to understand how these ideas arise from the ready visualization and media portrayals of pollution (Finn and O'Fallon 2017). Second, people widely believe that chemicals and products must be tested for safety before they are sold in the U.S. They think that the federal government oversees consumer products similarly to the evaluation of drugs and food safety, so things on the store shelf can be assumed to meet minimum health-protective standards.

Participants vividly describe the shift in mental model when personal results show chemicals from consumer products are in their blood or urine, house dust, or indoor air. For example, the participants quoted below highlight shifts in understanding that the sources of chemicals can come from everyday products in the home, and that despite efforts to avoid chemicals, exposures are ubiquitous:

> *I mean I'm surprised that they can find that many things by looking at your dust and looking at your air, I mean that's amazing to me that they can actually find chemicals in your air in any amount whatsoever. (Altman et al.* 2008*)*

> *I really did not expect to have any [target chemicals]. Or if I had any, it wasn't going to be high levels of anything because I actually believed that I was living a really healthy life. And that I was lucky enough not to live next to a landfill, and be in an inner-city, in a ghetto, where a lot of families are exposed to these types of things.*

> *Chemicals! I have residues from insecticides and disinfectants and wood preservatives and mothballs…I never, never saw it coming for me. I always have lived a healthy life, lifestyle. And this just shocked me. (Dunagan et al.* 2013*)*

Study results generated a "curiosity gap" (Loewenstein 1994) that led people to reconsider what they thought they knew and open new lines of thinking about how exposures occur. For example, one participant said:

> *I did have some residual pesticides in my urine, and that surprised me because … whence might it have come? How long does it stay in your body? It's kind of a puzzle … I know that there is a company that comes around maybe once or twice a year as needed to take care of insect invasion around the base of the whole building…*

Sometimes study participants considered themselves well-informed about toxics in consumer products. These participants were particularly surprised that their careful consumer choices had limited effect, since the target chemicals are not always listed on product labels.

People also learned about the persistence of some chemicals:

> *The thing that I was really surprised about was that they found DDT here. In the home... that really surprised me because it was outlawed in 1972 I think, and we got this land in 1980 or 1982, and so that really surprised me. All over the Cape they had had a lot of farming here, so I guess it just hangs around for a long, long time. (Dunagan et al.* 2013*)*

Mental models *after* people receive their results in studies of EDCs typically included these ideas: Researchers found many chemicals in people (or homes); the chemicals come from ordinary products, known and unknown; and researchers are studying them because of concerns about health effects (Brody et al. 2014). These points are fundamentals for functional EHL about EDCs, providing good evidence that personal report-back can be a tool for improving understanding.

## Community Context

While fundamental shifts in mental models about EDCs are observed across many types of communities, some aspects of interpretation and response are rooted in the eco-social history of specific places and social groups. People look to past experiences with the environment and social power dynamics for cues to interpretation (Edelstein 2004; Kroll-Smith and Couch 1991).

For example, the California Household Exposure Study (CA HES) includes two communities that are very different in their environmental exposure profile and social context (Adams et al. 2011). Richmond, CA, is predominantly Latino; one fourth of the residents had incomes below the federal poverty level and half had incomes below 200% of the poverty line, according to the U.S. Census at the time of the study. The homes in the study are next to one of the largest oil refineries in North America, a marine port, and rail and highway corridors. The comparison community, Bolinas, is 20 miles west and predominantly white, middle class, and rural. The community is widely identified as liberal and environmentally conscious. Both communities have active environmental organizations: Communities for a Better Environment (CBE) in Richmond and Commonweal in Bolinas. Study results showed a greater number and higher concentrations of industrial and transportation pollutants in Richmond than Bolinas, and provided evidence of chemicals from the refinery penetrating inside homes (Brody et al. 2009). Both communities had very high levels of flame retardants, due to the state's flammability standard (Zota et al. 2008), which has since been modified, and the two communities had roughly similar levels of many other EDCs from consumer products used indoors (Rudel et al. 2010).

Before the study, Bolinas participants expected "a clean bill of health." In contrast, Richmond residents had some rational sense that outdoor polluters affect their lives, but still thought of their indoor environment as protected:

> *I just want to believe it's ok. But when I think rationally about our situation I'd have to say, "no, it's not surprising." But if I think on a feeling level, yeah. On a feeling level I was surprised. (Adams et al.* 2011*)*

Some Richmond participants noted the irony that the official policy for response to accidents at the refinery was "shelter-in-place," a process of sealing doors and windows to prevent emissions from entering, but homes were already violated by routine emissions. Residents in both communities rethought their understanding of "pollution" when they learned about consumer product chemicals and residues from banned pesticides still lingering in house dust and air. Seeing results for many chemicals, they also intuitively interpreted them in terms of cumulative impact, a concept that is becoming more central in environmental health science.

Residents of both communities viewed their results as resources for action. In Bolinas, the focus was on personal consumer choices and government regulation of EDCs in products. For example:

> *I'd like to see a ban on bisphenol A in products. And the whole PBDE thing, which I think is a little complicated because the politics around... fires and stuff in California... But I would like to see that people don't have to ... have their pillows doused in something toxic if they don't want it to be. (Adams et al.* 2011*)*

In Richmond, study participants were motivated to attend public hearings and rallies to oppose a specific proposal to expand the refinery, a plan expected to result in added emissions from dirtier petroleum sources than those currently in use. A resident who had not previously been engaged on this issue said:

> *I'm not very active, but I forced myself to go to the Chevron rally and march...I was so cold! And then I had to walk all the way home. It seemed like a worthwhile thing to do. I felt motivated to do that. (Adams et al.* 2011*)*

Another said:

> *At first I was thinking, "God, I wish I didn't know all this." But the more I think about it, the more I understand it, the more I feel like it helps me to, ... do whatever I can...if you know the information then you can't not participate in trying to make change. (Adams et al.* 2011*)*

The ongoing efforts of CBE made it easy for study participants to connect with environmental change efforts. Participants also came to community meetings to discuss action strategies with each other and propose new study hypotheses, for example to compare the effects of the refinery with a power plant in a nearby town. These excerpts illustrate how people absorbed the study results and were motivated to apply them within the eco-social context of their lives.

Economic connections between study participants, communities, and polluters can be important dimensions of community context that affect people's interpretation of opportunities for action. In Richmond, refinery employees had incomes sufficient to live farther from the refinery, so participants in our study, which focused on neighborhoods closest to the refinery, were not employed there and did not have a direct economic stake in the company. Conversely, in Little Hocking, Ohio, a community where drinking water was contaminated by fluorinated chemicals (PFOA, also known as C8) from DuPont, some residents perceived a conflict between economics and environmental health. Perhaps as a result, biomonitoring participants who learned their personal PFOA blood level results were more likely to take them to their doctor as a personal health concern rather than seeking system-change from industry (Judge et al. 2016). However, others in this community participated in a lawsuit against DuPont and the vast majority changed their drinking water source, as described below.

## Learning and Action

While the exposure experience and interpretation of results are filtered through local and personal eco-social histories, the PERE Study and others consistently find that people who receive their personal results are motivated to take action. As people read and interpret their results, they brainstorm about possible exposure sources and solutions, suggesting and evaluating alternative scenarios and seeking the strongest options. This phenomenon shows how personal report-back is effective in quickly advancing EHL beyond the "recognize" level of Bloom's taxonomy of knowledge to "apply, analyze, and evaluate," as described by Finn and O'Fallon (Finn and O'Fallon 2017).

Some EDC exposures are easier than others to avoid through simple changes in consumer behavior or housekeeping. Participants often turn first to these opportunities. They think about avoiding indoor pesticides, switching cleaning and personal care products, or discarding household items that may be exposure sources. Critics argue that focusing on changing consumer behavior can potentially create unreasonable burden and guilt, particularly for mothers who are already busy with shopping, cooking, housekeeping, and childcare, often on top of paid jobs (Mackendrick 2014, 2015). Further, exposure research shows that the effectiveness of behavior change is limited, for example by inadequate labeling (Dodson et al. 2012a, b), limited marketplace options, and the presence of legacy chemicals embedded in buildings and furnishings. We share the concern that the effectiveness of individual choices should not be over-sold without empirical evidence that they are practical and effective (Brody et al. 2007). However, empirically-based recommendations for personal exposure reduction continue to expand (e.g., (Adams et al. 2011; Harley et al. 2016; Rudel et al. 2011; Zota et al. 2017)).

Despite the limitations and caveats and the need for systemic reductions in pollutants, participants strongly prefer to have exposure reports that include options for simple actions to reduce exposure, and they pay attention to these messages. In interviews, study participants frequently report making changes that are reasonably expected to reduce exposures relevant to health. For example, this comment by a parent in the CDC Green Housing Study shows how the information in results reports generates learning as a basis for action:

*I thought that using spray chemicals to kill cockroaches was the most effective way to kill them, but now I have replaced it with cockroach sticky traps, decreasing the use of the spray chemicals. And moreover, after I learned from this research that products with fragrances can increase the chemical concentration in the house, I have decreased the use of products with fragrances. Take shower gel as an example, I now understand that those without fragrances would be better, and I have encouraged my kid to use products with no fragrance.*

When community-level or even national change is needed, researchers face challenges to provide direction. In some communities, such as Little Hocking and Richmond, industrial sources were obvious targets for action. In Little Hocking, the researchers made clear action recommendations for residents to reduce PFOA intake from drinking water by switching to other water sources, and the study resulted in DuPont providing free access to bottled water. As a result, it was relatively easy for residents to act on their new awareness and a large majority did so (Emmett et al. 2009). In Richmond, the local environmental justice organization, Communities for a Better Environment, provided an avenue for study participants and community members to take actions to influence refinery emissions. CBE hosted community meetings, alerted community members about opportunities for public testimony, organized rallies, and prepared documents for legal proceedings. These efforts contributed to a court victory that required Chevron to complete a cumulative health and environmental impact assessment, effectively stalling refinery expansion (Brody et al. 2009). However, CBE and other local organizations continue to press for environmental protections from Chevron and other local polluters, drawing strength in part from the study's success at raising EHL in this community.

The desire for collective action can be quite strong:

*So that's what I think biomonitoring is good for...personal choices, political action, public awareness... but also the creation of community. When you've got a community of people that trust each other and have moved through the grief, then you can start making changes. (Dunagan et al.* 2013)

In many studies, however, participants lack a clear avenue for collective action for various reasons. Participants in large cohort studies are not in touch with each other and may not be geographically localized, making collective action difficult. In other studies, the source may not be known or it may be due to national-level government and industrial policies, as is the case for personal care products or other consumer products. In these situations, personal exposure reports can encourage and help participants become engaged in civic life and connect with national efforts. Improving the efficacy of collective action is a major challenge in efforts to strengthen the ability of report-back to move people to the highest levels of EHL.

## Free-Choice Learning

The early report-back efforts for biomonitoring and personal exposure studies were motivated by CBPR and right-to-know ethics, but the learning outcomes of returning results in these studies highlighted the potential to more intentionally influence knowledge and behavior change. Reaching adults generally requires free-choice learning, which is driven by the interests and needs of the learner (Ramirez-Andreotta et al. 2016). The Metals Exposures in Homes (MESH) project adopted this framework in a study involving residents in Dewey-Humbolt, Arizona, near the Iron King Mine and Humboldt Smelter Superfund site and in a region with naturally high levels of arsenic and heavy metals in groundwater and soil. MESH collected biological samples (urine, toenails, and blood) and environmental samples (soil, dust, and water) to test for metalloid content. Participants were interviewed after they received results to see if they understood results and whether they had made informed choices to reduce risks. Participants reported adopting various strategies, including switching to bottled water, reducing soil and dust ingestion of metals by removing shoes at the door, washing hands, wet dusting, closing windows during wind events, and removing carpet. Some took actions such as planting ground-cover to reduce blowing dust. Personal results played a role in generating action, as one participant noted:

> *I had heard that the water was bad here prior to the warnings that we had gotten from the sheriff's department. But once you see it compared from home samples on your own land that you're living in … it strikes you deeper. (Ramirez-Andreotta* et al. 2016*).*

The action strategies showed understanding of risk and exposure routes. In addition, participants had many questions and sought further sampling, a health risk analysis, and government response to reduce contamination sources. (Ramirez-Andreotta et al. 2016).

## Building Trust

Research across many fields of communication reveals the importance of trust to how messages are received and understood. Report-back can be a key element in building trust between researchers and communities, improving the effectiveness of EHL messages and also strengthening the research by supporting recruitment and retention. Participants feel respected and are grateful for the time, attention, and honesty involved in preparing thorough reports. They simply appreciate that they are "not ignored" (Dunagan et al. 2013). The power of transparent reporting to strengthen trust was illustrated in the Growing Up Female cohort in the Breast Cancer and the Environment Research Centers. Researchers unexpectedly discovered high levels of PFOA in girls living in one part of the study area, probably due to contamination of the drinking water supply. Parents expressed their gratitude even though the source of exposure and health implications were not known at the time; as one commented:

> *Let me get this straight: You have found something, you do not know the cause or solution? Thank you for doing the right thing morally and ethically for sharing this information with us. (Hernick et al.* 2011*).*

This kind of disclosure can help to remedy past abuses when researchers failed to tell participants results that were relevant to their health. For example, in the prominent Kennedy-Kreiger case, a study of lead remediation in low-income housing, researchers failed to promptly tell study participants, predominantly African Americans, that their children's blood lead levels were rising due to one of the remediation strategies in the study, and failed to promptly end that part of the study (Mastroianni and Kahn 2002). This failure and others have understandably made it more difficult to enroll African Americans in studies, which, in turn, limits knowledge that could benefit them. Transparent report-back of exposure biomarkers can help to rebuild trust.

Report-back also strengthens relationships between researchers and participants by helping to equalize knowledge and create reciprocity. As a CDC Green Housing Study participant said,

> *[I give] information to you guys who are doing the study and then get some information back for my own interest." Another said, "In the beginning I didn't like it [participating in sampling]... You get that feeling like when somebody's intruding ....But then, after seeing the result ... I felt more comfortable because this was for the benefit of my daughter and her asthma, her food allergies.*

As participants learn about the scientific process, they understand and feel pride in their contribution to knowledge and the health of their family and community.

While report-back strengthens perceptions of researchers as allies, conversely, results often are interpreted in the context of past distrust of industry and government that dampens the potentially empowering aspects of receiving results. Participants, across all types of communities, believed that government was not doing enough to protect people from harmful chemicals and industry would not act against its economic interests to protect health (Adams et al. 2011; Altman et al. 2008; Judge et al. 2016).

## Effects on Researchers

Interestingly, researchers as well as participants learn through the report-back process (Ohayon et al. 2017). They reported deeper engagement with study participants. By focusing on personal results, rather than central tendency results, their attention is drawn to unusually high exposures and to patterns of exposure across multiple chemicals in the same person's sample. Through this process, report-back has led to the discovery of previously unrecognized chemical sources, such as high PCBs resulting from a wood floor finish used in the 1950s and 1960s (Rudel et al. 2008).

## Methods for Returning *Personal* Results

Returning personal exposure results, as practiced in the studies just described, means much more than telling people a single number, like the number of nanograms per liter of an unpronounceable chemical in their blood. Meaningful results reports provide extensive contextual information about what the numbers mean, including information about how the person's results compare to others, what is known and not known about health effects, where the chemicals come from and how to reduce exposures, when that is possible. The Personal Exposure Report-back Ethics (PERE) Study prepared a handbook for researchers that guides development of reports and includes examples of text and graphs (Dunagan et al. 2013). We developed the Digital Exposure Report-Back Interface (DERBI) to make it efficient to produce high-quality reports in studies of any size (Boronow et al. 2017). Examples of several reports are accessible online to show the range and flow of information (Silent Spring Institute 2017). DERBI also includes researcher tools that allow scientists to readily identify outliers and examine mixtures of chemicals, supporting the opportunity described above for researchers to learn through report-back and identify new hypothesis.

Table 6.1 shows how data from the study and from scientific literature can be combined to answer people's typical questions. Knowledge from multiple scientific fields goes into the reports, including epidemiology, exposure science, toxicology,

**Table 6.1** Personal exposure reports can combine study data with broader scientific knowledge about chemicals and health in order to answer typical participant questions about their results

| Questions | Responses |
|---|---|
| What did you find? How much? | Results can list the chemical's technical name and a common language functional name. For example, polybrominated diphenyl ethers (PBDEs) can be described in common language as fire retardants. Participants often want to know the numerical value of their result, but often need contextual information to understand units of analysis. |
| Is that high? Is it safe? | In cases where an up-to-date government health guideline exists, it can be a useful benchmark for participants to evaluate whether their result is "high" and whether it is safe. |
| | Participants are also interested in how their own result compares with others in the study and with national benchmarks. When a chemical level for one individual or subgroup is much higher than most others, results can stimulate exposure reduction. |
| | When health effects are uncertain, researchers can explain the possible health implications that led to the study. Uncertainty can be acknowledged without offering false reassurance. |
| Where did it come from? How was I exposed? | Information about the likely sources of chemical exposures and how they get into people's bodies helps build a mental model of exposure and forms a basis for people to brainstorm about exposure reduction that is relevant to their own life. |
| What can I do? | Participants are keenly interested in opportunities to reduce their exposure. Both individual-level and community-, policy-level strategies should be shared. Exposure reduction tips should be evidence-based and acknowledge limitations. |
| What did the study learn? | Presenting study-wide results can emphasize the most important messages about health and exposure reduction, adding to EHL. Participants take pride in learning how their data contributed to knowledge to benefit the health of their community. |

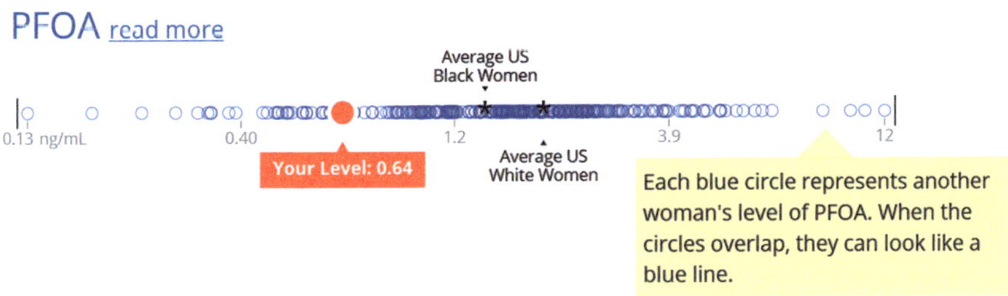


**Fig. 6.2** Strip plots, such as this one from CHDS, show people how their result compares with others in the study and benchmarks from other studies, such as NHANES

and communications. Because different people have different learning and communications styles, including both text and graphs is helpful.

Graphs are particularly useful for showing relationships within the data, since people can distinguish larger-than, smaller-than, and above-below relationships (Few 2004; Kosslyn 2006). These are skills that don't rely on literacy or numeracy. Figure 6.2 from the Child Health and Development Studies shows an example strip plot for the study participant's urine result in relation to others in the study and similar people in NHANES (Boronow et al. 2017). Since half the women in the study were African-American and half were not, the graph includes NHANES benchmarks for these groups separately. This type of graph has been usability-tested by the PERE Study, Health Research in Action at the University of California Berkeley (Brown-Williams et al. 2009), and the Three Generations Study (Judd and Plumb 2012). Even participants who lack confidence or "don't like" graphs generally are able to report accurately on the main messages when queried. Other types of graphs are useful for reporting study-wide results. Bar graphs can be effective for showing differences between groups, and line graphs are effective for showing changes over time.

As one example, when we have consulted with community research partners and held local focus groups, community members in Puerto Rico and Latina communities in the contiguous US have told us that reports on fragrance chemicals and phthalates should talk specifically about Fabuloso cleaner, a commonly used brand in Latina communities, and address the idea that chemical fragrances don't mean "clean." Other important messages in personalized chemical exposure reports can teach about environmental health fundamentals. For example, parents can learn through exposure reports that children are particularly vulnerable to the adverse effects of environmental chemicals because their bodies are growing and some compounds affect the developing brain, programming of bodyweight, changes at puberty, and other organs. Participants can also learn about routes of exposure – how chemicals enter the body in food and drinking water, ingested house dust, breathed air, and through the surface of the skin. Figure 6.3 shows how study-wide results in the Green Housing Study educate about fragrance as an avoidable asthma trigger.

In studies of endocrine disrupting chemicals, reports explain how laboratory studies in animals and cells lead to hypotheses about effects in humans. This infor-

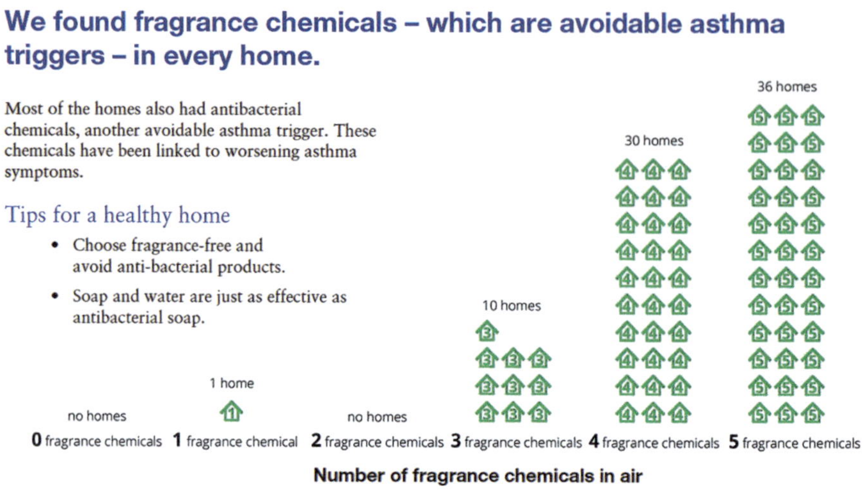


**Fig. 6.3** This pictograph of study-wide results from the CDC Green Housing Study visually shows that fragrance use was ubiquitous in homes. The text links the results to actions that can reduce asthma symptoms for children in the study

mation also makes clear that there is scientific uncertainty about how certain chemicals affect humans because the body of evidence is diverse and evolving. For example, the Child Health and Development Studies (CHDS) report says, "There are different degrees of evidence for links between chemicals and health. Some chemicals have known health effects, while others are suspected to cause certain health problems. When chemicals cause health effects in animals or human cells, scientists often suspect they will cause health problems in people, too. Since we don't do experiments on people, we often learn how chemicals may affect health by testing in animals or cells in a laboratory, similar to the way we test new drugs for safety." (CHDS (Child Health and Development Studies) and Silent Spring Institute 2017). Reports also explain that an individual's exposure results usually cannot be connected directly to a specific illness. The CDC explains, "Detecting levels of an environmental chemical in a person's blood or urine does not necessarily mean the chemical will cause health effects or disease." (HHS (Department of Health and Human Services) 2017b).

Results reports can also teach participants about the classes of chemicals included in a study. Biomonitoring California is one good resource for emerging and evolving information about the health effects and sources of environmental chemicals (https://biomonitoring.ca.gov/chemicals/fact-sheets), and DERBI reports also are based on libraries of carefully vetted information (Boronow et al. 2017). Scientific consensus statements help clarify whether the scientific evidence is sufficient to conclude that many biomonitored chemicals are harmful to health and if efforts to reduce exposures are warranted (Bennett et al. 2016; Di Renzo et al. 2015; Gore et al. 2015; UNEP/WHO 2013).

To make sure exposure report content is appropriate, researchers can work with community research partners, create community advisory councils, and convene focus groups to get input for tailoring reports. Once a prototype report is created, one-on-one usability testing is a good method to check how people understand the report and navigate through it. The rough-draft prototype can be tested, revised, and tested again until observations show that it is working well for the study population.

Beyond developing the content of personal results reports, research teams, in consultation with community members, need to decide on protocols for returning results. We recommend beginning the report-back process during the informed consent process before samples are collected. At that point, the study team can set expectations for participants about what the results will and won't be able to tell them, and give them the choice to receive their personal results or not. When results are ready, one or more members of the study team will need to be designated to respond to participant questions. Other protocol decisions include determining whether results will be returned in person, at a community meeting, or online, and related decisions about whether they will be print or digital (computers or smartphone platforms). Digital reports have the advantage of layering information, so even large reports appear more manageable and people can navigate to the content that interests them most. Using digital tools, such as DERBI, to generate reports – returned in print or online – also makes it easier to personalize reports to draw people's attention to the most important results for them. Using DERBI, researchers can develop decision rules to automate the process of creating personalized headlines to direct participants to notable results and to actions specifically relevant to those headlined findings.

Sometimes developing action recommendations can be challenging because the sources of exposure may be unknown or difficult for individuals or local communities to modify. U.S. law allows manufacturers to sell products without disclosing their ingredients, so the specific sources of chemicals detected in people or homes can be unclear (Dodson et al. 2012a, b). Laws, such as flammability standards, and marketplace practices can confuse and limit consumer choices of toxic-free products. For example, recent studies document how state-driven phase-outs of brominated flame retardants decreased the levels of brominated flame retardants but increased exposures to substitute chemicals used to meet the state's flammability standard (Dodson et al. 2012a, b). Since then, an update to the California rules prompted manufacturers to market flame-retardant-free furniture and children's products, providing consumers with choices they previously did not have.

Similarly, legacy contamination from persistent organic pollutants that have been banned for decades, such as organochlorine pesticides like DDT or flame-retardant compounds like PCBs, can be difficult and costly to remediate. In some situations, less-toxic options, such as eating more organic food, can be costly, thus limiting access for low-income people. Low income residents also are more likely to be affected by industrial pollution, often from major employers, which requires legal or social action to create change. Researchers studying hard-to-avoid toxics face

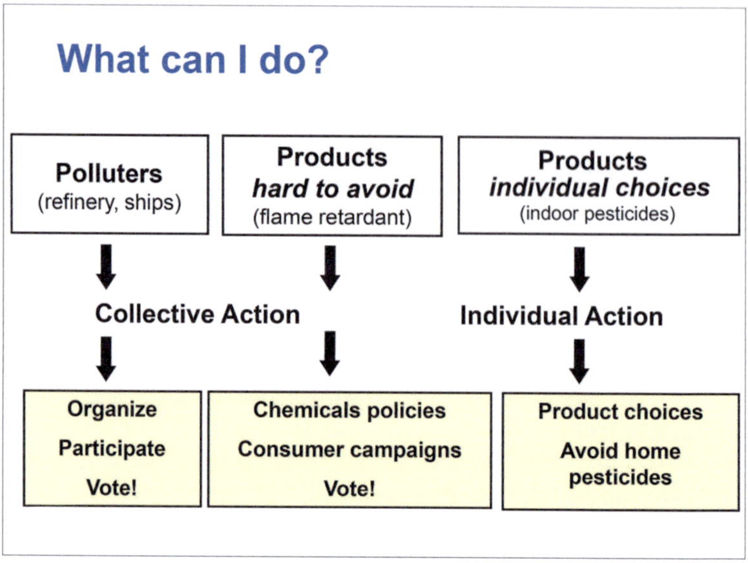


**Fig. 6.4** Biomonitoring reports should present options for reducing exposures from industry, transportation, and hard-to-avoid products as well as practical consumer choices

challenges to inform and support participants' efforts to act on their study results and reduce exposures. Figure 6.4 shows a conceptual map of how report-back messages can be tailored to realistically describe options for exposure reduction in light of the challenges outlined above.

## Community Level EHL

In place-based studies, communicating results at a community meeting can raise EHL for both study participants and the broader community. Researchers can present community-wide findings that put personal results in context, and they can directly answer questions. Several major studies, including CHAMACOS and the CDC Green Housing Study, have returned personal results to participants at community meetings. This strategy has several advantages. For participants who attend, it ensures that they personally receive their results, and the researcher is present to directly answer questions. The setting can be a festive celebration of participants' contributions to their community's health coupled with serious discussion of the overall study results and implications for health. IRBs are sometimes reluctant to allow meetings of study participants or return of results at a meeting, because of concerns about protecting participants' confidentiality in terms of their participation in studies. However, community members often have different values, giving lower priority to privacy (Brown et al. 2010). When results are returned at community meetings, participants can seek help and support from each other in reading the results, and they can brainstorm together about actions to reduce exposure. In communities with effective environmental advocacy organizations, community meetings can connect individuals with opportunities for collective action.

Ted Emmett and his colleagues have developed an approach, called "Community First" Communication, for building trust, effective communication, and public health action in a community affected by industrial pollution (Emmett et al. 2009). The study began with collaboration between researchers based at the University of Pennsylvania School of Medicine and the Decatur Community Association to address community concerns in Little Hocking, on the Ohio River near a fluoropolymer production facility that resulted in drinking water contamination with perfluorooctanoic acid (PFOA). They began by forming a Community Advisory Committee (CAC) composed of community organization leaders, town representatives, state and US Environmental Protection agency staff, and local physicians who were trained in environmental health. The CAC developed communications guidelines with several notable provisions even before the data collection had begun. For example, CAC principles stated that results would be released as soon as the investigators were comfortable that they were ready, rather than waiting for the often-slow full peer-review and publication cycle. Individuals would receive their results before they would be published in the scientific literature. Communications would maximize constructive response, a guideline that encourages researchers to articulate recommendations, and they would minimize "pointless" concern, recognizing that some concern may be warranted and motivating. After individual results were returned, overall community results were presented at an open meeting in a location pre-determined by the CAC. This careful roll-out of results combined with the study data documenting high blood levels of PFOA in community residents resulted in the company providing alternative bottled water and 78% of community members adopted alternative water supplies, either from the company or by using filters or other sources.

Community-level report-back of aggregate results fosters an interactive milieu in which both participants and non-participants living in the study community can share details and responses. In our community-level report-back in Richmond, there was a strong sense of project "ownership," which residents expressed at a very early phase by volunteering to participate, and later, by active participation at community meetings and public hearings. One of the things that surprised and delighted the research team was the response of people suggesting additional research questions. For example, one person suggested that we study another industrial community for comparison, and another argued for using the Environmental Impact Report to see which pollutants were predicted to rise with refinery expansion, and then see if those were the same chemicals we detected, in accordance with the California Environmental Quality Act of 1970. Another said we should look at chemicals that are now at or near EPA and Cal-EPA standards, to predict if refinery expansion would lead to exceedances. These examples of hoped-for future research indicate that residents' increased EHL led them to participate in discussions of scientific research design (Brown et al. 2012).

## Future Needs

The authors of this chapter are optimistic about the opportunities for personal exposure report-back to become a powerful tool for increasing EHL and hopeful that researchers will continue to develop best practices. Currently, report-back in environmental health is not as well developed as in medicine and genetics research, so we recommend the development of guidance documents by NIEHS and professional societies, such as the International Society for Environmental Epidemiology, to establish return of environmental results as the ethical norm, consistent with HHS guidance for human research. Researchers also need access to models and tools, such as DERBI, to make report-back practical. These tools can automate the process of personalizing results as well as provide resources such as libraries of vetted information in lay language (Boronow et al. 2017). Developing the content for reports requires specialized, multi-disciplinary knowledge from toxicology, exposure science, epidemiology, and communications, and individual researchers should not be expected to cover all of these fields anew for each study. Instead, they can draw on past reports for the same compounds and, in concert with representatives of their study community and participants, modify content and presentation to their own setting. New applications of report-back can also creatively attack resistant problems, such as better ways to visualize log-scale distributions. Studies of baseline EHL can help to highlight the core messages that need to be communicated along with study results. Both qualitative and quantitative assessments of how report-back influences EHL, behavior change, and community engagement will continue to improve practices and make report-back more effective for advancing environmental public health.

## Conclusion

Exposure biomonitoring – the measurement of chemicals in biological samples of blood, urine, and other tissues – and environmental sampling in household air and dust and other personal spaces have become a cornerstone of environmental health science. With these personal measurements comes an ethical responsibility to report back to people about their own exposures, both giving people their personal results and giving communities their aggregate results. Beyond just an ethical responsibility, this is a powerful opportunity to expand EHL. Going even further, we can see that the ethical issues and EHL issues are completely intertwined, combining to make a major advance in environmental health science.

The ethics framework for reporting personal results draws on principles of community-based participatory research, including concepts of co-ownership of data, co-learning by experts and lay people, and knowledge-sharing to equalize power and support autonomy. A "research right to know" allows people the choice to receive their results or not. When results are returned with contextual information about chemical sources, health effects and uncertainties, and strategies to reduce exposure, people feel respected and do not become overly alarmed.

Instead, evidence from studies in many different types of communities shows that returning personal results can quickly advance EHL. People are highly interested in their own results, so report-back generates curiosity and motivates learning that changes people's mental models of toxic chemicals and health. Often, participants begin with a mental model of toxic chemicals coming from industrial smoke stacks, waste dumping, or military sites, and through the "exposure experience" of learning about invisible chemicals in their own body or home, they come to understand that some of the chemicals they encounter every day in air, drinking water, and consumer products may also affect health. The new information stimulates brainstorming about possible exposure sources and connections to their personal history. As participants learn about sources and pathways of exposure, they generally understand uncertainties in links to health. They are vitally interested in both individual and collective actions to reduce their exposures. By sharing knowledge through report-back, researchers can support communities in translating results into action to improve environmental public health. In addition, this dialogue can help increase the EHL of the researcher, too, by tapping into the local and cultural knowledge of the community partners.

To support public understanding, personal results reports include much more than a number. Study participants want to know what was found in their own sample and interpretive information about whether their levels are high in relation to other people and what is known about health effects. They particularly want to know about exposure sources and how to reduce them. Research teams need to carefully consider what options are available for their participants, considering the eco-social history and community context. By planning and pilot testing in collaboration with community advisers or others who are similar to study participants, researchers can make their reports convenient, accessible, and responsive to local concerns. Digital methods, such as DERBI – the Digital Exposure Report-Back Interface – streamline report preparation and allow researchers to draw on tested graph formats and libraries of evidence-based plain language about health effects and exposure reduction. Senior researchers on the study team will still need to develop main messages about their study-wide results, though DERBI provides models and guidance. Through this process, researchers themselves gain by increasing their own EHL, since developing reports can be considered an example of the "create" level in the EHL model (Finn and O'Fallon 2017).

By continuing to experiment with report-back methods and improve them, researchers can develop novel ways to advance EHL about personal exposures to toxic chemicals. These topics are vital to personal health and public health, as science reveals the complex ways in which low doses of certain chemicals can influence pathways to disease.

## References


[Anonymous]. (2011). Communicating individual-level results to participant families. In *Breast Cancer and the environment research program annual meeting*. Cincinnati.

Adams, C., Brown, P., Morello-Frosch, R., Brody, J. G., Rudel, R., Zota, A., et al. (2011). Disentangling the exposure experience: The roles of community context and report-back of environmental exposure data. *Journal of Health and Social Behavior, 52*(2), 180–196.

Altman, R. G., Morello-Frosch, R., Brody, J. G., Rudel, R., Brown, P., & Averick, M. (2008). Pollution comes home and gets personal: women's experience of household chemical exposure. *Journal of Health and Social Behavior, 49*(4), 417–435.

Barlow, J., & Kushi, L. (2011, November 16). Communicating individual-level research results to participant families. In *Proceedings of the BCERP annual meeting*. Cincinnati.

Becker, K., Seiwert, M., Casteleyn, L., Joas, R., Joas, A., Biot, P., et al. (2014). A systematic approach for designing a HBM pilot study for Europe. *International Journal of Hygiene and Environmental Health, 217*(2–3), 312–322.

Bennett, D., Bellinger, D. C., Birnbaum, L. S., Bradman, A., Chen, A., Cory-Slechta, D. A., et al. (2016). Project TENDR: Targeting environmental neuro-developmental risks the TENDR consensus statement. *Environmental Health Perspectives, 124*(7), A118–A122.

Boronow, K. E., Susmann, H. P., Gajos, K. Z., Rudel, R. A., Arnold, K. C., Brown, P., et al. (2017). DERBI: A digital method to help researchers offer "right-to-know" personal exposure results. *Environmental Health Perspectives, 125*(2), A27–A33.

Brody, J. G., Aschengrau, A., McKelvey, W., Rudel, R. A., Swartz, C. H., & Kennedy, T. (2004). Breast cancer risk and historical exposure to pesticides from wide-area applications assessed with GIS. *Environmental Health Perspectives, 112*(8), 889.

Brody, J. G., Tickner, J., & Rudel, R. A. (2005). Community-initiated breast cancer and environment studies and the precautionary principle. *Environmental Health Perspectives, 113*(8), 920.

Brody, J. G., Morello-Frosch, R., Brown, P., Rudel, R. A., Altman, R. G., Frye, M., et al. (2007). Improving disclosure and consent: "Is it safe?": New ethics for reporting personal exposures to environmental chemicals. *American Journal of Public Health, 97*(9), 1547–1554.

Brody, J. G., Morello-Frosch, R., Zota, A., Brown, P., Perez, C., & Rudel, R. A. (2009). Linking exposure assessment science with policy objectives for environmental justice and breast cancer advocacy: The northern California household exposure study. *American Journal of Public Health, 99*(Suppl 3), S600–S609.

Brody, J. G., Dunagan, S. C., Morello-Frosch, R., Brown, P., Patton, S., & Rudel, R. A. (2014). Reporting individual results for biomonitoring and environmental exposures: Lessons learned from environmental communication case studies. *Environmental Health, 13*, 40.

Brown, P. (2007). *Toxic exposures: Contested illnesses and the environmental health movement*. New York: Columbia University Press.

Brown, P., & Mikkelsen, E. J. (1997). *No safe place: Toxic waste, leukemia, and community action*. Berkeley: University of California Press.

Brown, P., Morello-Frosch, R., Brody, J. G., Altman, R. G., Rudel, R. A., Senier, L., et al. (2010). Institutional review board challenges related to community-based participatory research on human exposure to environmental toxins: A case study. *Environmental Health, 9*, 39.

Brown, P., Brody, J. G., Morello-Frosch, R., Tovar, J., Zota, A. R., & Rudel, R. A. (2012). Measuring the success of community science: The northern California household exposure study. *Environmental Health Perspectives, 120*(3), 326.

Brown-Williams, H., Graham, C., & Oliver, M. (2009). *Report on CYGNET study focus groups: Communicating results to study participants*. Berkeley: Health Research to Action, University of California, Berkeley School of Public Health.

Cordner, A., Mulcahy, M., & Brown, P. (2013). Chemical regulation on fire: Rapid policy advances on flame retardants. *Environmental Science & Technology, 47*(13), 7067–7076.

Day, B., Langlois, R., Tremblay, M., & Knoppers, B. M. (2007). Canadian health measures survey: Ethical, legal and social issues. *Health Reports, 18*(Suppl), 37–51.

Delbanco, T., Walker, J., Darer, J. D., et al. (2010). Open notes: Doctors and patients signing on. *Annals of Internal Medicine, 153*(2), 121–125.
Department of Health Education and Welfare. (1979). The Belmont report: Ethical principles and guidelines for the protection of human subjects of research. *Federal Register, 44*(76), 23192–23197.
Di Renzo, G. C., Conry, J. A., Blake, J., DeFrancesco, M. S., DeNicola, N., Martin, J. N., et al. (2015). International Federation of Gynecology and Obstetrics opinion on reproductive health impacts of exposure to toxic environmental chemicals. *International Journal of Gynaecology and Obstetrics, 131*(3), 219–225.
Dodson, R. E., Nishioka, M., Standley, L. J., Perovich, L. J., Brody, J. G., & Rudel, R. A. (2012a). Endocrine disruptors and asthma-associated chemicals in consumer products. *Environmental Health Perspectives, 120*(7), 935–943.
Dodson, R. E., Perovich, L. J., Covaci, A., Van den Eede, N., Ionas, A. C., Dirtu, A. C., et al. (2012b). After the PBDE phase-out: A broad suite of flame retardants in repeat house dust samples from California. *Environmental Science & Technology, 46*(24), 13056–13066.
Dunagan, S., Brody, J., Morello-Frosch, R., Brown, P., Goho, S., Tovar, J., et al. (2013). *When pollution is personal: Handbook for reporting results to participants in biomonitoring and personal exposure studies.* Newton: Silent Spring Institute.
Edelstein, M. R. (2004). *Contaminated communities: Coping with residential toxic exposure.* Boulder: Westview Press.
Emmett, E. A., Zhang, H., Shofer, F. S., Rodway, N., Desai, C., Freeman, D., et al. (2009). Development and successful application of a "community-first" communication model for community-based environmental health research. *Journal of Occupational and Environmental Medicine, 51*(2), 146–156.
Few, S. (2004). *Show me the numbers: Designing tables and graphs to enlighten.* Oakland: Analytics Press.
Finn, S., & O'Fallon, L. (2017). The emergence of environmental health literacy—From its roots to its future potential. *Environmental Health Perspectives, 125*(4), 495.
Gibbs, L. M. (2011). *Love Canal: And the birth of the environmental health movement.* Washington, DC: Island Press.
Goho, S. A. (2016). The legal implications of report back in household exposure studies. *Environmental Health Perspectives, 124*(11), 1662–1670.
Gore, A. C., Chappell, V. A., Fenton, S. E., Flaws, J. A., Nadal, A., Prins, G. S., et al. (2015). Executive summary to EDC-2: The Endocrine Society's second scientific statement on endocrine-disrupting chemicals. *Endocrine Reviews, 36*(6), 593–602.
Harley, K. G., Kogut, K., Madrigal, D. S., Cardenas, M., Vera, I. A., Meza-Alfaro, G., et al. (2016). Reducing phthalate, paraben, and phenol exposure from personal care products in adolescent girls: Findings from the HERMOSA intervention study. *Environmental Health Perspectives, 124*(10), 1600–1607.
Hawthorne, M. (2017). *Federal panel votes to warn public about flame retardants in baby products, furniture.* Chicago Tribune (Chicago) 20 September 2017.
Hernick, A. D., Brown, M. K., Pinney, S. M., Biro, F. M., Ball, K. M., & Bornschein, R. L. (2011). Sharing unexpected biomarker results with study participants. *Environmental Health Perspectives, 119*(1), 1.
HHS (Department of Health and Human Services). (2017a). *NHANES health measurements.* Available: https://wwwn.cdc.gov/nchs/data/nhanes/2017-2018/documents/2017_health_measurements_list.pdf. Accessed 18 Oct 2017.
HHS (Department of Health and Human Services). (2017b). *National report on human exposure to environmental chemicals: Frequently asked questions.* Available: https://www.cdc.gov/exposurereport/faq.html. Accessed 23 Oct 2017.
Is it in us?. (2017). *Project results.* Available: http://isitinus.org/project.php. Accessed 10 Oct 2017
Judd, S., & Plumb, M. (2012). *Reporting personal exposures to environmental chemicals: Focus group feedback.* Oakland: Child Health and Development Studies, Public Health Institute.

Judge, J. M., Brown, P., Brody, J. G., & Ryan, S. (2016). The exposure experience: Ohio river valley residents respond to local perfluorooctanoic acid (PFOA) contamination. *Journal of Health and Social Behavior, 57*(3), 333–350.

Kosslyn, S. M. (2006). *Graph design for the eye and mind*. New York: Oxford University Press.

Kroll-Smith, J. S., & Couch, S. R. (1991). As if exposure to toxins were not enough: The social and cultural system as a secondary stressor. *Environmental Health Perspectives, 95*, 61–66.

Lazarus, R. S., & Folkman, S. (1984). *Stress, appraisal, and coping*. New York: Springer.

Loewenstein, G. (1994). The psychology of curiosity: A review and reinterpretation. *Psychological Bulletin, 116*(1), 75.

Mackendrick, N. (2014). More work for mother: Chemical body burdens as a maternal responsibility. *Gender & Society, 28*(5), 705–728.

Mackendrick, N. (2015). Protecting ourselves from chemicals: A study of gender and precautionary consumption. In D. N. Scott (Ed.), *Our chemical selves: Gender, toxics, and environmental health* (pp. 58–77). Vancouver: UBC Press.

Mastroianni, A. C., & Kahn, J. P. (2002). Risk and responsibility: Ethics, Grimes v Kennedy Krieger, and public health research involving children. *American Journal of Public Health, 92*(7), 1073–1076.

McKelvey, W., Brody, J. G., Aschengrau, A., & Swartz, C. H. (2004). Association between residence on Cape Cod, Massachusetts, and breast cancer. *Annals of Epidemiology, 14*(2), 89–94.

Mind Disrupted. (2017). *About the report*. Available: http://minddisrupted.org/about.php. Accessed 10 Oct 2017.

Minkler, M., & Wallerstein, N. (Eds.). (2008). *Community-based participatory research for health: From process to outcomes*. San Francisco: Jossey-Bass.

Morello-Frosch, R., Brody, J. G., Brown, P., Altman, R. G., Rudel, R. A., & Pérez, C. (2009). Toxic ignorance and right-to-know in biomonitoring results communication: A survey of scientists and study participants. *Environmental Health, 8*(1), 6.

Morgan, M., Fischhoff, B., Bostrom, A., & Atman, C. (2002). *Risk communication: A mental models approach*. Cambridge: Cambridge University Press.

National Research Council. (1989). *Improving risk communication, report of the committee on risk perception and communication, commission on Behavioural and social sciences and education, National Research Council*. Washington, DC: National Academy Press.

Nelson, J. W., Scammell, M. K., Altman, R. G., Webster, T. F., & Ozonoff, D. M. (2009). A new spin on research translation: The Boston consensus conference on human biomonitoring. *Environmental Health Perspectives, 117*(4), 495–499.

Ohayon, J. L., Cousins, E., Brown, P., Morello-Frosch, R., & Brody, J. G. (2017). Researcher and institutional review board perspectives on the benefits and challenges of reporting back biomonitoring and environmental exposure results. *Environmental Research, 153*, 140–149.

O'Rourke, D. (2005). Market movements: Nongovernmental organization strategies to influence global production and consumption. *Journal of Industrial Ecology, 9*(1–2), 115–128.

Quandt, S. A., Doran, A. M., Rao, P., Hoppin, J. A., Snively, B. M., & Arcury, T. A. (2004). Reporting pesticide assessment results to farmworker families: Development, implementation, and evaluation of a risk communication strategy. *Environmental Health Perspectives, 112*(5), 636.

Quigley, D. (2012). Applying bioethical principles to place-based communities and cultural group protections: The case of biomonitoring results communication. *The Journal of Law, Medicine & Ethics, 40*(2), 348–358.

Ramirez-Andreotta, M. D., Brody, J. G., Lothrop, N., Loh, M., Beamer, P. I., & Brown, P. (2016). Reporting back environmental exposure data and free choice learning. *Environmental Health, 15*(1), 2.

Rudel, R. A., Camann, D. E., Spengler, J. D., Korn, L. R., & Brody, J. G. (2003). Phthalates, alkylphenols, pesticides, polybrominated diphenyl ethers, and other endocrine-disrupting compounds in indoor air and dust. *Environmental Science & Technology, 37*(20), 4543–4553.

Rudel, R. A., Seryak, L. M., & Brody, J. G. (2008). PCB-containing wood floor finish is a likely source of elevated PCBs in residents' blood, household air and dust: A case study of exposure. *Environmental Health, 7*, 2.

Rudel, R. A., Dodson, R. E., Perovich, L. J., Morello-Frosch, R., Camann, D. E., Zuniga, M. M., et al. (2010). Semivolatile endocrine-disrupting compounds in paired indoor and outdoor air in two northern California communities. *Environmental Science & Technology, 44*(17), 6583–6590.

Rudel, R. A., Gray, J. M., Engel, C. L., Rawsthorne, T. W., Dodson, R. E., Ackerman, J. M., et al. (2011). Food packaging and bisphenol A and bis(2-ethyhexyl) phthalate exposure: Findings from a dietary intervention. *Environmental Health Perspectives, 119*(7), 914–920.

SACHRP (Secretary's Advisory Committee on Human Research Protections). (2016). Attachment B: Return of individual research results in July 21, 2016 SACHRP Letter to the HHS Secretary.U.S. Department of health and human services.

Saxton, D. I., Brown, P., Seguinot-Medina, S., Eckstein, L., Carpenter, D. O., Miller, P., et al. (2015). Environmental health and justice and the right to research: Institutional review board denials of community-based chemical biomonitoring of breast milk. *Environmental Health, 14*(1), 90.

Silent Spring Institute. (2017). *MyCHDSReport*. Available: http://45.33.82.210/. Accessed 18 October 2017.

State of California Senate. (2006). *California environmental contaminant biomonitoring program* (Senate Bill No. 1379). California Health and Safety Code § 105440–105444.

UNEP/WHO. (2013). *State of the science of endocrine disrupting chemicals - 2012*. Geneva: United Nations Environment Programme and the World Health Organization.

Wu, N., McClean, M. D., Brown, P., Aschengrau, A., & Webster, T. F. (2009). Participant experiences in a breastmilk biomonitoring study: A qualitative assessment. *Environmental Health, 8*, 4.

Zikmund-Fisher, B. J., Turkelson, A., Franzblau, A., Diebol, J. K., Allerton, L. A., & Parker, E. A. (2013). The effect of misunderstanding the chemical properties of environmental contaminants on exposure beliefs: A case involving dioxins. *Science of the Total Environment, 447*, 293–300.

Zota, A. R., Rudel, R. A., Morello-Frosch, R. A., & Brody, J. G. (2008). Elevated house dust and serum concentrations of PBDEs in California: Unintended consequences of furniture flammability standards? *Environmental Science & Technology, 42*(21), 8158–8164.

Zota, A. R., Singla, V., Adamkiewicz, G., Mitro, S. D., & Dodson, R. E. (2017). Reducing chemical exposures at home: Opportunities for action. *Journal of Epidemiology and Community Health, 71*(9), 937–940.

# Chapter 7
# Strengthening Environmental Health Literacy Through Precollege STEM and Environmental Health Education

**Nancy P. Moreno**

**Abstract** High quality science, technology, engineering and mathematics (STEM) education provides a strong foundation for developing the environmental health literacy of students in grades pre-Kindergarten through 12 (pK-12). STEM learning inside and out of school develops the critical thinking, communications and numeracy skills, along with deep knowledge of systems and processes, that are necessary for the informed evaluation and use of environmental health information in everyday life.

**Keywords** Problem-based learning · STEM education · Environmental health education · Teaching · Teacher · Curriculum · Interdisciplinary · Precollege · Afterschool · Grades K–12

STEM (science, technology, engineering and mathematics) learning experiences develop foundational knowledge and skills that support environmental health literacy of students in grades pre-Kindergarten through 12 (pK-12). STEM educational programs, however, do not necessarily address environmental health topics, nor do they explicitly require students to apply critical thinking and problem-solving skills to health- or environmental health-related challenges. Across the landscape of STEM teaching and learning for precollege youth, there is room for more environmental health content and examples in all aspects of program and curriculum design and delivery. Many current approaches shortchange students in terms of building their understandings of how human health and quality of life are inextricably linked with local and global environmental processes and systems. This deficit represents a missed opportunity, because topics that are personally relevant to students, such as asthma resulting from indoor air pollution, can be powerful catalysts for the development of a range of STEM-related knowledge and skills, including aspects of environmental health literacy.

N. P. Moreno (✉)
Baylor College of Medicine, Houston, TX, USA
e-mail: nmoreno@bcm.edu



S. Finn, L. R. O'Fallon (eds.), *Environmental Health Literacy*,
https://doi.org/10.1007/978-3-319-94108-0_7

When environmental health concepts explicitly are woven into STEM education, teachers and students increase their knowledge of complex environments and ecosystems, sources and applications of environmental health information, individual variability in disease susceptibility and informed approaches to decision-making. These key science and health understandings form part of the framework of environmental health literacy and contribute to improvements in the health outcomes overall for individuals and communities (Finn and O'Fallon 2017).

Environmental health content and examples are an obvious fit for science classes. By focusing on human-induced changes in indoor and outdoor environments and how these changes affect disease risk factors, teachers can make abstract concepts related to microorganisms, climate or earth systems immediately relevant to students' everyday experiences. Since environmental health also is part of the larger fabric of STEM education, examples based on safe drinking water, pesticide presence in foods or extreme weather, for example, can enrich instruction in any of the STEM areas across all grade levels while also raising students' environmental health literacy.

STEM learning, including experiences related to environmental health literacy, takes place inside and outside the classroom (see Chap. 12). Research internships in universities, government agencies, communities and private companies enable students to participate in science investigations or engineering design experiences, and give students a taste of possible future careers. Other opportunities exist through afterschool programs, and informal out-of-school programs at museums, science centers, and youth organizations, to name a few. Teachers also benefit from these programs, in addition to reaping rewards from ones that specifically aim to increase teachers' disciplinary knowledge or teaching repertoire. Most successful educational approaches for any of the STEM fields, can be adapted and applied in the context of building environmental health knowledge and skills–and ultimately, literacy across a range of environmental health concerns.

This chapter examines STEM education and its relationship to development of environmental health literacy, current approaches to teaching STEM using examples from environmental health, strategies for stimulating student interest and learning, and STEM education models and programs that have used environmental health themes to foster related literacy skills of students and their teachers. These topics are presented with the aim of inspiring the application of models from across the spectrum of best practices in STEM education to the development of students' environmental health literacy.

## What Is Science, Technology, Engineering and Mathematics Education?

The acronym STEM (science, technology, engineering and mathematics) became common in the 1990s through applications in a variety of contexts at the National Science Foundation (Bybee 2010a). In practice, "STEM education," frequently is applied to educational programs, interventions, teaching activities or curriculum resources that involve any of the four component subject areas. However, STEM also can be thought of as a meta-discipline that combines elements of all four of the constituent subject areas.

It is difficult to find agreement on a single meaning of STEM education. The following definition provides a starting point and reflects many current practices (Tsupros et al. 2009; Brown et al. 2011): *STEM education is an interdisciplinary approach to teaching and learning in which science, technology, engineering and mathematics concepts and skills are presented or taught holistically, without separating discipline-specific content, and make connections among individuals, school, community and the world through authentic problems or topics.*

When used in this sense, STEM education implies the use of active, problem-solving approaches to learning, rather than traditional teacher-centered instruction in which students are passive recipients of information from textbooks or lectures. Effective STEM education immerses students deeply in a topic and develops a range of critical thinking, collaborative and communications skills that are essential components of environmental health literacy, even when the skills are not developed in the explicit context of environmental health competencies. Importantly, an interdisciplinary STEM approach prepares students to be informed citizens and consumers, who can make sense of information and healthy decisions for themselves, their families and their communities. This approach is compatible with, and supportive of, the development of environmental health literacy.

The recent adaptation of Bloom's taxonomy as a construct for six educational stages of environmental health literacy, building from recognition to new applications of knowledge, highlights the overlap of STEM education and development of students' environmental health literacy (Finn and O'Fallon 2017). In both contexts, it is important for students to be able to develop knowledge, skills and understandings that apply in everyday life. Thus, a STEM literate individual is able to evaluate the quality of science or health information in social media posts, accurately estimate risks associated with behaviors or exposures, and interpret and apply information to make informed decisions for her- or himself, a family or the community (NAS 1996).

The skills taught in the STEM disciplines are particularly critical, because health topics, including environmental health, sometimes are left out of STEM teaching and learning in both formal and informal settings. Thus, students must be able to apply the same problem-solving and analytical skills in the potentially unfamiliar context of environmental health. Ecology and earth systems figure prominently in most state and national standards, for example, but required teaching standards vary in the degree to which they cover risks, health impacts of environmental change or health-related decision-making.

In many cases, it is up to the teacher to provide examples of how a particular science topic relates to health. For example, lessons for third grade students on weather-related hazards, could be expanded in various ways to include grade-level appropriate environmental health concepts. These ideas might include exploring how flooding leads to health risks from water- and mosquito-borne diseases or chemical contaminants, or how sun screen and other protections shield against damaging UV radiation in warm, sunny climates. The challenge for teachers is locate appropriate reference materials or lessons to make these connections, and to weave these ideas into the content they are required to cover. One way to support teachers in these tasks is to strengthen their own environmental health literacy through professional development related to the environmental science, earth and space science, nutrition and other topics. In addition, the general development of teachers and their students as critical thinkers with strong baseline STEM knowledge, will take them well along the road toward environmental health literacy.

## A New View for Science and STEM Education

Expectations of STEM learning have changed over time. In this section, we explore how the current vision for STEM education began more than 20 years ago. The discussion is relevant to the teaching of environmental health literacy, because many of the same competencies and kinds of disciplinary knowledge are needed both for literacy in environmental health and STEM.

In the 1990s, low performance of United States students on international assessments of science and mathematics learning, as compared to students in other nations, caused growing consternation among educators and policy makers. Analysis of fourth and eighth grade student achievement from the 1996 Third International Mathematics and Science Study (TIMSS) (Beaton et al. 1996a, b) led to the characterization of US curricula in both subject areas as being "a mile wide and an inch deep," with few opportunities for students to develop conceptual understanding (Schmidt et al. 2002). Access to high quality teachers and well-designed curricula also were identified as essential for effective learning by students. Shortly thereafter, international assessments of 15-year-old students in 57 countries by the Programme for International Student Assessment (PISA) found US students placing below the international average in scientific literacy. The test examined students' capabilities to explain scientific phenomena and apply scientific knowledge, in addition to assessing for science knowledge, applying science in range of contexts (including health, resources, environments, hazards and frontiers of science and technology) and attitudes (OECD 2007). These outcomes fueled an already ongoing national dialogue about how to teach science more effectively (Bybee 2010b). These concerns about general "science literacy," parallel the gradual identification of knowledge and skills that contribute to environmental health literacy, as discussed throughout this volume.

Concerns about workforce development also drove interest in the quality of STEM education. US students were enrolling in STEM degree programs in lower numbers than students in other countries that compete against American companies in the global economy. Development of a highly capable STEM workforce for the twenty-first Century was viewed as essential for the continued vitality of the US economy (Committee on Prospering in the Global Economy 2007). Consequently,

STEM literacy and skills development by future workers and citizens became not only educational imperatives, but were identified as crucial to maintaining national prosperity and security.

Key initiatives by the American Association for the Advancement of Science (AAAS) and the National Academies already were underway by the time results from the TIMMS and PISA studies were communicated widely. Both organizations spearheaded complementary, multidisciplinary programs to develop a new vision for STEM education by focusing deeply on fewer topics and systems, and engaging students in authentic scientific processes of inquiry. Project 2061 developed by AAAS advocated for STEM literacy for all and outlined an ambitious agenda in *Science for All Americans* (Rutherford and Ahlgren 1991). The project recommendations focused on the science learning needs of all children and emphasized that "race, language, sex or economic circumstances must no longer be permitted to be factors in determining who does and who does not receive a good education in science, mathematics and technology" (Rutherford and Ahlgren 1991, Chap. 14). Subsequently, AAAS released a comprehensive set of learning outcomes, *Benchmarks for Science Literacy*, for students of all grade levels encompassing the nature of science, mathematics and technology, physical setting, living environments, the human organism, human society, the designed world, the mathematical world, historical perspectives, common themes and habits of mind (AAAS 1993).

The National Academies concurrently developed a set of *National Science Education Standards* (NAS 1996) that encompasses not only learning goals for students, but standards for science teaching, professional development for teachers of science, assessment, science content, science education programs and sciences education systems. Importantly, the Standards offered a new vision for STEM learning by asserting that all students, not just those who are aiming for science careers, should have opportunities to learn science by engaging in authentic scientific practices, in other words, inquiry. The call to action issued in 1996 by the Standards continues to ring true today in a world in which science and environmental health illiteracies paralyze the discussion on key issues, such as global climate change, vaccination and genetically modified organisms. In today's milieu, environmental health literacy should be considered a key component of general science and STEM literacy.

> In a world filled with the products of scientific inquiry, scientific literacy has become a necessity for everyone. Everyone needs to use scientific information to make choices that arise every day. Everyone needs to be able to engage intelligently in public discourse and debate about important issues that involve science and technology. *National Science Education Standards*, p 1.

Both the *Benchmarks for Science Literacy* and the *National Science Education Standards* included fundamental health concepts in their expectations for student learning. The Standards proposed Content Standards on Science in Personal and Social Perspectives as part of the science curriculum. This subset of standards covered many aspects of human health and disease, and outlined environmental health themes such as the following.

- Grades K–4 (Changes in Environments): Changes in environments can be natural or influenced by humans. Some changes are good, some are bad, and some are neither good nor bad. Pollution is a change in the environment that can influence the health, survival or activities of organisms, including humans.
- Grades 5–8 (Risks and Benefits): Risk analysis considers the type of hazard and estimates the number of people that might be exposed and the number likely to suffer consequences. The results are used to determine the options for reducing or eliminating risks.
- Grades 9–12 (Science and Technology in Local, National, and Global Challenges): Humans have a major effect on other species. For example, the influence of humans on other organisms occurs through land use–which decreases space available to other species–and pollution–which changes the chemical composition of air, soil, and water.

The *Benchmarks for Science Literacy* included sections on The Human Organism, Human Society, The Designed World and The Nature of Technology. These sections touched on themes that are consistent with a broad range of environmental health topics, such as population growth, trade-offs and decision-making, energy uses and applications of technology to solve human problems. Teaching these topics in the context of inquiry and problem-solving are aligned with and support the development of students' literacy in environmental health. Importantly, both documents continue to serve as useful guides about ways to integrate STEM topics into instruction in science and other classes. Many state education standards and assessment guidelines for science still are based on the approaches and concepts outlined in these initial sets of recommendations, particularly those of the *National Science Education Standards* (NAS 1996).

Thus, teachers and instructional designers can continue to use the *Benchmarks for Science Literacy* and the *National Science Education Standards* as resources to identify grade levels and science disciplines where they can insert environmental health examples or problems into ongoing instruction for students.

The *National Science Education Standards* and *Benchmarks* specifically targeted science teaching and learning, with connections to mathematics and technology. These documents, however, made little mention of the importance of engineering design, practices and problem-solving to modern life, nor did they reflect research that was emerging on the science of learning. Two influential publications, *A Framework for K-12 Science Education* (National Research Council 2012) and the *Next Generation Science Standards* (NGSS) (NGSS Lead States 2013) incorporated this information into an updated vision for STEM learning. Importantly, the *Next Generation Science Standards* explicitly call out key understandings, major ideas that cut across all sciences, and essential skills that are needed for fluency in science. Not surprisingly, as outlined in the next section, these components include knowledge and skills essential for environmental health literacy.

## Next Generation Science Standards and Environmental Health

The *Next Generation Science Standards* (NGSS) (NGSS Lead States 2013) expanded upon earlier recommendations and incorporated new research on how people learn into a detailed set of educational outcomes for students in grades K-12 (Bransford et al. 2000). Development of the NGSS followed a set of guiding principles and key understandings related to preparing students for future careers and as educated citizens. These guiding principles are outlined below (National Research Council 2012, pp. 24–25).

- **Children are born investigators**. Children naturally seek to understand and influence the world around them, and their early ideas about how things work can be used as foundation for teaching science.
- **Focusing on core ideas and practices**. A framework for science learning should focus on key (or, core) ideas, and on the basic practices and approaches used by scientists and engineers in the real world.
- **Understanding develops over time**. Core ideas should be developed over time through learning progressions, which provide instructional supports to help students make progress toward mastery.
- **Science and engineering require both knowledge and practice**. Scientists use inquiry and problem-solving approaches to develop explanations and predict future outcomes. Engineers aim to solve problems related to a human problem or need.
- **Connecting to students' interest and experiences**. Science learning should connect to students' everyday lives, and link to future potential careers.
- **Promoting equity**. All students should have access to quality teaching, resources and time to learn science effectively, and engage in activities that inspire further participation in science.

The same guiding principles apply to the teaching of environmental health concepts at any grade level, because they reflect current understandings of how students learn STEM and prepare for employment. Importantly, the NGSS document offers various entry points for environmental health science content into K–12 science teaching, because it provides student learning outcomes by grade level within a framework of three different strands or "dimensions" of science learning. The dimensions, which are Practices, Crosscutting Concepts and Disciplinary Core Ideas, reflect not only essential science knowledge, but the ways in which scientists and engineers solve problems and seek explanations. Practices refer to the approaches used by scientists or engineers to answer a question or solve a problem. Crosscutting Concepts are themes with broad applicability across all sciences, such as patterns in the natural world, energy and matter, or cause and effect. Disciplinary Core Ideas are major science ideas that build sequentially as students' progress through school Table 7.1.

**Table 7.1** Dimensions of the *Next Generation Science Standards (NGSS Lead States 2013)*

| **Dimension 1: Practices** |
|---|
| Describes eight major practices used by scientists to investigate, and build models and theories about the world; and key engineering practices used by engineers to design and build systems. |
| 1.Asking questions and defining problems |
| 2. Developing and using models |
| 3. Planning and carrying out investigations |
| 4. Analyzing and interpreting data |
| 5. Using mathematics and computational thinking |
| 6. Constructing explanations and designing solutions |
| 7. Engaging in argument from evidence |
| 8. Obtaining, evaluating, and communicating information |
| Students are expected to participate in these practices, NOT learn about them secondhand through lectures or reading from textbooks. |
| **Dimension 2: Crosscutting Concepts** |
| Consists of seven unifying processes and ideas, which are common themes across all domains of science: patterns; cause and effect; scale, proportion, and quantity; systems and system models; energy and matter in systems; structure and function; and stability and change of systems. |
| Concepts are intended to help students make sense of disciplinary content across grade levels and promote systems thinking. |
| **Dimension 3: Disciplinary Core Ideas** |
| Set of essential ideas from four major domains: physical sciences; life sciences; earth and space science; and engineering, technology and applications of science. |
| Core ideas meet at least two (and usually three or all four) of the following criteria. |
| 1. Have broad importance across multiple sciences or engineering disciplines or be a key organizing principle of a single discipline. |
| 2. Provide a key tool for understanding or investigating more complex ideas and solving problems. |
| 3. Relate to the interests and life experiences of students or be connected to societal or personal concerns that require scientific or technological knowledge. |
| 4. Be teachable and learnable over multiple grades at increasing levels of depth and sophistication. That is, the idea can be made accessible to younger students but is broad enough to sustain continued investigation over years. |

Like the *National Science Education Standards*, the NGSS Practices, Crosscutting Concepts and Disciplinary Core Ideas do not constitute curricula or lessons. Instead they reflect what a student should know or be able to do. Our current understanding of environmental health literacy is that it also relates to stages of knowing as well as to action. Thus, the three NGSS dimensions apply equally to the development of environmental health literacy. Just as the concepts contained within the NGSS are intended to build progressively and coherently across grades K–12, while focusing on building deep understanding of science content and processes–so should students have opportunities to develop concurrent skills and understandings related to the environment and impacts on health in the context of core ideas that are taught with increasing depth over years.

Importantly, all three dimensions of the NGSS standards are nested under performance expectations, which outline how students should demonstrate their skills and understandings at different grade levels. The performance expectations provide a coherent story line to guide curriculum planning. For example, students in grade four are expected to "obtain and combine information to describe that energy and fuels are derived from natural resources and that their uses affect the environment" (4-ESS3-1). This performance expectation provides teaching opportunities related to the practices of obtaining evaluating and communicating information, and constructing explanations and designing solutions. It encompasses core ideas related to national resources and natural hazards–in addition to the crosscutting concept of cause and effect. This example does not directly link environmental impacts to health concerns. In this case, it would be up to the classroom teacher to extend her or his teaching to explore health impacts of fossil fuel induced air pollution or worldwide changes in global climate. The performance expectation provides an opening to introduce ideas related environmental health, but the ideas are implicit and rely on the teacher's own expertise to make the necessary connections.

Other sections in the NGSS, however, are more explicit regarding connections to health. A later set of middle school performance expectations focuses on human impacts on climate, earth systems and local environments. Students are expected to "apply scientific principles to design methods for monitoring and minimizing a human impact on the environment" (MS-ESS3-3) and "construct an argument supported by evidence for how increases in human population and per-capita consumption of natural resources impact Earth's systems" (MS-ESS3-3). Importantly, these explanations and design challenges are linked explicitly to society, "all human activity draws on natural resources and has both short- and long-term consequences, positive as well as negative, for the health of people and the natural environment" (MS-ESSe-4). The notion of weighing benefits and risks is clear in this expectation for middle school students and contributes directly to the development of environmental health literacy.

These two examples illustrate the challenge of leveraging NGSS recommendations to raise environmental health literacy of students and their teachers. In the second example, the connections between health and environment, risk and consequences are made explicit for teachers, curriculum designers and textbook publishers–and we have some assurance that essential linkages to human health will find their way into teaching. In the first example, however, a teacher might focus on how production and use of fossil fuels affect the global carbon cycle or influence climate change, without exploring how these practices and outcomes impact water supplies, crops, patterns of disease or cancer risk. This knowledge gap can only be filled by providing needed information to teachers through professional development, by promoting the use of curricular materials and resources that make the connections for them, through partnerships with science organizations or individual science and health professionals and developing connections with textbook publishers and curriculum developers.

The NGSS recommend that students actively develop and practice a range of skills and be able to demonstrate the depth of their understanding. This represents another formidable challenge for educators. Active learning implies involvement of students in experimentation and inquiry, engagement with data and other opportunities to construct their own knowledge. These approaches are not necessarily part of K–12 science teaching, which still sometimes relies on textbooks and multiple-choice tests. Achieving the vision of the NGSS and improving US science (and STEM) education will therefore require large-scale involvement of teachers in existing or new professional development programs, as well as consistent instruction in similar teaching approaches as part of undergraduate and graduate K–12 teacher preparation and certification programs. In addition, when teachers are weaving environmental health concepts into STEM lessons or units, they must reach beyond expectations of state or national standards to include ideas related to individual susceptibilities, environmental exposures and risk. Even though these topics link naturally to disciplinary core ideas, such as genetics and human impacts on the environment, the connections are not laid out explicitly for teachers. Thus, supplementary curriculum programs and teacher professional development become essential if teachers are to embed environmental health concepts into their ongoing science or other STEM instruction with the aim of developing students as citizens who are literate across a range environmental health topics and concerns.

## Integration across Subject Areas

The *Next Generation Science Standards* (NGSS Lead States 2013) and preceding standards focus on STEM education. However, science is only one aspect of the complex set of requirements related to teaching and instruction that vary from state to state. Responsibilities for interpreting and implementing required or recommended standards lie with state, county or city education agencies, districts, schools, administrators, curriculum specialists and individual teachers.

Science topics typically compete for instructional time with other required areas of instruction such as reading/language arts, mathematics or social science. One approach to managing a crowded school day is to combine or integrate two or more subjects into the same lesson. Integrating science teaching with reading or social studies, for example, enables teachers or schools to meet both sets of standards for student learning simultaneously. Examples provided later in this chapter demonstrate the practicality and value of this approach, which is particularly viable for elementary school teachers. Many elementary teachers still teach all subjects to the same group of students, so using environmental health themes to integrate reading/language arts, science and mathematics represents an efficient way to address required education standards as well as introduce knowledge and develop skills needed for environmental health literacy.

Ultimately, individual classroom teachers are charged with delivering the educational experience to students. This instruction does not happen in isolation. Instead, teachers daily navigate a complex landscape of layered requirements and regulations, while handling issues such as large class sizes, family factors like income, language or caregiver status, violence or student health. Thus, any discussion of how to provide science or other STEM learning experiences to students must also be sensitive to community, school and social concerns.

These same concerns, however, provide another entry point for discussing and teaching about environmental issues in the contexts of health and social justice. Because of its immediate connections to individuals and populations, environmental health provides a context for using local events and news as the basis for case-based learning, student research projects or in-class activities. These kinds of learning activities develop both science disciplinary knowledge and literacy in environmental health. Helping students identify and understand health risks and hazards in their own communities can contribute to better decisions for their own health and that of their families, in addition to promoting informed advocacy, which are key elements of environmental health literacy (White et al. 2014).

Most teachers, however, do not have sufficient time or adequate resources to develop detailed and scientifically rich cases or interactive activities for students. Instead, they often rely on curriculum models and teaching materials created by university-based or commercial developers to be able to provide instruction on environmental health (or any other specialized science topic). Helping teachers find and learn how to use these materials, however, remains a challenge across all STEM fields. Stakeholders must continue to focus on dissemination and engagement with educators at all levels to encourage use of environmental health topics as vehicles for STEM teaching and learning. In addition, new materials that meet the latest state education standards and recommendations of the NGSS while developing students' environmental health literacy are needed.

Many schools still adhere to traditional scheduling with 50-minute (or so) class periods, which are not conducive to student investigations or laboratory experiences. Grades K–5 offer more opportunities for integrated approaches to STEM education, because many teachers still teach the same students throughout the entire day in self-contained classrooms. Or, when elementary teachers specialize in teaching one subject area, such as reading, they are more likely to collaborate with other team members and coordinate instruction. STEM topics, including environmental health, are suitable themes for integration into formal elementary classrooms. In addition, elementary teachers focus much of their efforts on building students' skills in reading and English language arts. This emphasis on communication provides a natural opening for developing skills necessary for environmental health literacy. Examples of these cross-over communications and analytical skills include inquiry and literature research, argumentation, and integration of information from oral, visual, quantitative, and media sources.

One framework that has guided a number of curriculum and teacher professional development programs is the Environment as an Integrating Context (EIC) model. This model uses locally relevant environmental issues as unifying themes to create seamless instruction across several subject areas. The EIC approach offers a solution to teachers who would like to expand their teaching beyond the traditional fragmented approach to teaching the STEM disciplines. The model has been implemented in a variety of settings and has inspired several curriculum programs.

Without guidance, it can be difficult for teachers to integrate content and instruction across disciplinary boundaries during the school day on their own. Teachers may lack specialized science knowledge, access to appropriate lessons or experience with leading student inquiry. The EIC model addresses these challenges and, also contributes to development of teachers' own environmental health literacy by immersing them in the study of local environmental challenges along with their students.

EIC is aligned with current standards, such as the NGSS, and is based on six key educational strategies (Lieberman 2013):

- Integrated interdisciplinary instruction that crosses traditional boundaries among disciplines;
- Community-based investigations as learning experiences for students, through activities, investigations or service learning;
- Collaborative instruction shared by teachers, students and community members;
- Learner-centered approaches to teaching, in which the teacher acts as a facilitator and students have opportunities to construct their own knowledge and develop skills;
- Combinations of independent and cooperative learning;
- Local and natural community surroundings as the "context" for teaching and learning.

The EIC model has been shown to promote science learning by students from a wide range of backgrounds using a range of implementation strategies from individual teacher-developed units to large scale curriculum development and implementation projects (Lieberman and Hoody 1998). Improvements are reported in student academic achievement related to state standards, engagement of students in instruction and readiness for post-secondary education or careers–when classroom teaching is integrated around environmental themes (Lieberman 2013).

The potential for increased student engagement is a particularly valuable aspect of the EIC model and other curricular modalities that let students explore pertinent community-based issues. Many students fail to keep up in school because they do not perceive their studies as relevant, nor do they have exposure to professional role models or mentors. EIC is activity- and inquiry-based, and helps students and teachers cross traditional boundaries between disciplines. Results of EIC-based instruction were documented in 40 schools across the US. The study team found that environmental settings encompassed by EIC programs spanned across classrooms, laboratories, developed areas of school campuses, undeveloped school properties, and even off-site study areas. Students in 36 of the 40 study schools outperformed traditional students in reading, writing, mathematics, science and social studies (Leiberman and Hoody 1998). The EIC approach has clear benefits for improving the environmental health literacy of both teachers and students.

The EIC model was a foundation for nine environmental health education projects funded by the National Institute of Environmental Health Science (NIEHS) beginning in 2000 under its Environment Health Science as an Integrated Context (EHSIC) for Learning program. The initiative built on previous funding opportunities by the same agency for the development of teaching materials on environmental health for students in grades K-12 and related teacher professional development. The EHSIC-funded projects applied EIC principles to a range of integrated educational curricular approaches and collectively produced more than eighty individual educational units (NIEHS 2017). Some of this work has been continued and extended through many of the Community Engagement Cores at the various Centers funded by NIEHS.

Outcomes from NIEHS-funded projects demonstrated that integrating science instruction around environmental health sciences themes increased students' knowledge of environmental health or toxicology concepts, such as dose-response, risk assessment or carcinogens–while learning life and physical science, as well as deepening their mathematics, language and social science knowledge. Although the term, environmental health literacy, was not used at that time, these outcomes clearly demonstrate the value of the integrative approach to building pertinent skills and understandings (Hursh et al. 2011). Three examples from the funded projects are described below.

- Under the Environment as a Context for Opportunities in Schools (ECOS) project, scientists and educators at Baylor College of Medicine developed and piloted integrated instructional units focused on food production and safety, water pollution, indoor air pollution and global atmospheric change with elementary school students. The units consist of adventure storybooks, guided science and health investigations (designed to be taught during science class), related language arts activities and mathematics exercises. A study of second grade students in three urban schools found that students' content knowledge of science concepts related to photosynthesis and plant growth, food use by organisms, food chains, nutrition and food safety increased statistically significantly, after completion of the first edition of the integrated unit entitled, *The Science of Food* (Moreno and Tharp 2011). In addition, students' open-ended writing samples provided evidence of improved use of science vocabulary words in context, and more detailed descriptions of scientific processes, such as "cause and effect." (Moreno et al. 2006).
- The Hydroville curriculum developed at Oregon State University for high school students aims to enhance students' science content knowledge, their inquiry capacity and disciplinary literacy (reading, writing and speaking in the context of science) and awareness of environmental health issues (Oregon State University 2017). The *Hydroville Water Quality* module, for example, engages students in finding the source of water contamination impacting an imaginary town. Using a quasi-experimental design, the problem-based module was compared to a more traditional approach, with different classes of students. The students using the integrated problem-based curriculum performed better than the comparison

group in terms of posing scientific questions, and generating hypothesis-drive approaches to investigate their questions. In addition, the students in the problem-based group increased their content knowledge statistically significantly from pretest to posttest over a ten-week period (Kang et al. 2012).

- The ToxRap™ project at the University of Medicine and Dentistry of New Jersey focuses on grades K-9 with a cross-curricular approach based on concepts from toxicology and environmental health risk assessment. Units for different grade level bands enable students to identify a problem, gather information, form a hypothesis, conduct experiments, collect and analyze data and formulate and communicate their conclusions. The program has been disseminated to 23 states, and actively is used with students through Environmental Health Science Centers, such as the ones at University of Arizona and Oregon State University.

## Best Practices for Teaching and Student Engagement in STEM

How students learn science is an ongoing area of investigation–and discussion–for all grades and educational levels, including undergraduate and graduate or professional training. The previous examples, which used environmental health themes as the basis for integrated instruction, all provided evidence in support of active learning by students. In these models, teachers are facilitators of student science learning, rather than instructors at the front of the classroom.

However, traditional teacher-directed instruction–in which teachers transmit information and students are relatively passive recipients–continues in many classrooms from elementary school through advanced programs in science, engineering or medicine. Long lectures, student memorization, and rote homework generally are less effective than experiences in which students are asked to challenge their own misconceptions, make meaning of new information or observations and reflect on their experiences (NRC 2000). Nonetheless, the teacher as "teller," still prevails as the model of teaching STEM disciplines in many settings, despite growing evidence to the contrary regarding its effectiveness as an instructional model (Freeman et al. 2014).

Impactful approaches to STEM learning immerse students in significant content and questions–a process that leads to deep understanding of content and development of a range of critical thinking and problem-solving skills. Teaching situations, in which students build knowledge through questioning, problem-solving, and personal experiences are characterized as constructivist. Constructivist pedagogy has its roots in psychological theories of learning, which suggest that learners build meaning around observable phenomena and that the actual construction of new knowledge depends on prior experience or background, and the learner's existing knowledge or misconceptions (Richardson 2003; NRC 2000). Using environmental health themes as the basis for student investigations, as advocated by the EIC model, is aligned with constructivist approaches for STEM learning.

Another way to think about STEM teaching and learning is by looking at whether the teacher or the student actively is participating in the process, or "doing the work." Strategies in which the teacher provides information through lectures or other didactic approaches to students, who are limited to note-taking or answering occasional questions, are considered "teacher-centered." Contrast this approach to a "learner-centered" active process, which requires students to work in small groups, solve problems, and share ideas with peers or the class. Learner-centered instruction gives value to the various backgrounds, perspectives, learning needs and interests that individual students bring to the educational process (McCombs and Miller 2007, p 16). Learner-centered instruction, as noted above, also is "constructivist," because students are building and applying knowledge.

"The one who does the work does the learning." (Doyle 2011, p 7)

Hursh et al. (2011) offers a simple example of a constructivist approach to teaching environmental health. Students are presented with a problem–well water with a foul smell. They investigate and gather information from reliable sources, such as the US Environmental Protection Agency, to examine possible causes of the odor and potential risks from different possible contaminants, such as pesticides. Finally, students come up with an explanation for the offensive odor, and propose policy changes and an educational component to inform the fictional community. All of this is conducted in the context of a realistic scenario originating from a fictional letter from a community member. Through immersion in the problem, students learn key environmental health and science concepts–related to water chemistry, common pollutants, health impacts, and risk–because they need to apply the information to solve the mystery. They also learn how to find and evaluate scientific information, develop teamwork competencies, and improve their written and oral communication and presentation skills. In other words, their environmental health literacy is strengthened. They have gone from the simple understanding of an environmental health concept to the application of that knowledge.

Aligning science and STEM teaching to the NGSS recommendations requires shifting away from the traditional approaches, already mentioned, that rely on textbooks, lectures, worksheets and memorization. Instead, as in the example above, students are engaged in developing explanations, systems thinking, and conducting investigations or solving problems. These practices also provide opportunities for students to read and write in a variety of contexts, and to engage in teacher-facilitated discussions (NRC 2015b). Effective learner-centered STEM instruction typically includes some or all of the following elements, which can be used to evaluate the quality and potential usefulness of environmental health or other STEM programs and curriculum resources. When selecting or creating a program to develop students' environmental health literacy in the context of STEM teaching and learning, users or developers should look for the following features.

- **Active learning**. Students are immersed in single lessons, cases or significant problems that are drawn from examples in the real world, and have opportunities to use observations, numerical data and other resources to answer questions, conduct investigations, synthesize information and communicate their findings.
- **Collaborative grouping**. Students work in teams, typically of 2–4 students, with assigned roles and responsibilities. These experiences potentially develop desirable workforce skills, such as leadership and written or verbal communication and feedback.
- **Assessment as part of the learning process**. Instead of a single test at the end of a chapter or unit, students self-assess their progress–which helps them take responsibility for the learning process. In addition to grading students, teachers use assessments to tailor or modify instruction to meet individual or group needs.
- **Attention to students' misconceptions**. Overcoming persistent erroneous beliefs is essential to helping students learn new STEM concepts and skills (NRC 2000). Experienced teachers identify students' misconceptions about phenomena or processes though assessment and bridge the knowledge gap by providing students' with first hand experiences that promote understanding.
- **Teachers as facilitators**. Teachers serve as coaches and guides, while students actively conduct their own investigations or engage in designing solutions to a problem.
- **Learning cycle**. Students become engaged in a topic through a real-world focus or challenge designed to stimulate their interest, followed by a sequence of learning activities that enable them to explore the idea further, develop explanations, evaluate their knowledge and elaborate or apply the knowledge to a new situation. The 5E learning cycle is a common model: Engage, Explore, Explain, Evaluate and Elaborate (Bybee 2015).
- **Interdisciplinary approach**. Students use technology and mathematics seamlessly with science and engineering problem-solving–in ways that are similar to how STEM professionals address challenges in the workplace. Other subjects, including reading and writing, social studies and health also are integrated into student learning experiences.

Learner-centered teaching, based on the features above, is effective using environmental health topics. For example, environmental health scientists and collaborating educators at the University of Wisconsin-Milwaukee have used zebrafish embryos and minnows of other species as model systems for student investigations of the impacts of environmental exposures on health. One module, for example, enables students to learn about normal development and evaluate how various environmental toxicants affect development of zebrafish embryos. The module was evaluated with 37 teachers and more than 850 high school students enrolled in a range of biology and advanced life science courses (such as advanced placement environmental science or anatomy) during 2011–2012 and 2012–2013. Teachers received preparation in environmental health science concepts and the module's experimental methodology, including care of zebrafish, during a one-week summer institute. They also received all necessary materials and opportunities to interact with the project team during the school year. As part of the module, students gathered and analyzed their own data from control and experimental embryos, and prepared written scientific communications or papers with their findings. The culminating activity was a student science research conference held in the spring with oral and poster presentations. Pre-test/post-test results from student assessments found statistically significant positive changes in students' opinions about environmental health and questions about environmental toxicants. For example, on the post-test, significantly more students agreed that "Seeing how an environmental agent affects fish helps me understand that those same agents can also affect me" (Tomasiewicz et al. 2014). Another module developed by the same team enables students to investigate how lead exposures affect behaviors of fathead minnows, and contributes to their understanding of negative consequences of chemical exposures (Weber et al. 2013).

*Chemicals, the Environment, and You* is an example of a standards-based curriculum module that follows a learning cycle approach. Students explore the relationships between chemicals in the environment and the health of living organisms, including humans, by progressing through all stages of the 5E model (BSCS 2012; Bybee 2010a). Developed by scientists, including subject area experts from NIEHS, medical experts, educators and curriculum designers, the unit incorporates actual case studies into classroom activities, and consists of six lessons that are mapped to the 5E learning cycle. Students begin their explorations through an "engage" activity in which they learn that everything in the environment is make of chemicals. Next, they "explore" the effects different doses of chemicals on seed growth and are introduced to concepts of toxicology. Further activities in the "explain" and "elaborate" phases further develop students' knowledge and skills as they use dose-response curves and learn about individual and community exposures to chemicals. In the culminating activity ("evaluate"), students apply what they have learned to solve a fictitious problem in which participants complain of headache and nausea after returning from a museum field trip. After examining evidence from the case, students are able to conclude that carbon monoxide exposure from a faulty exhaust system in the bus used for field trip transportation led to the observed illnesses. This module develops environmental health literacy, because students apply new knowledge and skills to solve problems related to chemical exposures, risk and health impacts on individuals.

**The Five E Learning Cycle Model**

Engage
Explore
Explain
Elaborate
Evaluate

The Science Take-Out project represents a different approach, which removes barriers to teaching environmental health and other science topics by providing simple, low cost kits with all of the materials and instructions to complete a single investigation. The kits are designed for use in situations where access to lab facilities or equipment is limited. Individual kits developed by investigators and educators at the University of Rochester enable teachers to explore important environmental health concepts, such as dose-response relationships, effects of environmental agents on body systems, changes in susceptibility to environmental agents at different life stages, and prevention of environmental exposures. Importantly, Science Take-Out kits have utilized the Small Business Technology Transfer grant mechanism of the National Institute of Environmental Health Sciences and National Institutes of Health to scale-up a research-based curriculum development project into a commercial company that creates, manufactures and markets science teaching materials (URMC 2017). Environmental health and educational expertise for the project is provided by the Community Outreach and Engagement Core (COEC) of the University of Rochester Environmental Health Sciences Center. This innovative model ensures that accurate and up-to-date science and health information is made available in an easy-to-teach format at low cost to teachers and community educators.

An even shorter format has been successful for the Sunwise Program, which demonstrates that not all effective environmental health or STEM units require several days or weeks for implementation. The Sunwise Program is an example of a set of successful, environmental health teaching units for different grade level bands that require a minimal time commitment and can be integrated with existing curricula. Developed in 1998 by the Environmental Protection Agency and now distributed by the National Environmental Education Agency, the program aims to educate and children and adult caregivers about health risks associated with UV radiation and how to protect against excessive UV exposure (EPA 2017a). By 2015, the program was estimated to protect against more than 11,000 cases of skin exposure. Between 2000 and 2015, more than 58,000 teachers and 34,000 schools registered for Sunwise and received the downloadable Sunwise Tool Kit to teach students in grades K-8 about the health risks of UV overexposure (EPA 2017b).

Use of the program led to statistically significant increases from pretest to posttest in students' knowledge about sun safety, and some improvements in children's perceptions of the healthiness of a tan and their intentions to play in the shade (Geller et al. 2003). Widespread implementation of the program is attributed to broad recruitment of schools through teacher and school nurse conferences and organizations, referrals from partner organizations, such as the American Cancer Society and alignment with state and national standards for science and health education. Despite its short duration, the Sunwise Tool kit exemplifies several elements of effective learner-centered STEM instruction. Students are engaged actively in investigations, including use of a UV meter to measure radiation; they work in collaborative groups on a variety of challenges; teachers serve in the roles of facilitators; and the approach is interdisciplinary with numerous linkages to reading, writing, speaking, and applications of technology. Thus, the units achieve both STEM and environmental health literacy goals.

## STEM Teachers and Environmental Health Literacy

Learner-centered educational approaches require teachers to develop different, and often new, sets of facilitation and classroom management skills. As noted, STEM teaching that is aligned with the NGSS requires teachers to know how to guide independent or small-group work by students. In addition, teachers must possess deep content knowledge, know how to convey complex concepts effectively to students and be able to address common misconceptions (Schulman 1986, 2000). Building this practical knowledge, referred to as pedagogical content knowledge, is an important part of teacher development and is essential for effective STEM education (Van Driel et al. 2001).

Given that the US continues to face regional shortages of STEM teachers–and teachers of some subjects, such as physics, chemistry and mathematics, are in perpetual short supply (Marder 2016)–providing professional development to teachers across all of the STEM disciplines remains a priority. In addition, elementary school teachers typically are less prepared to teach science and few of them have deep disciplinary knowledge in STEM or health-related fields. In order to teach environmental health ideas effectively, teachers need background in areas related to chronic and infectious disease, toxicology, risk and common environmental challenges to health related to chemical pollution, animal and insect pests, radiation, noise or climate. In other words, teachers also need to build their own environmental health literacy as a prerequisite to instilling the same knowledge and applicable skills in their students.

Professional development programs for teachers who already are employed in school (known as in-service professional development) are especially critical to improving the quality of STEM teaching consistent with the vision of the NGSS. Effective professional development enables teachers to learn new science content, develop experience with hands-on lessons and work with peers to share effective instructional approaches. Sustained programs, rather than single-day workshops, are more effective in building teachers' content knowledge, skills and beliefs about their abilities as teachers–particularly when the programs provide opportunities for teachers to enact new strategies and reflect on their own teaching (Van Driel and Berry 2012; Roberts et al. 2001).

Job-embedded forms of professional development, such as coaching, team-teaching, or lesson study (an approach developed in Japan, in which teachers collaboratively develop, evaluate and revise an actual classroom lesson, called a "research lesson"), also are effective in building discipline-specific teaching skills (pedagogical content knowledge) and science teaching efficacy beliefs (Lewis et al. 2004; Weiss and Pasley 2009). Direct partnering of scientists with teachers, either through summer research experiences or as part of curriculum development, further deepen teachers' knowledge and enhance their understanding of the scientific enterprise (Moreno 2005; Moreno et al. 2001).

Five general characteristics of effective STEM teacher professional development have been identified from the plethora of teacher development options available in schools, districts and organizations across US. These approaches are: focusing on specific content to encourage depth of knowledge; engaging teachers in the same active learning modalities they will implement with their students; enabling teachers to work collectively and collaboratively to examine their practices, curricula and planning for improvement; coherence with other school policies and practices; and sufficient duration in terms of intensity and contact hours (Wilson 2013). In general, professional development that extends over multiple sessions during the summer and school year leads to better teacher learning and sustained implementation of new practices. Opportunities to practice new instructional approaches and receive peer feedback also contribute to gradual and enduring changes in classroom instruction. However, results–even from large scale studies of innovative science curriculum programs with intensive teacher professional development are uneven–in terms of the ultimate development of students' scientific practices and acquisition of deep content knowledge (Grigg et al. 2012).

Project EXCITE (Environmental health science eXperiences through Cross-disciplinary and Investigative Team Experiences) is an example of a teacher professional development program that enabled middle school teachers to develop, implement, and revise problem-based, interdisciplinary curricula focusing on locally relevant environmental health issues. The program, conducted by Bowling Green State University, involved 18 middle school teachers in summer institutes and monthly meetings, in addition to collaborating with relevant community agencies, local scientists and education faculty members (Haney et al. 2007). The program exemplifies the characteristics of effective teacher professional development. It engaged teachers in a conceptual model of an environmental health system and promoted development of deep content knowledge through teacher institutes and year-round activities. Teachers were engaged actively in their learning and collaboratively worked to develop and refine problem-based learning modules for students. The program was aligned with local and state standards, and involved teachers over a two-year period. As a result, teachers' beliefs about their own efficacy in teaching science using constructivist approaches improved. In addition, teacher content knowledge and students' skills related to science inquiry increased with statistical significance from pre- to post-assessments (Haney et al. 2007).

The NIEHS supported Project EXCITE and has funded a wide range of teacher professional development programs conducted by individual organizations or administered by Community Engagement Cores since 1999 (O'Fallon 2017). These programs include: short programs and multi-day workshops either focused on environmental health modules or specific themes; summer institutes; summer research experiences for teachers and distance learning courses. Many of these experiences, such as those conducted by the University of Wisconsin-Milwaukee, are designed to develop teachers' abilities to use particular curricular materials. Others, such as the Environmental Health Sciences Summer Institute, enable teachers to participate in an extensive variety of professional development workshops focused on the connections between health and the environment over the course of a week. Currently conducted by Texas A&M University Health Science Center, the program provides opportunities for teachers to discuss environmental health issues that directly impact people's lives, participate in hands-on, classroom ready learning activities, share ideas with scientists and professionals, and discover how environmental health curricular materials contribute to student achievement (TAMU 2017).

To address shortages of expert teachers within certain subject areas, the UTeach program at the University of Texas-Austin enables students to simultaneously obtain a teaching credential while earning a degree in engineering, science or mathematics. This approach is developing more teachers with formal backgrounds in STEM subject areas. The program has been replicated widely and is available at 44 universities in 12 states. UTeach programs are expected to graduate more than 9000 STEM teachers by 2020 (Backes et al. 2016). While UTeach does not explicitly target environmental health as a disciplinary area, the program is addressing general needs for developing STEM content experts as well-prepared teachers.

## Informal and Out-of-School STEM Education

Formal education occurs in classroom settings during the regular school day, usually under the guidance of a trained teacher. In most states, STEM education during the school day occurs within each subject area domain–for example, in mathematics, science, computer coding or health classes. Informal education happens outside of organized classrooms. This kind of learning experience can occur in museums, science centers, clubs, summer programs, schools outside of normal classroom hours, parks and other settings, which present exciting possibilities for engaging youth with interdisciplinary themes related to environmental health.

Afterschool settings enable students to engage with peers and STEM professionals in challenging and exciting activities related to real world situations (Phillips and Miller 2014). Afterschool programs (and programs that occur before school or during the summer) offer opportunities to engage students from populations that traditionally are underrepresented in STEM fields. Girls, for example, attend afterschool programs at similar rates to boys, and African-American and Hispanic children are twice as likely to participate in afterschool programs than students from other groups (Afterschool Alliance 2015). Further, afterschool programs create opportunities for young people to build identity, see themselves as future STEM professionals, develop skills, enhance their academic performance and embark on STEM pathways (Krishnamurthi et al. 2014).

Afterschool programs also build students' interests and help them develop academic and social skills needed for success in STEM career pathways. Interest is an important motivator for students. Sadler et al. (2012) reported that the key factor predicting STEM career interest at the end of high school was interest at ninth grade. In addition, students' initial STEM career interests influenced their retention in STEM areas in college. Thus, programs that stimulate and nurture students' interest contribute to their ultimate persistence in STEM careers.

Afterschool programs tap into valuable community resources that include designed settings, such as schools, clubs or museums; naturalistic settings provided by parks or arboreta; organizations and networks of people, consisting of practices STEM professionals, educators, hobbyists, and business leaders; and everyday exposures to STEM ideas on the Internet, television and other media. The rich interplay of interactions of students with individuals across contexts (school, neighborhoods, out-of-school experiences) in combination with broader cultural influences has been described as a STEM learning ecosystem (NRC 2015b). The STEM learning ecosystem includes all of the various influences, contexts and learning modalities that influence students over time.

Well-designed STEM informal and out-of-school programs build skills and knowledge that contribute to environmental health literacy in areas such as communications, problem-solving, teamwork, career awareness and community involvement, even when the programs are not focused explicitly on environmental health science themes. High quality informal and out-of-school STEM learning opportunities have three design features in common as identified by the National Research Council (2015a). Productive programs:

- Engage young people by providing first-hand experiences, sustained STEM practices and a supportive community;
- Respond to young people's immediate interests by being socially meaningful and culturally responsive, while promoting development of their collaborative and leadership skills;
- Connect learning across the STEM learning ecosystem by leveraging community resources and partnerships.

First hand experiences are powerful motivators for students. Afterschool activities that enable students to work with peers to solve challenges were found to predict content knowledge gains related to life science and epidemiology in a Houston afterschool program, *Think Like a Microbiologist* (Newell et al. 2015). In addition, the afterschool program boosted students' perceptions of their own willingness to work hard on difficult topics.

Garden Mosaics youth activities developed by Cornell University, and implemented in urban community gardens across the country, is an example of a culturally responsive out-of-school initiative that involves a wide range of organizations. The program involves students in exploring the gardening assets within their communities, interviewing gardeners about their planting practices or cultural traditions, conducting a community garden inventory and development of Action Projects, in which they apply what they have learned to enhance their neighborhoods or local gardens. Over 4 years of the program, approximately 1200 educators, 14,000 youth and 2500 gardeners participated in Garden Mosaics activities (Krasny and Tidball 2016). When interviewed after the pilot phase of the project, student participants most frequently mentioned benefits related to gardening skills, learning from and developing relationships with elder gardeners, academic/research skills, such as writing and measuring, teamwork and appreciation for the value of gardens (Krasny and Doyle 2002).

Community-engaged research and other community-focused programs provide other vehicles for meaningful out-of-school learning. Community-engaged research (CEnR) requires students to work side-by-side with members of neighborhoods to identify and define problems, collect information and answer research questions. The Health Science and Technology Academy (HSTA) in West Virginia has created a successful STEM pipeline model for underrepresented minority students across the entire state. Approximately 800 high school students (grades 9–12) participate each year in HSTA, which is led by faculty at the University of West Virginia and local governing boards comprised of community members. Students enter the program in ninth grade, attend two summer camp experiences, additional activities during the school and complete 75 h of community service. Over 4 years, the students engage in increasingly rigorous activities, including summer research experiences and community-related projects, including community-based participatory research on disease prevention (Branch and Chester 2009). Importantly, HSTA has improved high school graduation rates, college attendance and graduation in health science and STEM fields of underrepresented minority students–96% of HSTA graduates attend college, 90% graduate with a four-year degree or better, and 65% major in a health or STEM subject, with 59% choosing one of the health professions (McKendall et al. 2014).

Student laboratory or other research experiences place students in the roles of investigators and research assistants. These programs, which occur in the field, laboratory or communities, let students understand that STEM fields, including environmental health, are collaborative endeavors with real-world application. By working with a range of STEM professionals engaged in environmental health research, students gain insights into possible career pathways and develop key professional and social skills, such as written and verbal communication, that contribute to success in any field. The Life Sciences Learning Center at the University of Rochester School of Medicine and Dentistry investigated the impacts of its Summer Science Academy for high school students. The program includes guided and independent laboratory experiments, bioethics discussions, computer workshops, library research and field trips. The program enrolled between 20 and 39 students each year from 1996 through 2002. Long term follow-up with former participants found that the program had a positive influence on their performance in advanced science courses, as well as their decisions to participate in other science-related programs and possible career pathways (Markowitz 2004).

High quality STEM afterschool and out-of-school programs share a suite of characteristics. These best practices, as exemplified by projects described in this section, should be applied to any informal learning experiences that aim to develop students' environmental health literacy, regardless of whether the program is related to a particular topic, such as asthma, or a skill, such as community advocacy.

## Environmental Health Literacy and Closing the Achievement Gap

Environmental health topics bring immediate relevance to classroom STEM activities. Twenty percent of US students are part of families living in poverty (NCES 2017)–and many of these students are at high risk for disengaging from school, dropping out or not having access to post-secondary opportunities. At the same time, environmental risks are disproportionately high for many low income and underrepresented minority students. These exposures occur both in homes and neighborhoods. Children in rural areas may be close to industrial sites, mines, agricultural chemicals or agricultural or industrial waste. Urban children may be exposed to lead, poor water quality, or indoor air pollutants–to name a few hazards. In addition, smoking rates continue to be higher among low income families (Gochfield and Burger 2011).

In addition to greater exposure to environmental hazards, low income students or those from underrepresented minority groups often have less access to early learning opportunities and are more likely to attend schools with fewer high quality teachers, fewer opportunities for advanced coursework and unsupportive school environments. These effects are cumulative over a lifetime of attending school. By the time many students reach high school, they already are behind in mathematics, reading or science, and are less likely to be prepared for or persist in STEM course sequences or pathways that lead to STEM careers (Darling Hammond 2010).

Despite efforts to improve test scores in reading, writing, mathematics, science and social studies by all students–United States Hispanic, African American or economically disadvantaged students perform below Asian, white or non-economically disadvantaged students at all grade levels on many state, national and international assessments. The income achievement gap already is present when students enter kindergarten, and persists as students continue through school (Reardon 2013). Disparities in students' test scores between white and non-white students continue to be reflected in national assessments, such as the National Assessment of Educational Progress (NAEP). Based on 2012 mathematic data, for example, the average scores for Hispanic and African American fourth grade students were below the average score of white students (Musu et al. 2016).

Closing the achievement gaps between high and low-income students, and among white or Asian students and Hispanic or African American students at all grade levels is a national priority. Relevant instruction, well-prepared teachers, and a classroom climate that cultivates and supports students' social identities and sense of belonging are key contributors to student success and persistence (Cohen and Garcia 2008). In fact, successful learning and development of a professional identity (e.g., feeling like a scientist) are mutually reinforcing experiences–which, in turn, increase confidence and motivation to pursue science (or other areas). These elements have been described as the "persistence framework" (Graham et al. 2013). In the framework, student confidence and motivation essential for identity development are fostered through strategies such as active learning, being part of a community, acquiring relevant knowledge and skills, and engaging in authentic problems. It should not be surprising that the same elements described within the context of the persistence framework have been outlined elsewhere in this chapter as components of effective STEM education and contribute to raising students' environmental health literacy. The challenge, however, once again falls on teachers and instructional designers to develop and implement programs of instruction that are relevant to students, while building essential knowledge, skills, behaviors and attitudes.

## Summary

STEM (science, technology, engineering and mathematics) education and efforts to develop the environmental health literacy of elementary and secondary school students share an emphasis on critical thinking, problem-solving, teamwork and informed decision-making in the context of real-world problems. STEM implies the use of active, interdisciplinary approaches to teaching and learning in which concepts and skills are taught holistically and make connections among individuals, school, community and the world through authentic problems. Yet, even though environmental health topics are engaging to students and form the basis of many effective formal and out-of-school programs for students–they are not always part of STEM education. More explicit connections to pertinent skills and environmental health knowledge are needed as part of formal and informal program development, curricula and textbooks, and teacher professional development, in order to raise environmental health literacy of students, teachers and even families.

The Next Generation Science Standards (NGSS) present a way to build students' environmental health literacies through practices, crosscutting concepts and disciplinary core ideas related to essential science understandings. Environmental health topics and themes are reflected in the standards, and provide appropriate cases and topics for investigations by students. However, more explicit connections between earth/space, life and physical science teaching are required if current needs for environmental health literacy are to be met. These connections can be made through greater dissemination of existing curricular materials and through more incorporation of environmental health topics and literacy skills into commercial textbooks.

High quality school-based STEM education depends on skilled teachers. STEM teachers need professional development that focuses on fostering deep knowledge; engages them in active learning and collaborative work; is coherent with other school policies and practices; and has sufficient duration in terms of intensity and contact hours to drive lasting improvements in instructional practices. Environmental health education programs, usually conducted in affiliation with universities, research organizations or museums and science centers, have made important contributions to STEM teacher development, and should be expanded to meet needs for teacher literacy in environmental health.

Environmental health science education programs across the country have contributed to closing achievement gaps between high and low-income students, and among white or Asian students and Hispanic or African American students at all grade levels. In addition, successful models for a variety of STEM programs help develop problem-solving, teamwork and communications skills that are components of environmental literacy. These programs should be continued or expanded, so that all students have access to the STEM learning experiences that serve as a springboard to environmental health literacy.

## References

[AAAS]American Association for the Advancement of Science. (1993). *Benchmarks in science literacy* (p. 418). New York: Oxford University Press.

[EPA]Environmental Protection Agency (2017a). *Sunwise: A Program that Radiates Good Ideas*. Accessed online, 30 Oct 2107. https://www.neefusa.org/resource/sunwise-tool-kit.

[EPA]Environmental Protection Agency (2017b). History of the Sunwise Program at EPA. Accessed online, 30 Oct 2107. https://www.epa.gov/sunsafety/history-sunwise-program-epa.

[NAS]National Academy of Sciences. (1996). *National science education standards*. Washington, DC: National Academy Press.

[NCES]National Center for Educational Statistics (2017). Characteristics of Children's families. The Condition of Education. Accessed online, 30 Oct 2017, https://nces.ed.gov/programs/coe/indicator_cce.asp

[NIEHS]National Institute of Environmental Health Sciences (2017) *Environmental Health Science Education*. Accessed online, 30 Oct 2017, https://www.niehs.nih.gov/research/supported/translational/ehsic/index.cfm

[NRC]National Research Council. (2000). *Inquiry and the National Science Education Standards: A guide for teaching and learning*. Washington, DC: The National Academies Press.

[NRC]National Research Council. (2012). *A framework for K-12 science education: Practices, crosscutting concepts, and core ideas. Committee on a conceptual framework for new K-12 science education standards. Board on science education, division of behavioral and social sciences and education*. Washington, DC: The National Academies Press.

[NRC]National Research Council. (2015a). *Guide to implementing the next generation science standards. Committee on guidance on implementing the next generation science standards. Board on science education, division of behavioral and social sciences and education*. Washington, DC: The National Academies Press.

[NRC]National Research Council. (2015b). *Identifying and supporting productive STEM programs in out-of-school settings. Committee on successful out-of-school STEM learning. Board on science education, division of behavioral and social sciences and education*. Washington, DC: The National Academies Press.

[OECD]Organisation for Economic Co-operation and Development. (2007). *PISA 2006: Science competencies for tomorrow's world: Volume 1: Analysis*. Paris: OECD Publishing. https://doi.org/10.1787/9789264040014-en.

[TAMU] Texas A&M University (2017). *The environmental health sciences Summer Institute*. Accessed online, 30 Oct 2017, http://www.k12summerinstitute.org/index.php

[URMC] University of Rochester Medical Center (2017). *Grant helps Build Understanding of Environmental Health with Hands-on Science Kits*. Research@URMC, June 2017. Accessed online 15 Feb 2018): https://www.urmc.rochester.edu/research/blog/june-2017/grant-helps-build-understanding-of-environmental-h.aspx?feed=blogs

After School Alliance (2015). America After 3 pm: Full STEM Ahead. Accessed online, 13 Jun 2018. http://www.afterschoolalliance.org/AA3PM/

Backes B, Goldhaber D, Cade W, Sullivan K, Dodson M (2016) Can UTeach? *Assessing the Relative Effectiveness of STEM teachers*. CALDER Working Paper No. 173 December 2016. Accessed online, Oct 4, https://eric.ed.gov/?id=ED573187.

Beaton, A. (1996a). *Mathematics achievement in the middle school years: IEA's third international mathematics and science study (TMSS)*. Chestnut Hill: TIMSS International Study Center, Boston College.

Beaton, A. (1996b). *Science achievement in the middle school years: IEA's Third International Mathematics and Science Study (TMSS)*. Chestnut Hill: TIMSS International Study Center, Boston College.

Branch, R., & Chester, A. (2009). Community-based participatory clinical research in obesity by adolescents: Pipeline for researchers of the future. *Clinical and Translational Science, 2*, 350–354. https://doi.org/10.1111/j.1752-8062.2009.00149.x.

Bransford, J., Brown, A., & Cocking, R. (Eds.). (2000). *How People Learn: Brain, Mind, Experience, And School*. Washington, DC: National Academy Press.

Brown, R., Brown, J., Reardon, K., & Merrill, C. (2011). Understanding STEM current perceptions. *Technology & Engineering Teacher, 70*(6), 5–9.

BSCS (2012) *Chemicals, the Environment, and You: Explorations in science and human health*. Colorado Springs, CO: BSCS. Accessed online, 30 Oct 2017. https://science.education.nih.gov/supplements/ChemicalsEnvironmentAndYou.pdf

Bybee, R. (2010a). Advancing STEM education: A 2020 vision. *Technology and Engineering Teacher, 70*(1), 30–35.

Bybee, R. (2010b). *The teaching of science: 21st century perspectives* (p. 95). Arlington: NSTA Press.

Bybee R (2015) The 5E instructional model: Creating teachable moments. Arlington, VA.

Cohen, G., & Garcia, J. (2008). Identity, belonging, and achievement. *Current Directions in Psychological Science, 17*(6), 365–369.

Committee on Prospering in the Global Economy of the 21st Century: An Agenda for American Science and Technology, National Academy of Sciences, National Academy of Engineering & Institute of Medicine. (2007). *Rising above the gathering storm : Energizing and employing America for a brighter economic future*. Washington, DC: National Academies Press.

Darling-Hammond, L. (2010). *The flat world and education: How America's commitment to equity will determine our future*. New York: Teachers College Press.

Doyle, T. (2011). *Learner-centered teaching: Putting the research on learning into practice*. Sterling Virginia: Stylus Publishers.

Finn, S., & O'Fallon, L. (2017). The emergence of environmental health literacy—From its roots to its future potential. *Environmental Health Perspectives, 125*(4), 495–501

Freeman, S., Eddy, S., McDonough, M., Smith, M., Okoroafor, N., Jordt, H., & Wenderoth, M. (2014). Active learning increases student performance in science, engineering, and mathematics. *Proceedings National Academy of Sciences, 111*(23), 8410–8415.

Geller A, Rutsch L, Kenausis K, Selzer P, Zhang Z (2003) Can an hour or two of sun protection education keep the sunburn away? *Evaluation of the Environmental Protection Agency's Sunwise School Program*. 2:13, Accessed online, https://doi.org/10.1186/1476-069X-2-13

Gochfield, M., & Burger, J. (2011). Disproportionate exposures in environmental justice and other populations: The importance of outliers. *American Journal of Public Health, 101*(Suppl 1), S53–S63.

Graham, M., Frederick, J., Byars-Winston, A., Hunter, A., & Handelsman, J. (2013). Increasing persistence of college students in STEM. *Science, 341*, 1455–1456.

Grigg, J., Kelly, K., Gamoran, A., & Borman, G. (2012). Effects of two scientific inquiry professional development interventions on teaching practice. *Educational Evaluation and Policy Analysis, 35*(1), 38–56.

Haney, J., Want, J., Keil, C., & Zoffel. (2007). Enhancing teachers' beliefs and practices through problem-based learning focused on pertinent issues of environmental health science. *The Journal of Environmental Education, 38*(4), 25–33.

Hursh, D., Martin, C., Trush, M., & Davis, H. (2011). *Teaching environmental health to children*. New York: Springer.

Kang, N., DeChenne, S., & Smith, G. (2012). Inquiry learning of high school students through a problem-based environmental health science curriculum. *School Science and Mathematics, 112*(3), 147–158.

Krasny, M., & Doyle, R. (2002). Participatory approaches to program development and engaging youth in research: The case of an Inter-Generational Urban Community Gardening Program. *Journal of Extension, 40*(2), Article 5FEA3 Accessed online. http://www.joe.org/joe/2002october/a3.php.

Krasny, M., & Tidball, K. (2016). Community gardens as contexts for science, stewardship, and civic action learning. In J. Blum (Ed.), *Urban horticulture: Ecology, landscape and agriculture* (pp. 267–290). Waretown: Apple Academic Press.

Krishnamurthi, A., Ballard, M., Noam, G. (2014). Examining the impact of afterschool STEM programs. Paper commissioned by the Noyce Foundation. Accessed online, 30 Oct 2017. http://www.afterschoolalliance.org/ExaminingtheImpactofAfterschoolSTEMPrograms.pdf

Lewis, C., Perry, R., & Hurd, J. (2004). A deeper look at lesson study. *Educational Leadership, 61*(5), 6–11.

Lieberman, G. (2013). *Education and the Environment: Creating Standards-Based Programs in Schools and Districts*. Cambridge, MA: Harvard Education Press.

Lieberman, G., & Hoody, L. (1998). *Closing the achievement gap: Using the environment as an integrating context for learning*. San Diego: State Education and Environment Roundtable.

Marder M. (2016). Is *STEM Education in Permanent Crisis*. Education Week. Oct 26, 2106. Accessed online 4 Oct 2016, http://www.edweek.org/ew/articles/2016/10/26/is-stem-education-in-permanent-crisis.html

Markowitz, D. (2004). Evaluation of the long-term impact of a university high school summer science program on students' interest and perceived abilities in science. *Journal of Science Education and Technology, 13*(3), 395–407.

McCombs, B., & Miller, L. (2007). *Learner-centered classroom practices and assessments: Maximizing student motivation, learning and achievement*. Thousand Oaks: Corwin Press.

McKendall, S., Kasten, K., Hanks, S., & Chester, A. (2014). The health sciences and technology academy: An educational pipeline to address health care disparities in West Virginia. *Academic Medicine. Journal of the Association of American Medical Colleges, 89*(1), 37–42.

Moreno, N. (2005). Science education partnerships: being realistic about meeting expectations. *CBE Life Sciences Education, 4*, 30–32.

Moreno, N., Chang, K., Tharp, B., Denk, J., Roberts, J., & Rahmati, S. (2001). Teaming up with scientists. *Science and Children, 39*, 42–45.

Musu-Gillette L, Robinson J, Mc Farland J, Kewal Ramani A, Zhang A, Wilkinson-Flicker S (2016) Status and Trends in the Education of Racial and Ethnic Groups 2016 (NCES 2016-007). U.S. Department of Education, National Center for Education Statistics. Washington, DC. Accessed online, 30 Oct 2017. http://nces.ed.gov/pubsearch.

Moreno, N., & Tharp, B. (2006). How do students learn science? In J. Rhoton and P. Shane (Eds.), Teaching Science in the 21st Century. National Science Teachers Association and the National Science Education Leadership Association. Arlington, VA: NSTA Press. pp 292–305.

Moreno, N., Denk, J., & Tharp, B. (2006). An interdisciplinary approach to teaching food and nutrition in elementary classrooms. *CESI Science, The Journal of the Council for Elementary Science International, 39*(1):15–21.

Newell, A., Zientek, L., Tharp, B., Vogt, G., & Moreno, N. (2015). Students' attitudes toward science as predictors of gains on student content knowledge: Benefits of an after-school program. *School Science and Mathematics, 115*(5), 216–225.

NGSS Lead States. (2013). *Next generation science standards: For states, by states*. Washington, DC: The National Academies Press.
Oregon State University (2017) Welcome to Hydroville, USA. Accessed online, 30 Oct 2017. http://blogs.oregonstate.edu/hydroville/
O'Fallon, L. (2017). Environmental Health Science Education. Accessed online, 13 Jun 2018. https://www.niehs.nih.gov/research/supported/translational/ehsic/index.cfm
Phillips, A., & Miller, J. (2014). Career interests of students in a STEAM camp. In M. Searson & M. Ochoa (Eds.), *Proceedings of society for information technology & teacher education international conference 2014* (pp. 2277–2282). Chesapeake: AACE.
Reardon, S. (2013). The widening income achievement gap. *Educational Leadership, 70*(8), 10–16.
Richardson, V. (2003). Constructivist pedagogy. *Teachers College Record., 105*(9), 1623–1640.
Roberts, J., Henson, R., Tharp, B., & Moreno, N. (2001). An examination of change in teacher self-efficacy beliefs in science education based on duration of inservice activities. *Journal of Science Teacher Education, 12*(3), 199–213.
Rutherford J, Ahlgren A (1991) Science for all Americans. New York: Oxford University Press, ISBN 9780195067712. pp 272.
Sadler, P., Sonnert, G., Hazari, Z., & Tai, R. (2012). Stability and volatility of STEM career interest in high school: A gender study. *Science Education, 96*, 411–427.
Schmidt, W., McKnight, C., Cogan, L., Jakwerth, P., & Houang, R. (2002). *Facing the consequences: Using TIMSS for a closer look at US mathematics and science education*. New York: Kluwer Academic Publishers.
Schulman, L. (1986). Those who understand: Knowledge grown in teaching. *Educational Researcher, 15*, 4–14.
Schulman, L. (2000). Teacher development: Roles of domain expertise and pedagogical knowledge. *Journal of Applied Developmental Psychology, 21*, 129–135.
Tomasiewicz, H., Hesselbach, R., Carvan, M., Goldberg, B., Berg, C., & Petering, D. (2014). Zebrafish as a model system for environmental health studies in the grade 9–12 classroom. *Zebrafish, 11*(4), 384–395. https://doi.org/10.1089/zeb.2013.0881.
Tsupros N, Kohler R, Hallinen J (2009). *STEM education in southwestern Pennsylvania: Report of a project to identify the missing components, intermediate unit 1 and Carnegie Mellon, Pennsylvania*. Accessed online, 30 Oct 2017. https://www.cmu.edu/gelfand/documents/stem-survey-report-cmu-iu1.pdf
Van Driel, J., & Berry, A. (2001). Teacher professional development focusing on pedagogical content knowledge. *Educational Researcher, 41*(1), 26–28.
Weber, K., Hesselbach, R., Kane, A., Petering, D., Petering, L., & Berg, C. (2013). Minnows as a classroom model for environmental health. *The American Biology Teacher, 75*(3), 203–209.
Weiss, I., & Pasley, J. (2009). *Mathematics And Science For A Change: How To Design, Implement And Sustain High-Quality Professional Development*. Portsmouth, NH: Heinemann.
White, B. M., Hall, E. S., & Johnson, C. (2014). Environmental health literacy in support of social action: An environmental justice perspective. *Journal of Environmental Health, 77*(1), 24–29.
Wilson, S. (2013). Professional development for science teachers. *Science, 340*(6130), 310–313.

# Chapter 8
# Health Professionals' Environmental Health Literacy

**Phil Brown, Stephanie Clark, Emily Zimmerman, Maria Valenti, and Mark D. Miller**

**Abstract** Health care providers play an important role in communicating health risks to patients, community residents, and public health agencies. For their own patients, the role of the provider could include taking an environmental health (EH) history, answering patient questions about exposures, providing anticipatory guidance to prevent exposures, and remaining alert to the possibility of toxicants or other environmental influences causing acute or chronic illness. However, few clinicians feel confident in discussing environmental risks, although patients rate them of high concern. There is ample evidence from clinician surveys and case reports of EH trainings to show that health professionals are not sufficiently literate in EH. This is also evident in community-engaged research projects. Lack of training and knowledge on the links between environment and health and lack of available clinical tools for both provider and patient education contribute to the inattention to this topic in the clinical, public health, and community settings. Health professionals likely have no experience in dealing with local officials and the general public in

Research reported in this publication was supported by the National Institute Of Environmental Health Sciences of the National Institutes of Health under Award Number T32ES023769. The content is solely the responsibility of the authors and does not necessarily represent the official views of the National Institutes of Health.

This publication was supported by the cooperative agreement award number U61TS000238-04 from the Agency for Toxic Substances and Disease Registry (ATSDR). I was also supported by the US National Institute of Environmental Health Sciences (NIEHS) (grants P01 ES018172 and P50ES018172) and the USEPA (grants RD83451101 and RD83615901), as part of the Center for Integrative Research on Childhood Leukemia and the Environment (CIRCLE). Its contents are the responsibility of the authors and do not necessarily represent the official views of ATSDR, EPA, or NIEHS.

P. Brown (✉) · S. Clark · E. Zimmerman
Northeastern University, Boston, MA, USA
e-mail: p.brown@northeastern.edu

M. Valenti
Commonweal, Bolinas, CA, USA

M. D. Miller
Western States Pediatric Environmental Health Specialty Unit, University of California, San Francisco, CA, USA



S. Finn, L. R. O'Fallon (eds.), *Environmental Health Literacy*,
https://doi.org/10.1007/978-3-319-94108-0_8

affected communities, nor with outside agencies, yet they may be called upon to give information and guidance to those parties that extend beyond their own patients. In this sense, they become part of a broader public health network in which the community is the patient. This chapter reviews health professionals' current environmental health literacy (EHL) from the academic to community arenas, and focuses on examples of success and opportunities for expansion.

**Keywords** Physicians · Nurses · Speech-language pathologists · Environmental health literacy · Environmental factors in health · Medical education · Health education · Health communication · Toxic chemicals · Health policy

## Theoretical Frameworks

This chapter is guided in part by a "right-to-know" approach that holds that people are entitled to receive complete information on EH exposures and hazards. Right-to-know is enshrined in a variety of legislative mandates and scientific practices. Labor-environmental coalitions succeeded in passing the 1981 Philadelphia right-to-know legislation, based on concerns about workplace toxics exposures, and in the 1983 Worker and Community Right to Know Act in New Jersey, based on concern about the state's many chemical plants (Mayer et al. 2010). At the federal level, the Emergency Planning and Community Right-to-Know Act (EPCRA), which re-authorized the Superfund Program, was passed in 1986, by only one vote. The EPCRA was passed in response to the 1984 disaster in Bhopal, India, caused by a massive release of methyl isocyanate that killed or severely injured more than 2000 people. Based on these environmental disasters, Congress imposed requirements for federal, state and local governments, tribes, and industry, that covered emergency planning and "Community Right-to-Know" reporting on hazardous and toxic chemicals, in order to increase public knowledge and access to information on chemicals at individual facilities, their uses, and releases into the environment. This legislation also required facilities handling or storing any hazardous chemicals to submit Safety Data Sheets (SDSs) to state and local officials and local fire departments. While EPCRA is not likely known to the public, the public is relatively familiar with one thing that EPCRA created – the Toxics Release Inventory (TRI), a requirement for facilities to report on over 600 chemicals that are released in high volumes (EPA.gov/EPCRA). Many people have used TRI data to learn of local hazards and press for corporate reform and EPA pressure, and environmental research has employed TRI data as well (Fung and O'Rourke 2000).

From the research perspective, right-to-know has increasingly been transformed into the "research right-to-know," as situated in report-back practices whereby EH data is shared with research participants (See Brody et al. Chap. 6). This method is central to the community-based participatory research (CBPR) approach that views such data to be owned by participants as it came from their homes and/or bodies. Besides a growing number of researchers who pursue these practices, (Quandt et al. 2004; Brody et al. 2007; Emmett et al. 2009), formal guidelines for such data sharing have been established by the National Academies of Science, California Biomonitoring Program, CDC National Conversation on Public Health and Chemical Exposure, the federal Interagency Breast Cancer and the Environment Coordinating Committee, Statistics Canada, and Consortium to Perform Human biomonitoring on a European Scale (Brody et al. 2014a, Brody et al. 2014b). This right-to-know background highlights the importance of EHL for health care professionals, while it simultaneously demonstrates the potential to strengthen EHL. Health professionals gain EH knowledge needed to answer questions from patients about their level of exposure and what it means for their health. Furthermore, health professionals learn from the populations exposed to toxic chemicals or "affected" communities. This experiential learning may be particularly relevant when media has covered a local issue such as the Flint, MI lead-contaminated water crisis in 2016.

The approach is also based on ethnographic work such as "contaminated communities" (Edelstein 1988) and "popular epidemiology" (Brown and Mikkelsen 1990) in order to show the often, unmet needs for health professional awareness, and hopefully also involvement, in affected communities. In the wealth of such case studies that highlight community, rather than medical, discovery of contamination and/or health effects (Hoover et al. 2015), one learns that health professionals are largely unaware of environmental contaminations or how patients may be exposed to it, and despite frequent requests for advice, are unable to help affected residents understand their situation, recognize environmentally induced symptoms, or provide a specific diagnosis. Sometimes, professionals' lack of EH knowledge can be very detrimental, as noted by Balshem (1993), who showed health professionals in a cancer center looked at the hazard perceptions of people in a Philadelphia working class neighborhood, and used an individual-blaming approach that conflicted with the industry-blaming approach of sufferers. This shows how the absence of professional EHL reinforces a medical paternalism that creates an oppositional framework that blocks effective communication and practical resolution (e.g. through public health and industrial regulations and controls).

Given the large contribution of environmental factors to disease, it is in the general interest of all health professionals to know more. Still, it is not sufficient for health professionals to merely have knowledge of environmental contributors to disease. EH knowledge is not equivalent to EHL, as it is quite possible to have knowledge of environmental factors that impact diseases and conditions, yet not possess the skills or the resources to apply and transfer that knowledge in clinical encounters. This knowledge represents an early stage of EHL that occurs prior to the stage of EHL that leads to action in the clinic, in the local community, and in the policy arena.

The following levels of EHL are stages of awareness relevant to health professionals:

- Level 1 – Knowledge and understanding of environmental effects on human health. Similar, to what Finn and O'Fallon propose, this level is the foundation of EHL.
- Level 2 – Communicating and providing knowledge to patients. Examples of actions at this level could include, taking occupational and environmental histories, offering anticipatory guidance to patients to reduce exposures, or referring patients to environmental health experts such as the Pediatric Environmental Health Specialty Units.
- Level 3 – Communication to and engaging with community groups. At this stage, health professionals go beyond clinical communication knowledge and skills for individual patients, and engage in population health/public health approaches that reduce exposures to whole groups and that involve communities in discussing environmental hazards and how to respond to them. At this level, the healthcare professional appreciates the difference between individual and social determinants of health.
- Level 4 – Participation in professional societies and the policy arena. As mastery of knowledge and capacity to communicate increase, there are ways to engage in advocacy through professional societies and associations. Through these groups, professionals can take strong stances on environmental influences on health, such as advocating to make TRI and other relevant data more accessible to patients and community residents.
- Level 5 – Participation in public advocacy. In addition to health professionals getting their associations to take strong stances on policy, they may also act serve on task forces or work with other groups to influence policy. That could include supporting health protective legislative reforms such as improvements to the Toxic Substances Control Act (TSCA), or even assisting with evidence-gathering in states to help inform regulatory levels of toxicants (Fig. 8.1).

Taking into account these different levels, this chapter's central thesis is that EHL for health professionals means that they grasp the personal, community, and broader societal context of environmental exposures, and are able to engage on this awareness with people and groups at all levels.

EHL can be enhanced in many ways at any of the levels. Health care providers may initiate the process by providing advice to patients on how to prevent environmental exposures and maximize health through health protective actions. Patients and/or affected communities seeking professional help may begin the process, as in the example below of Testing for Pease, a community group dealing with per- and polyfluoroalkyl substances (PFAS, a class of man-made chemicals) contamination. Even at a higher level, professionals can take the lead, as in the example below of Hospitals for a Healthy Environment in Rhode Island. Finally, the process may be initiated from both the patient/public and health professional, as described in the example about local and national work on environmental factors in breast cancer.

**Fig. 8.1** Levels of EHL

## Scope of the Problem and the Need for Professional Environmental Health Literacy

Providers' lack of training and knowledge of environmental health is the most commonly discussed drawback to their EHL however one must also include the obstacles that affected communities face. In this section, both sets of issues are discussed with an emphasis on pediatric health care providers.

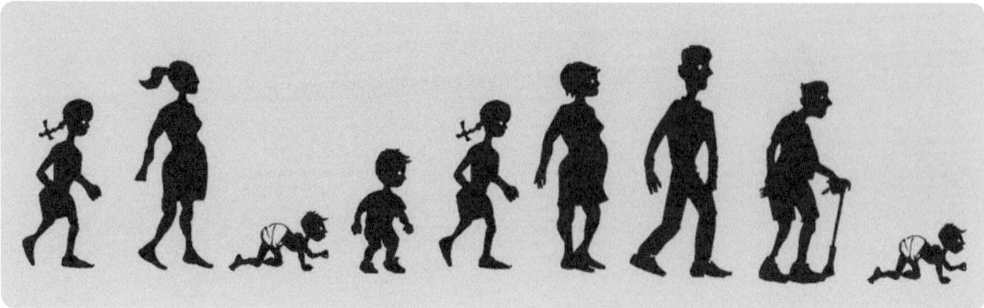

**Fig. 8.2** Health across the lifespan (Illustration: Stephen Burdick Design; Stein et al. 2008)

Environmental factors include all the aspects of the way humans live, work, play and socialize and how these factors influence physical and mental health throughout life. Factors include diet, exercise, the natural and built environments, familial and social interaction, and exposure to toxic substances. An ecological approach understands that all of these factors interact to help create the conditions, along with genetics, for health and wellbeing or disease. Because early life is so critical to health across the lifespan, (Miller and Marty 2012), obstetricians, pediatric care, and family health practitioners may be the most important health professionals to provide anticipatory guidance about environmental health threats and prevention strategies to patients, as well as to be advocates for protective health policies at all levels. There are many different windows of susceptibility across a person's lifespan, as Fig. 8.2 shows. For the purposes of this chapter, the focus is only on the most critical windows, pregnancy and early childhood. There is a clear future need to address this in geriatrics, where there has been little attention.

Beginning early in life and continuing throughout the life course multiple environmental factors are strong determinants of health, even decades later, and can include enduring impacts on brain aging and function. For example, maternal exposure to air pollution is associated with low birth weight and other problems. Air pollution also impairs lung development in children and increases the risk of respiratory tract infections, asthma, and cognitive behavioral problems. Low birth weight, followed by rapid catch-up growth, increases the risk of teenage and adult obesity. Low birth weight can also be a factor for adult cardiovascular disease, hypertension, and Type II diabetes. These diseases/disorders (especially diabetes and obesity) are associated with increased risk of Alzheimer's disease and dementia later in life.

## Providers: What's the Problem?

Nurses and medical doctors enjoy the highest ratings of honesty and ethical standards of all professions (Gallup 2014). They can play an important role in changing patterns of patient behavior as well as influencing public policies (Kreuter et al. 2000; Kovarik 2005). Health professionals' status and influence makes them a key stakeholder to educate about EH, because they are trusted messengers and make credible advocates for both patients and policymakers. However, their lack of knowledge of the subject is a major challenge when it comes to environmental health topics.

## State of Knowledge

To begin to address health professionals' knowledge and EHL, one first needs to know what their state of knowledge on EH issues is. Surveys of pediatricians, family health practitioners and obstetricians/gynecologists repeatedly show that providers typically do not address EH threats in discussions with patients. In multiple surveys, these medical professionals report lack of knowledge about taking the patients' environmental history, and also being concerned about their ability to follow-up on environmental issues that may be related to their patients' health, even though they are aware that patients are worried about these issues (AAP Council on Environmental Health 2011; Stotland et al. 2014; Trasande et al. 2006a, b, 2010, 2014; Zachek et al. 2015; AHRP/PSR 2007; Kilpatrick et al. 2002; Roberts and Gitterman 2003; Ibanez et al. 2015). In one survey of reproductive healthcare providers, the most important reason for not counseling patients on EH was time constraints, followed by lack of knowledge and then by lack of materials and evidence-based guidelines, followed by more pressing concerns (ARHP/PSR 2007). Another survey showed that routine guidance to patients on the health effects of environmental exposures was not a high priority, despite the fact that over three-quarters said that by counseling patients they could reduce patient exposures to hazards (Stotland et al. 2014).

A survey of pediatricians in Michigan reported high self-efficacy and knowledge of EH only in relation to lead and secondhand smoke (Trasande et al. 2010). Other common topics in pediatric EH (e.g. mercury and pesticides) elicited responses indicating much less confidence. In the wake of the wide-spread exposure to lead in the water supply in Flint, Michigan in 2014, local resident and faculty physicians at Hurley Medical Center felt ill prepared to handle the questions arising from exposure and the potential health outcomes (Taylor et al. 2017). In the Flint survey, less than 15% responded with strong ratings that they were comfortable answering patients' questions and confident in being able to accurately provide information. Only approximately one-third reported training in diagnosis and treatment of lead exposures. These results suggest that even in an area of medical knowledge such as lead poisoning, in which most medical trainees get at least some education and may rate themselves as knowledgeable and confident, when faced with actual patients with concerns, they realize that they are not confident in their knowledge base or experience.

In post-training surveys from the Pediatric Environmental Health Toolkit trainings (conducted by Physicians for Social Responsibility and the Pediatric Environmental Health Specialty Unit at UCSF in 2006), challenges described by individual pediatric respondents were varied but themes emerged. These include such practical issues as a lack of clinicians with expertise in EH to make referrals to if an issue arises, and lack of regulations and public health infrastructure that could be utilized to protect patients.

Other issues identified included the low confidence levels that physicians feel about discussing EH. One good reason may be that environmental medicine is "largely omitted in the continuum of U.S. medical education" (Gehle et al. 2011). Although, the National Environmental Education Foundation issued a position statement on "Health Professionals and Environmental Health Education" that was endorsed by sixteen health organizations and professional societies, a survey of environmental medicine content in U.S. medical schools found that 75% of medical schools only require about 7 h of study in environmental medicine over 4 years. In addition, a survey of Migrant Clinician Network clinicians found that approximately half had not had any training or courses related to environmental and/or occupational health" (NEETF 2004). In a survey in 2010, over one-third of graduating respondents believed that curricula provided insufficient attention to EH (AAMC 2010). The Association of American Medical Colleges 2011–2016 Curriculum Report indicates that 125 of 142 colleges surveyed reported inclusion of required coursework on environmental health in 2015–16, but it is unknown how extensive that training is (AAMC 2018).

## Pathways to Change

Ultimately, clinicians want to address issues for which they can influence patient behaviors resulting in reduced risk and the resultant improved health. The integrated model of behavioral prediction suggests that health behaviors are a function of attitudes, self-efficacy, and perceived normative pressure, which is the influence of others and social norms (Mello and Hovick 2016). In a study of chemical exposures in new and expectant mothers, Mello and Hovick (2016) found that normative pressure was the best predictor of exposure reduction measures. The authors suggested that health communication messaging (and campaigns) should seek to reinforce or change perceived norms by using phrases like "Many pregnant women choose to eat organic produce" or "My pediatrician thinks that I should limit the use of plastic food packaging with my kids." They suggest that messages conveying a sense of normative behavior would be more effective than those focused on attitudes or self-efficacy. In addition to the need for clinicians to have a basic understanding of health impacts of environmental exposures, they need training in how to best communicate this information in practical ways that have a positive influence on behavior. As a trusted advisor, be it to an individual, a family, or a community group, the clinician may act as an educator, interpreting information, advising on where to get additional reliable information, or doing additional research.

Unlike governmental agencies and industry scientists, clinicians are more likely to assess exposure and hazards in a more qualitative way. Rather than take actual environmental measurements or develop mathematical models, physicians are more

**Table 8.1** Risk assessment as performed by risk assessors and clinicians

| Risk assessment step | Questions asked by risk assessor | Questions asked by clinician |
|---|---|---|
| Hazard assessment | What are the chemicals of concern, and what kind of harm are they known or suspected to cause? Which chemicals will we focus on? | What information do we have about an environmental problem, what chemicals were involved, and what sources of information are there? |
| Exposure assessment | What are the sources and duration of exposures? How many people are exposed? What do our monitoring or modeling data predict about the range of doses in the population? | Is there a chance that your child may be coming in contact with (breathing, touching, ingesting, etc) this source? How often and for how long? Is the source highly contaminated or only slightly contaminated? |
| Dose-response assessment | What effects are seen in animals or humans at different exposure levels? What are the doses at which cancerous and noncancerous effects occur? Is there a threshold below which no effect is expected? | (Review literature, consult with experts.) How do the levels at which effects have been demonstrated compare with levels in the community? Are these levels higher than regulatory limits? |
| Risk characterization | Given the above, what are the human impacts of current exposures? What is the population risk? Are there sensitive subpopulations? How confident are we in this analysis? | Are regulations based on effects in the fetus or child? Is there reason to be concerned that children may be at greater risk from exposure to this chemical than adults may? |
| Risk communication | Is the information relevant to the audience and understandable? Does it respond to public concerns? What are the limitations in this assessment? | Have I listened to the concerns presented, responded with compassion, and helped identify information needed and credible sources for obtaining it? What additional steps are needed? |

From Miller and Solomon (2003). Used with permission

likely to use approaches that involve questions about frequency and duration of exposure, which may be coupled with a rough assessment of the magnitude of the risk in order to determine the appropriate triage. Understanding this difference between researchers and healthcare professionals will be important for those hoping to improve environmental health literacy of the clinical community. Some of these differences, which may be valuable in working with the clinical community, are summarized in Table 8.1.

## Communities Relating Health Risks to Providers

Outside of regular clinical visits, there are few options for communities to voice their concerns about environmental hazards to health care providers. In addition, mechanisms for reporting environmental hazards in local communities fall under the governmental domain and such information is not relayed to healthcare providers. The lack of awareness of local environmental hazards was documented by Brown and Kelley (1996) who reported that physicians practicing in areas with a Superfund site, or other well-known contamination issues, were no more likely than those practicing in areas without a Superfund or other contaminated site to be aware of environmental determinants of health. Additionally, studies have documented that when patients do express concern about environmental causes of poor health outcomes, physicians often discount these or focus more on lifestyle factors such as alcohol consumption or smoking (Balshem 1993). This is not surprising given the lack of training on EH in medical education combined with the focus on individualized determinants of health.

Federally funded programs that require or encourage collaboration among health care professionals, communities and researchers is one approach to bring these groups together. Previous programs, such as the NIEHS-funded "Environmental Justice: Partnerships for Communication" and the EPA-funded Environmental Justice and Environmental Education programs, have demonstrated the success of these collaborations to increase the EHL of all partners. Even when funding has been available, the budget is often not sufficient given the costs of covering effort levels of physicians. EPA's Technical Assistance Grants (TAG) apply only to Superfund sites, and are typically used to cover the costs of having environmental engineers or site remediation professionals explain to affected residents the highly technical documents involved in cleanups. However, EPA also offers a Public Health TAG grant, that can be used in addition to a regular TAG grant, but it has not been publicized broadly and most community groups are not aware of it.

## What Works? – Examples of Progress

**Role of Providers** As discussed earlier, the first three levels of EHL for health care providers are to (1) improve their understanding of EH, then (2) strengthen their ability to translate that knowledge into actionable information and prevention strategies for their patients, and (3) further enhance their knowledge and skills to apply in the community setting. Recognizing the stated challenges, it would be helpful to see greater integration of required EH education into medical, nursing and public health schools, as well as during residency and other professional training. While such integration is not currently widespread, below are some examples of elective courses on EH issues directed to medical and other students and the EHL benefits these courses will provide for health care professionals.

The Icahn Medical School at Mt. Sinai in New York City has one of the more substantial programs incorporating EH into curriculum. EH is regularly included during students' first 2 years during required "In Focus" weeks devoted to human rights, advocacy, global health, and health disparities. These innovative sessions examine current EH topics of particular concern in a participatory fashion. EH fits comfortably with such community health topics as advocacy and health disparities, but it is important to also include them in traditional specialty-based clinical training. In the Pediatrics department, a Pediatric Environmental Health Specialty Unit (PEHSU) faculty member routinely helps conduct morning rounds and is able to include environmental exposures in the discussion of differential diagnoses. The PEHSU at Mount Sinai offers clinical rotations for medical students and residents, works with students on research products, and offers a Pediatric Environmental Health course in the MPH program. These activities are supported by staff from the PEHSU as well as the Occupational and Environmental Medicine department and assorted epidemiologists and other researchers at the institution.

The University of Texas Health Science Center at San Antonio's medical school offers the elective course "Medicine and the Environment." The course "covers the health impacts from chronic exposure to common chemicals and includes house calls for students to assess chemicals found in local residents' homes, such as cancer-causing compounds used in perfume fragrances, body lotions and plug-in air fresheners, as well as the threats posed by them. The students are then assigned to write an action plan for the families to reduce their exposure." (NYT Nov 21, 2017. https://www.nytimes.com/2017/11/21/opinion/flint-doctors-chemical-exposure.html).

Global health issues such as climate change may provide the breakthrough needed to better institutionalize environmental health throughout medical education and training. According to a 2017 commentary on the medical community and climate change in JAMA, Jay Lemery, an associate professor of emergency medicine at the University of Colorado School of Medicine, said students are "curious and still in a save-the-world mode. They're hungry for this information. There is no question the students want to know more. The big challenge, he said is how to get it into a crowded medical school curriculum" (Friedrich 2017). One effort that could help is the relatively new Global Consortium on Climate and Health Education at Columbia's Mailman School of Public Health, a global network of schools of public health, medicine and nursing organized to help educate health professionals on the effects of climate change (https://www.mailman.columbia.edu/research/global-consortium-climate-and-health-education). Fostering an interface between public health and traditional medicine, which have historically taken different tracks, could help develop a new cadre of environmentally savvy health care providers.

**Role of Government Agencies** Government agencies have played an important role to promote EHL. The U.S. Environmental Protection Agency (EPA), the Agency for Toxic Substances and Disease Registry (ATSDR), and the Centers for Disease Control and Prevention (CDC) are important players in developing materials and guidelines on environmental health issues that health professionals can access and use. For example, the PEHSUs, which are funded by ATSDR and EPA, provide information and training to health care professionals, as well as the general public, on protecting children and reproductive-age adults from environmental hazards. They also work with federal, state, and local agencies to address children's environmental health issues in homes, schools, and communities. The PEHSUs conduct conferences, provide online training and webinars, and publish peer-reviewed articles that raise environmental health literacy amongst clinicians and others (Burke and Miller 2011; Lai et al. 2015).

**Fig. 8.3** Graphic from "Rosa and Carlos get Married"

The National Institute of Environmental Health Sciences (NIEHS) is also a key player for both programs and funding. They have helped establish NIEHS/EPA Children's Environmental Health and Disease Prevention Research Centers whose discoveries have profoundly shaped how the impacts of environmental exposures and child and lifelong health are understood. Each center has a designated physician scientist to ensure that research is translated into practical information for health care providers. Several of the centers have collaborative relationships with PEHSUs to assist in outreach, including to the clinical community. For example, the NIEHS/EPA funded Center for Integrative Research on Childhood Leukemia and the Environment at UC Berkeley has worked with the Western States PEHSU to develop a program, "Improving Environmental Health Literacy of Young Adults". This program has produced innovative materials in English and Spanish (see Fig. 8.3) based on *A Story of Health* eBook to enhance adolescent and young adult EHL including a shadow puppet theater video and a comic book. These materials are being used by *promotoras de salud* (community-based lay health educators) and others to engage their audience in learning about environmental exposures, both prenatal and preconception, which may affect children's health.

**Role of Professional Societies** For years, pediatric professional societies have been at the forefront of EH counseling, providing practical tools, and championing for increased environmental health curricula in professional schools.

The American Academy of Pediatrics (AAP) first issued "Pediatric Environmental Health" (also known as the "Green Book") in 1999. Now in its

Third Edition, the book provides information on a wide range of EH hazards as well as practical prevention advice that doctors can give to parents. AAP has been at the forefront of environmental health policy statements as well, such as one on pesticide exposures and children (AAP Council on Environmental Health 2012). More recently, the American College of Obstetricians and Gynecologists and the American Society for Reproductive Medicine issued a committee opinion on "Exposure to Toxic Environmental Agents" calling for "timely action to identify and reduce exposure to toxic environmental agents" (ACOG and ASRM 2013). This was followed by The International Federation of Gynecology and Obstetrics (FIGO) opinion on the reproductive health impacts of exposure to toxic environmental chemicals with recommendations to prevent exposures to such toxic chemicals (Di Renzo et al. 2015).

Nurses are another group of health professionals who have been leading the way on EHL in the health professions. The American Nurses Association has a strong effort on EH, and some states have major environmental components in their nurses' associations, such as Rhode Island. Nurses played a leading role in establishing statewide collaborations such as Maryland Hospitals for a Healthy Environment and Hospitals for a Healthy Environmental in Rhode Island. The Alliance of Nurses for Healthy Environments (http://envirn.org) is an international network of nurses working to, among other things, integrate environmental health into nursing education, greening healthcare workplaces, incorporating environmental exposure questions into patient histories, and providing anticipatory guidance to patients about their exposures. Nurses represent the highest percentage of health professionals who have taken EH courses offered by ATSDR. For instance, of the over 8,000 who have registered for *A Story of Health* course, over 75 percent are nurses. The percentage of nurses registered for the Pediatric Environmental Health Toolkit course mirrors those numbers.

**Role of Community-Professional Collaborations** Another example of the powerful role of nurses is seen in an asthma education and intervention project. As part of a growing nationwide effort at highlighting environmental causation of asthma, a citywide program in Providence RI funded by the Robert Wood Johnson Foundation got 972 families to participate in the Providence School Asthma Partnership over 3 years. This included after-school asthma program at their child's school, with parents and children attending concurrent workshops. Additional evening programs were held for parent support, both in English and Spanish. In this low-income population, 74% of parents were Latino and 45% non-English speaking; 77% of the children were on Medicaid. The highly successful program greatly reduced asthma symptoms, use of oral steroids, emergency room visits, hospital admissions, and school absence. The deep involvement of all school nurses in the city and the leadership of Rhode Island Hospital's Pediatrics Department, combined with community-rooted work, resulted in health professionals new EH knowledge and showed them how strong parental involvement improved medical outcomes (DePue et al. 2007). Parental involvement is key for child health as the parents are in partial control of the environment that their child is exposed, and can advocate for preventive measures in school siting, construction, and choice of cleaning chemicals.

**Fig. 8.4** Healthy child graphic

The free, peer reviewed, Pediatric Environmental Health Toolkit app was created by the Pediatric Environmental Health Specialty Units (PEHSUs) and Physicians for Social Responsibility and endorsed by the American Academy of Pediatrics. It provides access to simple ways to incorporate anticipatory guidance on environmental health during well child visits. This web app covers:

- Brief overviews of environmental hazards to health found in the air, water, food, consumer products, and elsewhere. It includes sections on health effects, routes of exposure, and prevention strategies.
- Key concepts in children's EH inculding the unique vulnerabilities of children, how the chemical, built and food environments influence health, and environmental justice.
- Anticipatory Guidance keyed to age groups associated with routine visits from prenatal through the teenage years.

**Role of Nonprofit Organizations** Nonprofits such as Physicians for Social Responsibility (PSR) have been instrumental in building EHL by providing health professional trainings in environmental health. The PSR developed, in conjunction with the UCSF PEHSU (now Western States PEHSU at the University of California San Francisco, and the PEHSU network), developed the clinical tool *Pediatric Environmental Health Toolkit,* endorsed by the American Academy of Pediatrics (AAP). It provides reference and anticipatory guidance materials to both providers and patients including "Rx slips" for patients to take home with easy prevention tips. A Toolkit application ("app") followed from that (See Fig. 8.4.). PSR also regularly conducts both science and advocacy trainings for health professionals that enables them to bring their expertise to policymakers at the local, state, regional and federal levels and take the next step in EHL.

Fig. 8.5 A story of Health

A Story of Health eBook/online continuing education (CE) course focuses on multiple environmental contributors to health and how they interact with genetics across the lifespan. By grounding the science of health in stories of fictional people, their families, and communities, readers can explore risk factors for illnesses that are a serious problem for the health of our nation. *A Story of Health* examines how a wide range of factors have an impact on our health every day, including "the air we breathe, the water we drink, the food we eat and our social and economic circumstances." Some examples are: how early life stress can magnify the effects of exposure to air pollution and make asthma worse; how some genes can make children more susceptible to toxic substances like pesticides, which may contribute to the onset of developmental disabilities; how a common nutritional supplement can reduce the risk of childhood leukemia and other disorders: and how exposure to lead can affect fertility. Practical actions to promote health and prevent disease are features, from the personal to the policy levels. Illustrations, graphics, and videos, pop-ups with essential "key-concepts," and links to a wide range of hundreds of additional resources and scientific papers enrich each story. Current chapters, including those on asthma, developmental disabilities, childhood leukemia and infertility/reproductive health, are available to download for free on the web sites of the Western States PESHU and CHE. Health professionals can register for free CEs through the Centers for Disease Control and Prevention (CDC) with the Agency for Toxic Substances and Disease Registry (ATSDR) hosting the CE accreditation pages. The cooperation among federal and state government agencies, academics and nonprofit organizations has enabled them to draw on and leverage each other's expertise and resources during development and their different networks during the important outreach phase.

Another nonprofit that has been at the forefront is the Collaborative on Health and the Environment (CHE), a project of the nonprofit Commonweal education and research institute. CHE collaborated with the Western States PEHSU, ATSDR, the Office of Environmental Health Hazard Assessment, California EPA (OEHHA), and the Science and Environmental Health Network (SEHN), in the development of A Story of Health, a multimedia online eBook designed to harness the power of storytelling to increase EHL for health care professionals (see Fig. 8.5. Availability of free Continuing Education (CE) credits has encouraged over 8000 health care professionals to register for it. Evaluations have been overwhelmingly affirming. Responses from one quarter indicate that over 89 percent say they will apply the new knowledge to develop strategies and interventions in their practices.

**Role of Communities** Communities affected by environmental hazards have been a key source of inspiration and education for clinicians. Local residents in these communities are often the ones who identify disease excesses and clusters, point out local hazardous sites or exposures, and provide the felt experience of living in a contaminated community. Working collaboratively with community-based organizations has been shown to be a valuable way to build and raise EHL of healthcare professionals. Participation in CBPR projects, environmental justice organizations' outreach, environmental health research centers, mainstream health settings, and toxics-affected communities help to shed light on training needs and to provide experiential for healthcare professionals.

Environmental CBPR projects have a long tradition of involving health professionals, and some of NIEHS's early Requests for Proposals (RFA)s required projects to include health professionals on the team (Baron et al. 2009). Especially when doing report-back to communities and individuals, research teams have understood the need to have health professionals as part of their projects, to provide trusted responses to questions about the meaning of the research findings (Curtis and Wilding 2007). As part of the growing "advocacy biomonitoring" movement that centers around the "embodied experience" of the body burden, i.e. the contaminants found in their bodies (Altman et al. 2008), Physicians for Social Responsibility conducted a biomonitoring study of physicians and nurses aimed at helping health professionals grasp that aspect of toxic contamination (Wilding et al. 2009).

In the outreach work of environmental justice groups such as Alaska Community Action on Toxics (ACAT), health professionals have been an important audience. ACAT produced the Environmental Health Care Toolkit aimed at community health aides, who are often the only health professionals in remote Alaskan villages. ACAT launched a series of Bulletins that they distributed for free to all physicians in the state, one on *Diabetes: The Role of Persistent Toxic Chemicals in this Complex Disease* and another on *Body of Evidence: Reproductive Health and the Environment.* Most recently, Northeastern University's Superfund Research Program, Puerto Rico Test Site to Explore Contamination Threats (PROTECT) updated that Bulletin and added a Puerto Rican context to a third edition, for distribution to health professionals in Puerto Rico. Akwesasne Mohawks who played a central role in gathering data in a CBPR project on their tribal lands were able to show health professionals the importance of lay involvement and to teach them about aspects of EH that they were less aware of (Schell and Tarbell 1998). Further, the leadership role that Akwesasne midwife Katsi Cook played in disseminating the research findings provided a lesson on how non-physicians could take a leading role in improving community EH (Hoover in press).

Northeastern University's Children's Environmental Health Center, Center for Research on Early Childhood Exposure and Development in Puerto Rico (CRECE) and Superfund Research Program, PROTECT, collaborate on a variety of surveys to learn about EH knowledge, and on educational programs to expand EHL for physicians, nurses, and early intervention speech-language pathologists (the latter described below). They also collaborate extensively with March of Dimes on pre-term birth, providing education sessions on potential environmental factors such as phthalates. The clinic-centered nature of the PROTECT Cohort (over 1300 recruited, with goal of 1600) and CRECE cohort (over 120 recruited, with goal of 600) ties together regular clinical visits with research activities and thus allows for extensive engagement with health professionals who would not otherwise be exposed to EH research findings. Frequent research presentations to clinics and hospitals build on this to provide a wealth of education for providers (Velez Vega et al. 2016). While this example may seem more in the realm of provider-initiated efforts, it is largely a result of the March of Dimes' initial pressure for research on the extraordinarily high level of pre-term birth in Puerto Rico, and this work continues to be guided by a community advisory board that involves many community-based organizations.

Toxics-affected communities often find medical professionals lacking in specific knowledge of their local exposures of concern. While communities often get relevant information from local, state, and national EH and justice organizations (e.g. Toxics Action Center, Center for Health, Environment, and Justice), they sometimes mobilize local health professionals. Portsmouth NH residents learned of extremely high levels (up to ten times higher than NHANES) of per- and polyfluorinated compounds (PFAS) in well-water at the Pease Tradeport, formerly an Air Force base, and now home to 250 businesses, two daycare centers, and 9,525 employees. Activists formed Testing for Pease, which successfully lobbied the state health department to test blood of 1,575 people, yielding mean levels higher than NHANES (for one chemical, PFHxS the mean level was nearly four times the NHANES level). Activists found that medical professionals were largely unaware of potential health effects of these chemicals, but through such testing efforts, were able to educate some of them sufficiently so that they assisted in a research proposal to examine PFAS effects on children's immune function. The activists are working to provide health care professionals with information about potential health effects, which may improve patient care by alerting physicians about heightened risk for certain diseases (e.g., kidney cancer) or suppressed children's DTaP vaccination responses from their PFAS exposures. Activists plan to develop a fact sheet for doctors and nurses that can be used not only in Pease but in many PFAS-affected communities around the US and internationally (Amico et al. 2016).

Along the U.S.-Mexico border, a collaboration was formed between the University of Washington, the California Environmental Health Tracking Program, and Comite Civico Del Valle, a local social and environmental justice non-profit organization, to develop tools for real time monitoring of $PM_{2.5}$ and $PM_{10}$ in the air (particulate matter at, respectively, 2.5 microns and 10 microns diameter) in Imperial County, California (see Chap. 5). $PM_{2.5}$ and $PM_{10}$ are both fine inhalable particulate matter that can become trapped in the nose, throat, and lungs and cause adverse health outcomes (see Chap. 4). The project was developed in response to the poor air quality in the region, and high rates of poverty, asthma, and asthma related ER visits (California Environmental Health Tracking Program 2017). The California Environmental Health Tracking Program, a program developed by the Public Health Institute and the California Department of Public Health, created a website where community members can report environmental hazards and access real time air quality data. The website, IVAN (Identifying Violations Affecting Communities) contains a list and map of 40 air monitors in the community and allows users to select the monitor nearest them and, also provides an option to receive air quality alerts via email (ivan-imperial.org). The air monitoring system is beneficial and innovative because it allows for a rapid dispersion of data on EH hazards in Imperial county and equips both community members and stakeholders with data on $PM_{2.5}$ and $PM_{10}$, which exacerbates asthma. Such data allows residents and health practitioners to incorporate environmental data into health assessments.

## Joint Efforts: EHL at the Intersection of Professional Initiation and Public Initiation

An interesting example of strengthening EHL from both directions is seen in the efforts on environmental factors in breast cancer. From the patient/pre-patient/public end, a vibrant environmental breast cancer movement developed that sought prevention of environmental exposures (McCormick et al. 2003; Brown et al. 2006). From the health professional end came a growing number of environmental health scientists who developed a growing research program on this subject (Brody et al. 2014a; b; Rodgers et al. 2018). The intersection of the lay and professional worlds here is a special case of EHL that may be its most significant form.

## Bringing in a New Health Professional Group: Speech-Language Pathologists and Early Intervention Workers

Early Intervention (EI) Speech Language Pathologists (SLPs) can play a major role in educating families on EH hazards, as they often spend their sessions in the family's home or childcare facility, where the child could potentially be exposed to toxicants daily that can negatively impact their development. Therefore, SLPs are

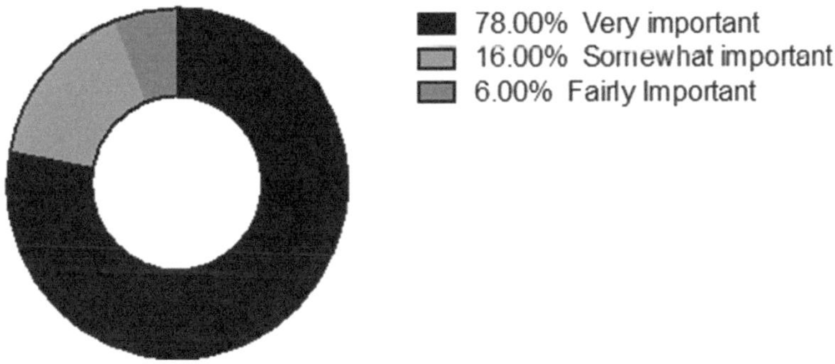


**Fig. 8.6** The importance of environmental influences on child development, and SLPs perspective

uniquely suited to educate both caregivers on the importance of EH and also treat the neurodevelopment outcomes that are evident in children exposed to various toxicants.

In order to understand how much SLPs working in EI know about EH, the Community Outreach and Translation Core of Northeastern University's Children's Environmental Health Center conducted a needs assessment by posting an EH link survey, custom tailored to SLPs working in EI, on the American Speech, Language, and Hearing Association's Early Intervention Community website (2500 members). One hundred fifty-eight participants completed the survey (Zimmerman et al. under review). Those surveyed had an average of 12.85 years of experience working as an SLP and 7.9 years working in EI. A majority (60%) of SLPs reported that they did not receive specific training regarding the effect of environmental exposures on child development.

Twenty-six percent of those surveyed reported seeing 2-to-4 children that, to their knowledge, have been affected by environmental exposures in the past year; 41% reported seeing 5 or more children affected. Regardless of training, a vast majority (78%) of the SLPs surveyed think that the role of environmental influences on child development is very important (Fig. 8.6).

Twenty-four percent of respondents always consider EH factors during EI assessment or treatment, 26% usually consider it, 10% consider it half the time, 32% consider it sometimes, and 7% never consider it. SLPs were also asked what amount of control they had over environmental health hazards; a majority reported thinking they had some control over EH hazards, see Fig. 8.7.

SLPs reported talking to parents/caregivers about diet/food choices, housing/home environment, school/childcare environment, and drugs. Smoking and lead exposure were overwhelmingly the two most prevalent specific environmental exposure issues that SLPs believed impact the clients they treat. Sixty-one percent of SLPs noted some level of dissatisfaction with their level of training about environmental exposures and their effect on child development. Sixty-four percent were dissatisfied regarding their feeling of preparation to be a resource in the

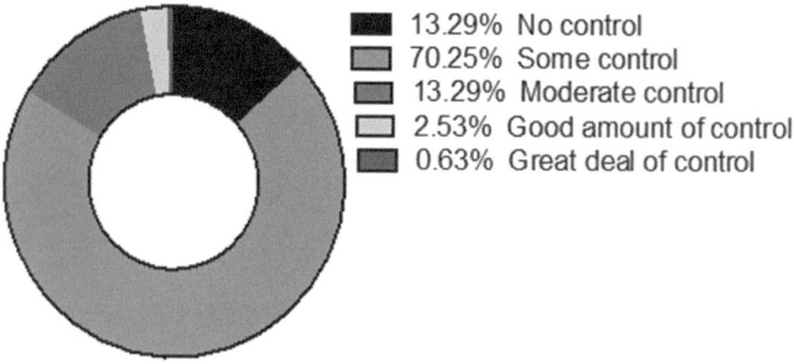


**Fig. 8.7** How much control do EI SLPs have over environmental health hazards?

community if a concern about environmental health were ever raised in the community in which they work. In fact, 56% of the SLPs do not feel as if they are prepared to be a health advocate about environmental exposure concerns within their community. Only half felt somewhat confident that they would be able to find resources about environmental health factors and outcomes. When given the opportunity to share what specific areas of environmental health they would like to learn more about, the answers included poor nutrition, pesticides, drug use, household cleaners and chemicals, air pollution, and GMOs.

Based on this survey, it was evident that there is a need for more formal education regarding EH and SLPs working in EI. To address this need, 26 SLPs and 22 master's students in SLP attended a CE event and took pre- and post-environmental knowledge tests to ensure that they gained a better understanding of the effect of environmental exposures on children as well as the role an SLP working in EI can play regarding environmental exposures by attending the CE event. All participants did significantly better on the test after the presentation, see Fig. 8.8 (Zimmerman et al. under review).

These findings indicate that the EI SLP community lacks an understanding about environmental exposures and their subsequent effects. However, based on the success of the CE event, it is evident that, given educational opportunities, SLPs can become more educated and further cognizant of the effects of environmental exposure, as well as their role as clinicians working in homes with families. Expanding an event like this to other allied health professions such as physical therapy, occupational therapy, and nursing would be beneficial since all provide key information and guidance to families. Given the success with the EI SLPs, the short pre- and post-test approach to assessing EH knowledge can be a useful model for these other groups as well. While nurses are often the most EH-aware, recent work in replicating the SLP education/evaluation with MS nursing students found low EH knowledge.

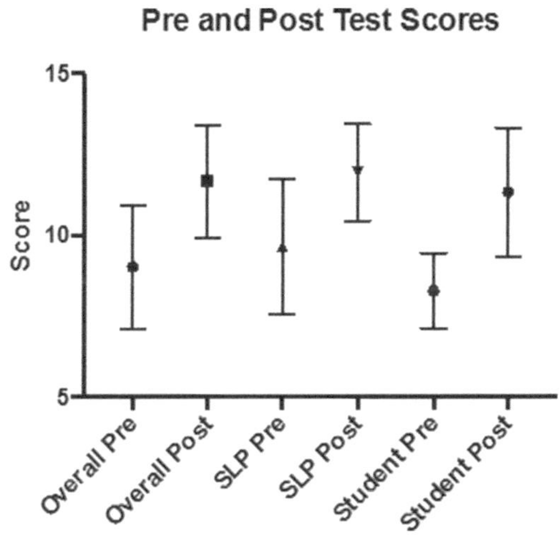


**Fig. 8.8** Pre- and Post-test scores of the entire group ($p < .001$), the SLP group ($p < .001$), and the SLP master's group ($p < .001$)

## Research to Real Life: What Health Professionals can Do

As proposed in the EHL model for health care professionals (page –), levels 4 and 5 involve stepping outside of clinical practice and patient advocacy to bring environmental health expertise to professional societies and other avenues that affect policy decision making. These activities require a higher level of EHL, which is exemplified by confidence to convey environmental health knowledge directly to policymakers. While daunting to many health care professionals, such efforts are important to advance public health. As discussed previously, health professionals are trusted messengers, and when they become involved in conveying environmental health threats to patients as well as to the public and policy makers it can have a profound impact.

One important example is the case of tobacco. When health professionals and scientists became involved, they helped influence both personal decisions and public health measures to mitigate tobacco smoking. The Surgeon General's report in 1964 set the stage for serious efforts to educate the public about the health dangers of cigarette smoking and a wide range of regulations to deter it (CDC History of SG report on smoking). According to the Massachusetts Department of Public health, "Each time a clinician intervenes with a patient who uses tobacco, that patient's likelihood of quitting increases by 30 percent" (MA DPH 2017).

Another example is the history of lead poisoning. In both Warren's (2000) "Brush with Death: A Social History of Lead Poisoning" and in Hernberg's (2000) "Lead Poisoning in a Historical Perspective," the importance of health professional involvement in turning the tide of scientific and public opinion about health threats from lead, a potent neurotoxin, and subsequent removal of it from gasoline and other products is well documented. More recently, the courage of pediatrician Dr.

Mona Hanna-Attisha to research and speak out about the health effects of lead leaching from city water pipes in Flint, Michigan helped call national attention to the crisis and eventually led to remediation efforts (CNN – Drs. Fight to expose Flint lead crisis). One person really can make a difference.

## Concerns about Environmental Health Advocacy

Concerns expressed by health professionals about engaging in EH advocacy include fear of appearing biased, fear of repercussions against their work and public attacks on their professional reputations, lack of knowledge about the topic, and inexperience speaking to the public and media. These concerns can prevent health professionals and scientists from engaging in advocacy. Indeed, the fear of publicly addressing inequities and environmental hazards are real. Herbert Needleman, a child psychiatrist, published the first study examining tooth lead as a marker of body burden in 1972 and subsequently its association with neurocognitive impacts (Needleman et al. 1972, 1979). As a result of his groundbreaking work, he noted that he was the subject of an intense public inquiry instigated by the National Institutes of Health Office of Scientific Integrity at the behest of industry, and subsequent derailing of a prominent career (Needleman 1992). When addressing environmental health in clinical practice, many clinicians may anticipate negative reactions from patients directed towards the clinician (Zachek et al. 2015). Though pediatricians have anticipated similar reactions to discussing parental smoking, the vast majority of smoking parents show strong support for addressing this issue at office visits (Cluss and Moss 2002; Moss et al. 2006). In fact, they believe that addressing these difficult issues is not only appropriate but a sign of a good clinician.

## Range of Activities

Health professionals can have such a profound impact on personal and public policies that it is essential to find ways to overcome these challenges. Education on EH issues, media and public speaking training and support from peer groups are some examples of ways to overcome the challenges. This may be particularly important as in recent years a surge of scientists and health care professionals are emerging from their laboratories and clinics to speak out in defense of evidence-based science in the face of science-denial around a range of environmental health issues.

EH advocacy can involve a range of activities from the personal to public policy arenas. However, each of these activities has the potential to both reinforce and strengthen the EHL of health professionals as well as policymakers. Below are listed and described activities for health professional involvement in advocacy.

- Professional association policies – Serving on committees that make environmental health policies or volunteering to start an environmental health committee can be an important way to contribute to policies that can raise the EHL of members and help encourage EH incorporation into practices and programs as well as EH public policies.
- Health care institutions and workplaces – Becoming involved with EH policies at health care institutions and private practices can help improve workplaces for practitioners and create healthier healing environments for patients. Health professionals involved with the organizations Health Care Without Harm and Practice Greenhealth have helped make enormous strides in "greening" hospitals, clinics and other places where health professionals work and care for patients. Health Care Without Harm's mission to "Transform health care worldwide so that it reduces its environmental footprint, becomes a community anchor for sustainability and a leader in the global movement for environmental health and justice," includes addressing medical waste, safer cleaning chemicals, green building construction and energy use, healthy food, green purchasing, and climate change effects on health.
- Policy making-entities – When decisions are made on rules, regulations and public policies related to the environment and health at the community, state, regional, national and international levels, they can have a significant impact on people's health. Scientific advisory committees to federal and state agencies, for example, can be influenced by the most prolific, prominent and loudest voices at the table. Corporate scientists and consultants are paid to sit on these committees. Public health scientists, health care professionals and other advocates often have to donate their time and expertise. The scale is often tipped in the direction of those with a profit-making interest, not those with the public health in mind. This makes it ever more critical that health professionals become involved. For example, Dr. Richard Clapp, the founding director of the Massachusetts Cancer Registry and Professor Emeritus of Environmental Health, Boston University School of Public Health, served on the Dioxin Reassessment Review Subcommittee of the EPA, which was charged with reviewing the potential health effects of a family of substances called dioxins. Dr. Clapp became concerned that the process of making final recommendations to the EPA Science Advisory Board (SAB) was being compromised by at least two industry-oriented members of the subcommittee. He was so concerned he wrote a letter to the SAB urging changes in the letter to be sent to the EPA director detailing what he described as an "abuse of process" that undermined the effort to bring thoughtful advice to the EPA." His intervention resulted in changes to the final letter sent to the EPA. This is another example of one person making a difference (Personal communication with Dr. Clapp & GBPSR news 200).
- Local and regional boards of health – Most communities or regions have boards of health that oversee local and/or regional regulations and activities to protect health. Health professional participation here can make a real difference in a community. For example, a Massachusetts physician with an interest in environmental health and who had attended a training on environmental threats to health

organized by PSR decided to serve on his local board of health where he subsequently was able to help change local regulations to prevent the spraying of pesticides in public areas in his community.

- Translation outreach from research centers – Many government-funded research centers are obligated by grant agreements to translate their research to make it relevant to the real world. They often work with community organizations and other nongovernmental organizations to conduct online or in-person sessions to help explain the relevance of the research to people's health, and often what the policy implications of the research findings might be. For example, how research on zebra fish can help us better understand the role of endocrine disrupting chemicals on human health. The NIEHS-funded Superfund Research Centers and the NIH-EPA-funded Environmental Health Disparities Research Centers regularly conduct outreach in cooperation with their research translation and community engagement cores. They may partner with nonprofits to share and discuss research findings via webinars, news feeds, community forums, and social media.
- Legislative arenas – Local, state, regional or county and federal government agencies conduct public hearings on issues relating to health and welfare of residents. This is an opportunity for experts in various fields to express concerns or support for proposed legislation, regulations, programs or public works. Health experts can often sway opinions, for better or worse. Local and state officials and legislators have recently played important roles in several states (New York, Massachusetts, New Hampshire, Vermont, Michigan) dealing with PFAS contamination. That includes working with health professionals to engage with ATSDR and the U.S. Air Force around research, alternative water sources, and remediation, and to help academic-community partnerships develop grant applications for CBPR projects on environmental health. Such actions provide a backdrop for health professionals to gain more legitimacy in speaking out about contamination.

Three Children's Environmental Health Research Centers (Columbia University, University of Washington, and UC Berkeley/CERCH) studied the health impacts of organophosphate pesticides on health. In part, based on the results of their studies, several state and local health laws and codes were strengthened to better protect children and workers. Similar results have occurred based on the research from other centers (Eskenazi et al. 1999).

- Community advocacy actions – Participating in local coalitions on such issues as sustainability, healthy housing, energy efficient transportation, and fresh and local food can benefit from an expert health voice as a member of the coalition. At Brown University's Superfund Research Program and Children's Environmental Health Center, a pediatrician who worked for the RI Department of Health played a central role in that broad range of topics, in collaboration with community groups such as Childhood Lead Action Project and Environmental Justice League of Rhode Island, enabling an expansion of EHL across many sectors.

- Social media, blogs, and other media – In the Internet age, it is ever more important that health professionals are able to communicate via social and other media that their patients and peers use to provide rational and reasonable analyses on important current events relating to environmental health. Most environmental health organizations and publications have Facebook pages, Twitter accounts, blogs and other social media outlets. Sharing important scientific and environmental health information this way can garner widespread readership. Commenting on blogs and other online postings can also provide an audience for environmental health issues that might not otherwise be heard. Creating one's own accounts and posting key information can be very valuable, such as the following examples in Fig. 8.9. For further details, see Chap. 12.

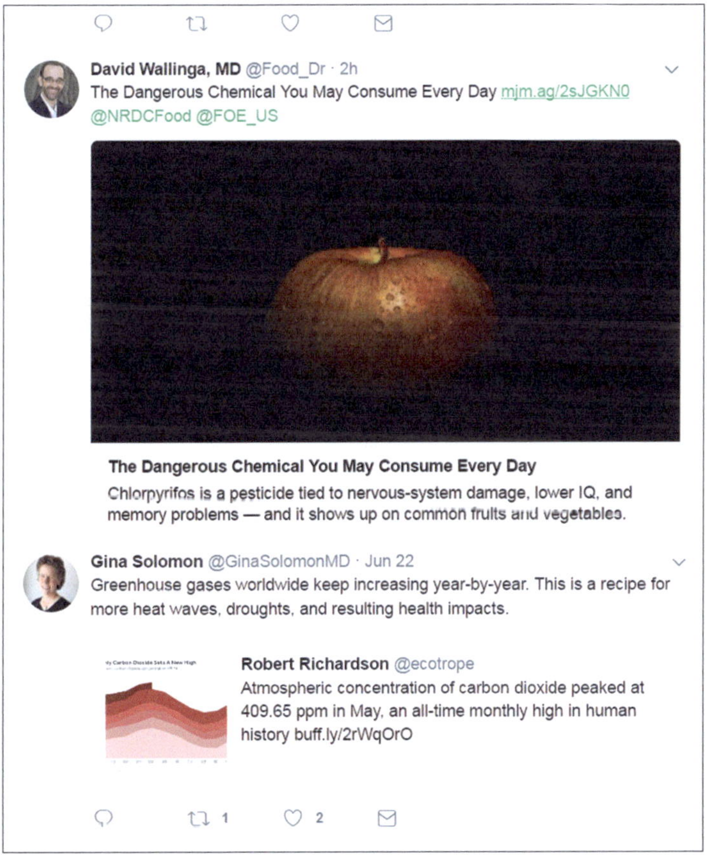


**Fig. 8.9** Twitter examples

## Challenges and Opportunities

**Challenges** In this chapter, several challenges to improving EHL among health professionals have been discussed. They include at the individual level time constraints, lack of knowledge, lack of supporting materials or ability to refer to proper experts, competing interests for time and medical education, fears about engaging in advocacy related to their professional standing and lack of communications/media skills. At the institutional level they include medical curricula issues, professional society positions, inadequate sharing of environmental information from government agencies and lack of funding. Some may feel they have no more time to take on new areas of knowledge, and others may think that it is unlikely they will encounter such issues. Some health professionals, like some EH scientists, are reluctant for many reasons to challenge local authorities and businesses in contaminated areas.

Affected communities typically face the problem of health professionals who lack EH knowledge. Even if some communities find some educated professionals, those professionals may lack the ability to effectively convey that information to patients or communities. Most affected communities lack connections to academic researchers, who may be gatekeepers to educating health professionals on the specific health implications of specific exposure and to getting them involved.

**Moving Beyond Challenges to Identifying Opportunities** Of the different challenges discussed above, lack of empirical evidence about environmental influences of health when counseling their patients on EH concerns is a big issue (Stotland et al. 2014). Evidence-based medicine (EBM) emerged in the 1980's and is widely accepted as the "gold standard" in the medical field. EBM is generally described as "the conscientious, explicit, and judicious use of current best evidence in making decisions about the care of individual patients" (Sackett et al. 1996). The reliance of EBM means that decisions and recommendations made by healthcare professionals must by supported by the scientific literature. In medical research, new drugs and therapies are tested on human subjects under strict regulations and are highly controlled. In these studies, a group of people are given a new drug or therapy while others are given a placebo and observations are made on the condition of the test subject. EBM treatment standards are based on such studies (Nardini 2014). Effects of environmental exposures cannot be tested in the same way that new pharmaceuticals are for multiple reasons. First, people are exposed to numerous chemicals and hazards at the same time and it is impossible to disentangle these exposures. Secondly, humans cannot be ethically exposed to environmental hazards in the same way because the is the aim is not therapeutic. As a result, there are relatively few empirical human health data on environmental hazards that can be applied in clinical settings the same way that medical studies are applied in clinical setting. To address the gap in EH literature and clinical practice, Woodruff et al. (2011) developed a methodology to evaluate the existing data on environmental toxicants and health outcomes. Widespread use of this methodology would allow medical professional organizations to make informed decisions about the potential risks of specific toxicants. Medical societies could then develop fact sheets and trainings to help professionals summarize risks with patients and how to mitigate them. This allows for better incorporation of EH into clinical practice and shifts away from the focus of individualized and lifestyle determinants of health while simultaneously addressing the concern that there is not definitive evidence of adverse health effects related to various environmental toxicants.

Communication skills are central to this model, and the examples demonstrate that research participants are appreciative of both individual and aggregate levels of report-back. They are able to use such information not only to learn about local exposures, but to bring that information to bear on government and industry to remediate and prevent contamination. This applies both to research participants and to other residents in the affected community who may not be in the sample (Brown et al. 2012). Another example is when communities learn of a contamination episode but are not given full and transparent communication about their health risks by officials. Complaints about the withholding of information are widespread. CBPR projects that report-back data to participants are an example of promoting EHL, as described in the chapter by Brody et al. In addition to using these metrics to learn about community response to health professionals' communication, the metrics developed to examine residents' EHL in such projects (Brown et al. 2012) can be modified to examine how health professionals learn from such activities. In a household exposure project in Richmond CA, criteria of effectiveness were developed that consisted of production of new science, methodological advances, support for the advocacy needs of community based organizations to generate health-protective action, strengthening collaboration between partners, carrying out multiple forms of dissemination and communication, and contributing to public participation (Brown et al. 2012). Modification of these criteria could include both how health professionals contribute to those criteria of effectiveness, but also how they learn more about those issues themselves. As with the formulation of EHL being broader than just communication, these criteria highlight those additional aspects of community and social levels of engagement.

In addition to the challenges for health care professionals in clinical settings, some hospitals may have occupational health clinics, and even though they are used to dealing with work-related exposures, they are likely to be board certified in Occupational and Environmental Medicine and prepared to address environmental exposures. If residents are involved in trade unions that have occupational health specialists, there is similar potential for improving EHL. Health professionals need continuing education credits, which offer valuable openings to include education and updates about environmental health.

## Recommendations

**Health Professional Education** Similar, to the recent inclusion of health social science on the Medical College Admission Test (MCAT), the authors urge that EHL material also be included. In addition, there is a need to develop models for improving the education of health professionals about EH (particularly in medical and nursing schools), and expanding options for CEUs. McCurdy et al. (2004) identified potential leverage points to incorporate environmental health into medical and nursing education. These included developing champions in medical-student organizations, on the faculty of schools of medicine, and in the Association of American Medical Colleges (AAMC), the Liaison Committee on Medical Education (LCME), and the National Board of Medical Examiners (NBME).

Incorporating information about environmental influences on health and how to take an environmental history in regular clinical training, including residency programs, can help health professionals bring this into their daily practices with patients. In addition, interactive learning programs can provide experiences to engage in environmental health advocacy in communities and elsewhere. For example, a PEHSU/University of California San Francisco (UCSF) community pediatrics rotation had residents develop a project where they taught a summer education course to high school students. The course included photography and environmental health. The high school students then undertook a community environmental photography project, using cameras provided by the UCSF residents. They photographed environmental health hazards in the neighborhood including traffic, abandoned buildings, etc. which they used to create a slide show that they presented at the community center. The residents added to the slide show and presented it to a university audience. Ideally, medical education would provide these kinds of active learning projects.

Medical and nursing schools (along with institutions training ancillary providers) should integrate EH from premed courses to specialty training. State medical societies and local and state health departments could dispatch health educators to work directly with health professionals to enable them to better communicate with patients and the public.

The increasing number of biomonitoring studies with report back to participants and communities underscores the need for outreach education for clinicians. Since many of the questions that inevitably arise from test results are referred to the subject's physician for advice, an introduction to biomonitoring, study specifics, and guidance on interpretation is critical for the clinical community. Resources for clinical support such as those developed by the Association of Occupational Environmental Medicine Clinics, the Pediatric Environmental Health Specialty Units, and ATSDR should be provided to clinicians as needed.

Developing models for community members and the public to share information with health professionals can improve community health studies. At Superfund sites with community engagement panels, and at sites with ATSDR health investigations, formal mechanisms in place for community engagement can be extended to include affected residents interacting with local health professionals. EPA and ATSDR could use material from those panels to develop model approaches and materials that can be adapted for other sites. Clinicians can be brought in by EPA and ATSDR at those sites.

If various teams can use both the same educational module and the same tests, a strong validated knowledge of what works would be accumulated. Educational and training activities will be best served if they can do such pre-and post-testing. When there is no ability to do pre-tests, brief questionnaires after activities will still be useful.

**Research** Most research on EH knowledge has been conducted on physicians, however there is a need to broaden the scope to other health professionals, especially nurses, nurse practitioners, public health practitioners, community health workers, physician assistants, psychologists, speech-language pathologists, and early intervention practitioners.

**Communications** Fostering cross-disciplinary communications/actions between EH scientists and health professionals can increase the body of knowledge on environmental health issues. Environmental health researchers do not often publish their key studies or reviews in journals that are commonly read by medical professionals (Zachek 2015). A corollary is that those journals may not identify environmental health research as of interest to their readers, and thus reject submissions, exacerbating a lack of communication between researchers and those in a position to help translate research to practice. Efforts to include more environmental health research and reviews in specialty and sub-specialty journals will help to engage this broader audience.

NIEHS's Partnerships in Environmental Public Health (PEPH) can help create and disseminate materials. PEPH has a Resource Center that is a repository of brochures, slide shows, pamphlets, reports, and other items that can be accessed by researchers and health professionals as well as the general public. A recommendation is to develop a dedicated section within the PEPH Resource Center on health professional EHL materials, that can be downloaded into printable versions.

**Public Policy** It is important to ensure that scientific evidence is included as a key element in decision making. One program that has responded to the need for more science-based input into federal agencies is the University of California San Francisco's Program on Reproductive Health and the Environment's "Reach the Decision Makers" fellowship which trains scientists, community members, clinicians and public health professionals to effectively promote science and health-based policies at the US EPA. More of these types of programs are needed. In addition, health professionals and scientists need to learn how to better communicate issues to policymakers, the media and the public. A recent collaboration between the International Society for Children's Health and the Environment (ISCHE) and the Boston University Superfund Research Program will develop a data base of researchers' expertise and advocacy groups needs for scientific guidance. This will help to match those with expertise with those who need guidance, and allow for translation of high quality scientific knowledge to those working to use that information to address policy needs.

## Conclusion

The role of environmental determinants on human health and health disparities has been well documented yet has been underemphasized in healthcare. Research consistently shows that medical students believe that their training on environmental health is lacking and physicians rarely take environmental health histories or discuss environmental exposures with their patients.

Physicians and health care professionals enjoy some of the highest professional rankings in the U.S. Their status and trusted community members' positions them as leaders for environmental health advocacy and training for individuals, families, and communities.

Throughout this chapter, primary issues have been identified that prevent EHL among health professionals (i.e. lack of training, time constraints), as well as pathways to change, examples of progress and recommendations for the future. In order to comprehensively train healthcare professionals on EH, there must be collaboration among the medical field, academics, governmental agencies, non-profits, and community members. Having a better grasp on EH could reduce exposure to environmental hazards and environmental health disparities.

**Acknowledgement** The U.S. Environmental Protection Agency (EPA) supports the PEHSU by providing partial funding to ATSDR under Inter-Agency Agreement number DW-75-95877701-4. Neither EPA nor ATSDR endorse the purchase of any commercial products or services mentioned in PEHSU publications.

## References

Altman, R., Brody, J., Rudel, R., Morello-Frosch, R., Brown, P., & Averick, M. (2008). Pollution comes home and pollution gets personal: Women's experience of household toxic exposure. *Journal of Health and Social Behavior, 49*, 417–435.

Amico, A., Davie, A., & Dalton, M. (2016). "The PFC Contamination at Pease: A Community Perspective." Presentation at Local environmental action conference March 13, 2016. Boston, MA, Northeastern University.

Association of American Medical Colleges (AAMC). (2010). 2010 medical school graduation questionnaire: all schools summary report: final. https://www.aamc.org/download/140716/data/2010_gq_all_schools.pdf

Association of American Medical Colleges (AAMC). (2018). Number of medical schools including topic in required courses and elective courses: environmental health. https://www.aamc.org/initiatives/cir/406462/06a.html

Balshem, M. (1993). *Cancer in the community: Class and medical authority*. Washington, DC: Smithsonian Institution Press.

Baron, S., Sinclair, R., Payne-Sturges, D., Phelps, J., Zenick, H., Collman, G. W., et al. (2009). Partnerships for environmental and occupational justice: Contributions to research, capacity and public health. *American Journal of Public Health, 99*, S517–S525.

Brody, J. G., Morello-Frosch, R., Brown, P., Rudel, R. A., Altman, R. G., Frye, M., Osimo, C. C., Perez, C., & Seryak, L. M. (2007). Is it safe? New ethics for reporting personal exposures to environmental chemicals. *American Journal of Public Health, 97*, 1547–1554.

Brody, J. G., Dunagan, S. C., Morello-Frosch, R., Brown, P., Patton, S., & Rudel, R. A. (2014a). Reporting individual results for biomonitoring and environmental exposures: Lessons learned from environmental communication case studies. *Environmental Health, 13*, 40.

Brody, J. G., Kripke, M. L., Kavanaugh-Lynch, M. H., & Forman, J. R. M. R. (2014b). Breast cancer and environmental research. *Science, 344*, 577.

Brown, P., & Kelley, J. D. (1996). Physician's knowledge, attitudes, and practices regarding environmental health hazards. *Industrial & Environmental Crisis Quarterly, 9*(4), 512–542.

Brown, P., & Mikkelsen, E. J. (1990). *No safe place: Toxic waste, leukemia, and community action*. Berkeley: University of California Press.

Brown, P., McCormick, S., Mayer, B., Zavestoski, S., Morello-Frosch, R., Gasior, R., & Senier, L. (2006). 'A lab of our own': Environmental causation of breast cancer and challenges to the dominant epidemiological paradigm. *Science, Technology, and Human Values, 31*, 499–536.

Brown, P., Brody, J. G., Morello-Frosch, R., Tovar, J., Zota, A. R., & Rudel, R. A. (2012). Measuring the success of community science: The northern California household exposure study. *Environmental Health Perspectives, 120*, 326–331.

Burke, M. G., & Miller, M. D. (2011). Practical guidelines for evaluating lead exposure in children with mental health conditions: Molecular effects and clinical implications. *Postgraduate Medicine, 123*(1), 160–168.

California Environmental Health Tracking Program. (2017). Imperial county community air monitoring project. Richmond, CA. http://www.cehtp.org/page/imperial_county

Cluss, P. A., & Moss, D. (2002). Parent attitudes about pediatricians addressing parental smoking. *Ambulatory Pediatrics, 2*, 485–488.

Council on Environmental Health. (2012). Policy statement: Pesticide exposure in children. *Pediatrics, 130*(6), e1757–e1763.

Curtis, K., & Wilding, B. C. (2007). "Is it in us? Toxic trespass, regulatory failure & opportunities for action." Commonweal biomonitoring resource center, coming clean body burden work group.

DePue, J. D., McQuaid, E. L., Koinis-Mitchell, D., Camillo, C., Alario, A., & Klein, R. B. (2007). Providence school asthma partnership: School-based asthma program for inner-city families. *Journal of Asthma, 44*, 449–445.

Di Renzo, G. C., et al. (2015). International federation of gynecology and obstetrics opinion on reproductive health impacts of exposure to toxic environmental chemicals. *International Journal of Gynecology & Obstetrics*. https://doi.org/10.1016/j.ijgo.2015.09.002

Edelstein, M. (1988). *Contaminated communities: The social and psychological impacts of residential toxic exposure*. Boulder: Westview Press/Perseus Books.

Emmett, E., Zhang, H., Shofer, F., Rodway, N., Desai, C., Freeman, D., & Hufford, M. (2009). Development and successful application of a 'Community-First' communication model for community-based environmental Health Research. *Journal of Occupational and Environmental Medicine, 51*(2), 146–156.

Eskenazi, B., Bradman, A., & Castorina R. (1999). Exposures of children to organophosphate pesticides and their potential adverse health effects. *Environmental Health Perspectives, 107*(Suppl 3), 409–419.

Friedrich, M. J. (2017). Medical community gathers steam to tackle climate's health effects. *JAMA, 317*(15), 1511.

Fung, A., & O'Rourke, D. (2000). Reinventing environmental regulation from the grassroots up: Explaining and expanding the success of the toxics release inventory. *Environmental Management, 25*, 115–127.

Gallup. (2014). Gallup poll. U.S. *Views on honesty and ethical standards in professions*. http://www.gallup.com/poll/180260/americans-rate-nurses-highest-honesty-ethical-standards.aspx

Gehle, K. S., Crawford, J. L., & Hatcher, M. T. (2011). Integrating environmental health into medical education. *American Journal of Preventive Medicine, 41*(4S3), S296–S301.

Hernberg, S. (2000). Lead poisoning in a historical perspective. *American Journal of Industrial Medicine., 38*, 244–254.

Hoover, E. (In press). *The River is in Us: Fighting Toxics in a Mohawk Community*. Minneapolis: University of Minnesota Press.

Hoover, E., Renauld, M., Edelstein, M., & Brown, P. (2015). Social science contributions to transdisciplinary environmental health. *Environmental Health Perspectives., 123*, 1100–1106.

Ibanez, G., Zabar, J., Cadwallader, J., Rondet, C., Lochard, M., & Magnier, A. M. (2015). Views of general practitioners on indoor environmental health risks in the perinatal period. *Frontiers in Medicine, 2*(32). https://doi.org/10.3389/fmed.2015.00032

Kilpatrick, N., Frumkin, H., Trowbridge, J., Escoffery, C., Geller, R., Rubin, L., Teague, G., & Nodvin, J. (2002). The environmental history in pediatric practice: A study of pediatricians' attitudes, beliefs, and practices. *Environmental Health Perspectives, 110*(8), 823–827.

Kovarik, W. (2005). Ethyl-leaded gasoline: How a classic occupational disease became an international public health disaster. *International Journl of Occupational and Environmental Health, 11*, 4.

Kreuter, M. W., Chheda, S. G., & Bull, F. C. (2000). How does physician advice influence patient behavior? Evidence for a priming effect. *Archives of Family Medicine, 9*, 426–433.

Lai, P. Y., Cottingham, K. L., Steinmaus, C., Karagas, M. R., & Miller, M. D. (2015). Arsenic and rice: Translating research to address health care providers' needs. *The Journal of Pediatrics, 167*(4), 797–803.

MA Department of Public Health. (2017). Make smoking history. http://makesmokinghistory.org/quit-now/for-providers. Accessed 25 May 2017.

Mayer, B., Brown, P., & Morello-Frosch, R. (2010). Labor-environmental coalition formation: Framing and the right-to-know. *Sociological Forum, 25*, 745–768.

McCormick S., Brown P., & Zavestoski S. (2003). The Personal Is Scientific, the Scientific Is Political: The Public Paradigm of the Environmental Breast Cancer Movement. *Sociological Forum, 18*(4), 545–576.

McCurdy, L. E., et al. (2004). Incorporating environmental health into pediatric medical and nursing education. *Environmental Health Perspectives, 112*(17), 1755–1760. https://www.ncbi.nlm.nih.gov/pmc/articles/PMC1253669/

Mello, S., & Hovick, S. R. (2016). Predicting behaviors to reduce toxic chemical exposures among new and expectant mothers: The role of distal variables within the integrative model of behavioral prediction. *Health Education & Behavior, 43*(6), 705–715. https://www.ncbi.nlm.nih.gov/pubmed/27179287

Miller, M. D., & Marty, M. A. (2012). Childhood - A time period uniquely vulnerable to environmental exposures. In R. H. Friis (Ed.), *The Praeger handbook of environmental health*. Santa Barbara: Praeger.

Miller, M., & Solomon, G. (2003). Environmental risk communication for the clinician. *Pediatrics, 112*, 211–217.

Moss, D., Cluss, P. A., Mesiano, M., & al, e. (2006). Accessing adult smokers in the pediatric setting: What do parents think? *Nicotine & Tobacco Research, 8*, 67–75.

Nardini, C. (2014). The ethics of clinical trials. *Ecancermedicalscience, 8*, 387. https://doi.org/10.3332/ecancer.2014.387.

Needleman, H. L. (1992). Salem comes to the National Institutes of Health: Notes from inside the crucible of scientific integrity. *Pediatrics, 90*, 977–981.

Needleman, H. L., Tuncay, O., & Shapiro, I. M. (1972). Lead levels in deciduous teeth of urban and suburban American children. *Nature, 235*, 111–112.

Needleman, H. L., Gunnoe, C., Leviton, A., et al. (1979). Deficits in psychological and classroom performance in children with elevated dentine lead levels. *The New England Journal of Medicine, 300*, 689–695.

NEETF – Health professionals and Environmental Health Education Position Statement. (2004). https://www.neefusa.org/resource/health-professionals-and-environmental-health-education-position-statement

Quandt, S. A., Doran, A. M., Rao, P., Hoppin, J. A., Snively, B. M., & Arcury, T. A. (2004). Reporting pesticide assessment results to farmworker families: Development, implementation, and evaluation of a risk communication strategy. *Environmental Health Perspectives, 112*, 636.

Roberts, J. R., & Gitterman, B. A. (2003). Pediatric environmental health education: A survey of US pediatric residency programs. *Ambulatory Pediatrics, 3*(1), 57–59.
Rodgers, K. M., Udesky, J. O., Rudel, R. A., & Brody, J. G. (2018). Environmental chemicals and breast cancer: An updated review of epidemiological literature informed by biological mechanisms. *Environmental Research, 160*, 152–182.
Sackett, D. L., Rosenberg, W. M. C., Gray, J. A., Haynes, B. R., & Richardson, W. S. (1996). Evidence based medicine: What it is and what it isn't. *British Medical Journal, 312*, 71–72.
Schell, L. M., & Tarbell, A. M. (1998). Commentary: A partnership study of PCBs and the health of Mohawk youth: Lessons from our past and guidelines for our future. *Environmental Health Perspectives, 106*(Supplement 3), 833–840.
Stein, J., Schettler, T., Rohr, B., Valenti, M. (2008). Environmental threats to healthy aging. Greater Boston physicians for social responsibility and science and environmental health network, www.agehealthy.org
Stotland, N. E., Sutton, P., Trowbridge, J., Atchley, D. S., Conry, J., Transande, L., Gerbert, B., Charlesworth, A., & Woodruff, T. J. (2014). Counseling patients on preventing prenatal environmental exposures- a mixed-methods study of obstetricians. *PLoS One, 9*(6), e98771.
Taylor, D. K., Lepisto, B. L., Lecea, N., Ghamrawi, R., Bachuwa, G., LaChance, J., & Hanna-Attisha, M. (2017). Surveying resident and faculty physician knowledge, attitudes, and experiences in response to public lead contamination. *Academic Medicine, 92*(3), 308–311.
Trasande, L., Boscarino, J., et al. (2006a). The environment in pediatric practice: A study of New York pediatricians' attitudes, beliefs, and practices towards Children's environmental health. *Journal of Urban Health, 83*(4), 760–772.
Trasande, L., Schapiro, M. L., Falk, R., Haynes, K. A., Behrmann, A., Vohmann, M., Stremski, E. S., Eisenberg, C., Evenstad, C., Anderson, H. A., & Landrigan, P. J. (2006b). Pediatrician attitudes, clinical activities, and knowledge of environmental health in Wisconsin. *WMJ, 105*(2), 45–49.
Trasande, L., Newman, N., Long, L., Howe, G., Kerwin, B. J., Martin, R. J., Gahagan, S. A., & Weil, W. B. (2010). Translating knowledge about environmental health to practitioners: Are we doing enough? *Mount Sinai Journal of Medicine, 77*(1), 114–123.
Trasande, L., et al. (2014). The environment and children's health care in Northwest China. *BMC Pediatrics, 14*, 82.
Vega, V., Milagros, C., Brown, P., Murphy, C., Figueroa, A., Cordero, J., & Alshawabkeh, A. (2016). Community engagement and research translation in Puerto Rico's northern karst region: The PROTECT superfund research program. *New Solutions, 26*, 475–495.
Warren, C. (2000). *Brush with death: A social history Of lead poisoning*. Baltimore: Johns Hopkins University Press.
Wilding, B. C., Curtis, K., & Welker-Hood, K. (2009). Hazardous chemicals in health care: A snapshot of chemicals in doctors and nurses. Physicians for social responsibility.
Woodruff, T. J., Sutton, P., & The Navigation Guide Work Group. (2011). An evidence-based medicine methodology to bridge the gap between clinical and environmental health sciences. *Health Affairs, 30*(5), 931–937. https://doi.org/10.1377/hlthaff.2010.1219.
Zachek, C. M., Miller, M. D., Hsu, C., Schiffman, J. D., Sallan, S., Metayer, C., & Dahl, G. V. (2015). Children's cancer and environmental exposures: Professional attitudes and practices. *Journal of Pediatric Hematology/Oncology, 37*(7), 491–497.
Zimmerman, E., Borkowski, C., Clark, S., & Brown, P. (n.d.) Under review. Education speech-language pathologists working in early intervention on environmental health.

# Part III
# Types of Communication Styles and their Effectiveness at Raising EHL

# Chapter 9
# Use of Traditional and Culturally Appropriate Modalities

**Matthew Dellinger and Jonathan Dellinger**

**Abstract** Researchers are specialists, but they must communicate their research to non-specialists to make their work meaningful. This tension exists from the inception to completion of a project, from finding support for an investigation, to obtaining informed consent, to reporting research findings. Monumental as it may be, this particular challenge of communication is often overlooked by researchers. Further complicating the situation is the influence of culture. As renowned anthropologist Edward Hall (1959) put it, culture hides more than it reveals, and it hides the most from its own members. More simply, culture is a force we take for granted, particularly because it describes our assumptions of what "normal" is and how it works. This holds true for the specialist and the layperson, alike.

**Keywords** Risk assessment · Mercury · Polychorinated biphenyls · Dose-response · Tribal research · Fish consumption · Health communication · Unimodel · Explication · Elaboration likelihood model

In 1989, researchers from the Arizona State University obtained informed consent from a population of Havasupai tribal members for use of genetic samples in an investigation of type two diabetes. The project was to determine if there was a hereditary factor contributing to the prevalence of the disease in that population, a relationship which was not found. Later uses for the DNA samples were varied, likely within the broad sweep of the informed consent, but problems arose when tribal members became aware that their DNA samples were used to investigate schizophrenia as well (Garrison 2012). In Havasupai culture, the topic of mental health is taboo. The tribal members sued the university, culminating in a $700,000 settlement and various other concessions, perhaps most importantly to the

M. Dellinger (✉)
Division of Epidemiology, Institute for Health and Equity, Medical College of Wisconsin, Milwaukee, WI, USA
e-mail: mdellinger@mcw.edu

J. Dellinger
Department of Communication, University of Wisconsin-Milwaukee, Milwaukee, WI, USA



S. Finn, L. R. O'Fallon (eds.), *Environmental Health Literacy*,
https://doi.org/10.1007/978-3-319-94108-0_9

Havasupai, the return of sample material (Garrison 2012). Two prominent cultural themes are evident in the Havasupai controversy: the previously mentioned taboo nature of certain research subjects, and, additionally the sacred nature of DNA. To the researchers who collected the samples, the DNA became property of the institution, to be used for scientific inquiry in the betterment of society, the furthering of knowledge, and to progress careers. To the Havasupai, however, the DNA belonged to the individuals from whence they originated, and the continued and unsanctioned use of the genetic samples was a violation of their essence as human beings.

This chapter focuses on the communication of research findings relating to environmental health science through culturally appropriate modalities. The above anecdote is not merely to illustrate the difficulty of cross-cultural, intercultural, and international research efforts. Rather, it is intended to demonstrate that even when researchers follow all established protocols, cultural and social considerations remain separate and enduring challenges. The participants described above were Native Americans with a vested interest in the subject matter (high prevalence of diabetes) and research findings, yet, their individual participation in subsequent uses of their genetic material was culturally inhibited. Culture shapes how people interpret what they see and hear, including new scientific discovery, news stories, and health interventions. Those seeking to bridge such knowledge gaps and to improve Environmental Health Literacy (EHL), must therefore simultaneously consider the relevant values, beliefs, taboos, and customs in specific subpopulations, not just what information these people are lacking. Generally, this requires the consideration of cultural norms and values be applied to the processes of translational science and risk communication. This can be accomplished by employing a formalized, iterative procedure in which members of each subpopulation or community involved in the research are consulted throughout the phases of a project.

## Background

Research findings derived from environmental science generate information that is relevant to public health. Researchers can leverage the techniques from disciplines such as health literacy, analytical chemistry, epidemiology, risk assessment, ecology, and toxicology into health recommendations. However, end users of data-driven advice interpret the data they receive through cultural and psychological filters. They may not share the expertise of researchers or statisticians. Recommendations undergo multiple iterations of development in the professional realm before reaching the public. The end users have not themselves selected and applied the quantitative techniques that generate estimated predictions and uncertainty ranges. Therefore, the intended interpretation of data-driven advice is often lost in translation to many lay communities. Science, by necessity and design, is often conducted in a culturally-sterilized manner that can be counterproductive to its purpose to share the results of discovery for the benefit of public health. This is because scientific research is often deliberately decontextualized to allow for the generalization and replicability of findings. This is a key barrier to improving EHL in diverse cultures which depends on its relevance in a particular cultural context. Professionals seeking to construct health advice often conflate sound methodology with the objective truth they seek. Sound methodology, as accepted by colleagues via peer review, approximates truth. Consequently, these producers of knowledge may overlook the fact that various cultures interpret information differently, not necessarily due to ignorance, but often because they hold a different perspective. Furthermore, a given population may know more about their context than the professional developing advice. Since the knowledge acquired via the scientific method remains an imperfect representation of truth, cultural context must be consulted when interpreting these approximations. These are subtle but important dynamics when translating EHL-promoting advice to other diverse cultures.

Although the focus of this chapter is on the specifics of culturally adapted risk messages, it is important to understand the underlying principles of risk communication. It is also crucial to note the differences between how experts understand their own assumptions versus how a given culture of laypersons would interpret the same processes and assumptions. All humans, including those who conduct scientific research, engage in motivated reasoning (Dieckmann et al. 2017; Kunda 1990; Scurich and Shniderman 2014) in which the interpretation of evidence is biased by existing goals and motives (Kunda 1990). The motivated reasoning of various audiences therefore presents a unique challenge to generating useable advice based on scientific findings. This understanding introduces two layers of interpretation: the researcher's and the intended audience's of their research findings. Empiricism, itself, is a response to the first challenge, attempting to remove as much subjectivity from the pursuit of knowledge as is possible. The implications of the second is that the health recommendations and research findings of environmental health science-related disciplines require additional translation, employing strategic communication and culturally relevant appeals, to adequately contextualize the knowledge gained from environmental health research. This may include the recognition of differing priorities as well as the use of artistic styles that are familiar to the culture of interest. Research in social influence and cultural relativity can also provide guidance for developing more meaningful and relevant translation of scientific knowledge to the needs of specific affected populations. Additionally, those seeking to translate their findings to various cultures require a thorough understanding of the multidisciplinary techniques leading to environmental health recommendations.

An overview of the history and methods of two important fields of study, Human Health Risk Assessment and Health Communication, can provide researchers the insights necessary to align underlying cultural values and assumptions with health advice. Human Health Risk Assessment is a formalized process to estimate the health risks from an environmental factor (usually exposure to a toxicant) to a given population or individual under assumed or assessed circumstances. Health communication theories include psychological models that attempt to explain and predict health behaviors. Lessons from these disciplines provide a sound foundation with which to promote environmental health literacy in culturally diverse populations. This will allow scientists a better chance of sharing important discoveries with the people that need that knowledge most. To assist in understanding the interplay between culture and scientific process, a case study of culturally-tailored fish consumption advisories is presented herein as an example. These fish consumption advisories are based on human health risk assessments for Methylmercury (MeHg) and Polychlorinated Biphenyls (PCBs) from a series of collaborations with Anisihnaabe tribes (Native Americans from the Laurentian Great Lakes).

## A Case Study of Cultural and Historical Influences on Anishinaabe Interpretations of Environmental Data

Fish, game, and wild edibles are more than just a healthy food source for the Anishinaabe. The tribes still rely upon fish, wildlife and plants to meet religious, ceremonial, medicinal, subsistence, as well as economic needs. As their traditional food culture declined in the twentieth century, the Anishinaabe likewise felt changes in other areas of their society. Historical distrust resulting from dispossession, threats to sovereignty, and threat of assimilation exert social forces on tribes that can interfere with cooperative progress with scientists (Burger and Gochfeld 2011; Hoover et al. 2012; Tobias and Richmond 2014). In a response to assimilating and colonizing pressure, the Anishinaabe tribes signed a series of treaties with the United States government in the early to mid-1800's. The ancestors of the Anishinaabe negotiated these treaties to protect their way of life and culture via hunting and fishing rights on the territories they ceded to the government. Today, many Anishinaabe families rely on income from harvesting of these fisheries (Taylor and Ferreri 2000).

From the perspective of the Anishinaabe, the treaty-ceded territories were established as part of unique rights indefinitely held by tribal members. The Great Lakes tribal fisheries are integral to the Anishinaabe right to pursue subsistence practices. These rights, and by extension their subsistence practices, were challenged in U.S.A. vs State of Michigan 1979. The federal government, initiated this litigation to protect the tribe's rights to fish in certain waters of the Great Lakes by virtue of the treaty's provisions related to aboriginal occupation and use. In its complaint, the United States asked that the State of Michigan be enjoined from interfering with the Anishinaabe's treaty-confirmed rights to fish in the Great Lakes. This lawsuit was filed in part because of the interruption of traditional practices since the nineteenth century as well as the modernization of food systems which out-compete subsistence foods. These factors interacted to shift the population away from traditional foods to create a nutritional health disparity that disproportionally impacts Native Americans (Ho et al. 2008; Johns et al. 2013; Schell 2012; Sheehy et al. 2015; Whiting and Mackenzie 1998). Although multifactorial in cause, the disparity most severely affects people living on rural Great Lakes reservations where healthy food alternatives to gas stations quickie marts and casino buffets are less abundant. Tribal members (even those with more food options) may feel a cultural aversion to follow any "official" advice warning them away from a subsistence practice which was previously threatened by legal, political, and social pressure.

A core motivation for the collaboration of inter-tribal consortia on projects such as those described here is to leverage an existing cultural interest in fish to promote health and wellbeing. There is also a strong motivation to support the subsistence practices that the earlier treaties were meant to preserve, thereby providing an additional sense of autonomy and self-determination.

Cultural differences, historic dispossession and issues of trust have created a persistent rift between these populations and contemporary ("Western") scientists. In addition, these cherished fishing traditions are undermined by anthropogenic contaminants which have affected traditional fish and led to the need for advisories. This creates a tension between environmental research and tribal communities. Many fish advisories were built with sport fishing in mind. At the very least, advice that assumes recreational fishing to be the norm is naive relative to the desired consumption rates of the local Anishinaabe. Adherence to sport fishing (i.e. occasional) consumption rates is considered by many Anishinaabe a violation of their culture and natural heritage. It is therefore critically important that attempts to raise EHL in this population consider the cultural and historical context, and input about the framing of risk messages be gathered from multiple local sources to ensure an effectively-translated EHL intervention. In addition to historical documents and anthropological literature, direct dialogue between researchers and the local population is recommended. This can be accomplished by applying participatory research principles (i.e. including community members in the research process) and by formalized qualitative techniques such as focus groups. Both of these routes were pursued with focus groups of Anishinaabe fish consumers, in which they were asked about their perceptions of local advisories (Dellinger et al. 2017). Stigmatizing fish consumption via risk assessment was a prominent concern in these discussions. Pollution of the lakes is common knowledge and discussed at length during these sessions and other interviews. When surveyed, many of the questionnaire responses (66%) reported that they do not limit consumption of wild caught fish due to contamination although 92% believed they are regularly exposed to contaminants. In these discussions, perceived, but perhaps inaccurate, sources of contaminants were a prominent concern: "freighters pull around by the Island and dump their waste." Someone noted that: "It'd make you not want to go in the water, let alone eat the fish." In tension with these concerns, another respondent worried that children were losing their cultural connection to fish: "We need to educate children, especially ages 9-12 or 2-3 grade, we need to teach them early, fresh out the gate. We used to be fishing families…we would harvest the fish."

**Fish Consumption and Natural Heritage**

"It [the advisory] needs to say, 'safe and healthy to eat fish' to focus on the healthy parts and not the contaminants of the fish. It's not the fish – [advisories] make it seem like the fish are bad, but it's not the fish, it's the environment." –Anishinaabe tribal member commenting on fish consumption advisories (Dellinger et al. 2017).

In addition to heritage, the economic harms to commercial tribal fishermen have been cited in multiple focus groups over the past decade (Dellinger et al. 2008, 2012; Dellinger 2007). During a 2016 survey of Anishinaabe fish consumers in the Upper Peninsula of Michigan (Dellinger et al. 2017), the notes from the accompanying focus group revealed that at least two of the participants held subsistence licenses, indicating that consumption of fish (contaminated or otherwise) is part of their livelihood. Approximately 10–20% of participants represent the subgroup of Anishinaabe who would not change their behaviors in response to fish consumption advisories due to the cultural and economic importance of fishing. These groups displayed a bias of mistrust towards governmental bodies and a motivation to continue their fishing practices. Both the lack of trust as well as the motivation to support economically and culturally important activities are understandable, but they are also good examples of the conditions leading to a motivated interpretation of scientific data regarding fish consumption. These dynamics present novel obstacles to improving EHL in Anishinaabe communities. This chapter will cover the process by which quantitative risk estimates are used to guide advisories such as those geared towards fish consumption and will describe the areas in which risk assessments are ill-equipped to grapple with cultural adaptation. The Anishinaabe example provides a successful model to meet these challenges and bridge the gap between science and culture thereby improving EHL.

## How Risk is Assessed Vs. How it is Understood

The concept of risk is one that laypeople are familiar with, but this public perception may differ from the technical definition used in professional risk assessment. The scientific process requires operationalization: the translation of abstract concepts such as risk into a discrete measurable variable for the purpose of testing and evaluation. This operationalization is based on statistical methods that scientists are familiar with but that remain abstract concepts to the public. Thus, it is the duty of a researcher to explain to research participants or affected community members how their investigation is evaluating risk through statistical analysis of the variables in question.

For scientists, the more abstract a concept is, the more important it becomes to justify the logical connection between the operationalized variable and the concept it represents. This justification is referred to as explication in the social sciences, and it is a critical step in the research process (Chaffee 1991). It allows logical connections and research methods to be scrutinized by the scientific community. In the process described below, risk is an operationalized concept, formalized through a rigorous process. However, the explication connecting the scientist's conception of risk is unlikely to match the popular conception of risk in the general population. Further translation is therefore needed to account for different contexts. For example, if a risk quantification was operationalized in a scientific study but will be communicated to the public, additional explanation of the process for establishing risk may be needed.

This chapter will explicate Environmental Risk Assessment as a scientific methodology that encompasses ecological and human health aspects. More specifically, Human Health Risk Assessment is the process of estimating the nature and probability of adverse health effects in humans who may be exposed to chemicals in contaminated environmental media, now or in the future. Human Health Risk Assessment adopts a toxicological focus to conceptualizing health risk and developing recommendations to avoid negative health outcomes. Accordingly, the focus of Human Health Risk Assessment is on estimating "safe" or "acceptable" levels of exposure for humans. This type of risk assessment is science-driven. It quantitatively evaluates the probability of risk. As such, risk assessment is mostly removed from subjective factors that influence risk perception. Much community-based research has been conducted since the original definitions were established by the EPA for hazard identification and exposure assessment. This has led to a shift in risk assessment paradigms where analytical characterization of risk and uncertainty is influenced by deliberation among relevant stakeholders. This approach to risk assessment, where stakeholders are engaged in the characterization of risk, ostensibly increases transparency thereby heightening knowledge and trust of the public. This integration of community engagement principles is an important step and it provides the opportunity for cultural beliefs and constraints to shape the risk assessment process and thereby improve its reception in the target population.

Human Health Risk Assessment consists of the following steps: (1) hazard identification, (2) dose-response evaluation, (3) exposure assessment, and (4) risk characterization. Each of these steps and their inherent assumptions are not well-understood by end-users.

**Hazard Identification** A hazard is a factor or condition that could lead to an undesirable or harmful outcome. In risk assessment, this relationship is probabilistic and therefore causal relationships are inferred through statistical associations with varying degrees of certainty. An environmental hazard may be identified from historical events, such as the unusually prolific birth defects in association with Methylmercury exposure in 1950's Japanese fishing communities (Harada 1967), or from concerns over a new consumer product that chemically or structurally resembles a hormone. Thus, techniques for identifying hazardous factors may be proactive or reactive and are dependent on the nature of the hazard.

In this chapter, the example of methylmercury toxicity (MeHg, found in fish and seafood) as an environmental hazard for Great Lakes tribal communities will be used. This hazard was initially identified from an MeHg outbreak in Minamata Japan, in which thousands of neurologically impaired infants were documented as resulting from eating contaminated seafood. Minamata syndrome symptoms included ataxia, numbness in the hands and feet, general muscle weakness, loss of peripheral vision, and damage to hearing and speech. The congenital form of the disease is of particular concern since fetuses in the womb are more sensitive to the effects of MeHg. This occurrence obviated the hazard to humans and is therefore relevant to all seafood/fish eating cultures. The Minamata outbreak occurred due to extraordinary circumstances in which a chemical company was dumping MeHg directly into Minamata Bay, thereby contaminating the aquatic food chain. Normally, MeHg contamination of the food chain relies on the slower process of atmospherically-deposited elemental Hg, which is gradually transformed by microorganisms into MeHg (its bioavailable form). In the case of Minamata, the production and dumping of MeHg greatly accelerated this process. The local citizens depended on fishing for subsistence and their exposure was considerable. This hazard has also been well-documented in Anishinaabe tribes (Great Lakes Region Native Americans) in Ontario, another subsistence fishing culture. Great Lakes Native Americans (Anishinaabe) maintain a traditional fishing culture which is similar to the Japanese in Minamata and are similarly exposed to MeHg.

**Dose Response** Dose Response requires understanding of carefully selected assumptions (Box 1) as well as technical/statistical knowledge to grasp, and therefore represents a greater burden for translation to the public than hazard identification. Since these assumptions vary depending on the context, improving the EHL of culturally-diverse data end-users presents challenges that must be addressed for each specific culture. The Anishinaabe, for example, exhibit varying fish consumption patterns throughout the year such as spear fishing, ice fishing, and netting. Due to the cultural importance of these activities, the assumptions used to describe "acceptable risks" may not accurately reflect the subjective concept of acceptability in practice. Each cultural situation is different, and these differences are also compounded by factors such as geography and history. For example, many tribal populations, like the Anishinaabe, hold unique hunting and fishing rights through treaties with the US Government. These rights have been challenged and defended in the courts (see for example United States v. Michigan, 471 F. Supp. 192 (W.D. Mich. 1979), aff'd 653 F.2d 277 (6th Cir. 1981), cert. denied, 454 U.S. 1124 (1981). Many of these legal battles represent hard-fought efforts to maintain traditional ways of life, including fish consumption. Therefore, certain circumstances in which high-mercury fish such as walleye are consumed under sanctioned cultural practices may warrant special considerations regarding "acceptable" levels of risk. Specifically, many Anishinaabe tribes practice a spring spear fishing season of spawning walleye. During the spring spear-fishing harvest (April–May) average fish consumption is elevated (15 g/day). The summer months (June–September) present a four-month window that is representative (9 g/day) of overall fish consumption (8.2 g/day); this is consistent across years (DeWeese et al. 2009; GLIFWC 2003). If dose-response assumptions are used to estimate an amount of fish/day, these variations, as well as the cultural importance of temporarily elevating fish consumption, should be integrated with advice that typically assumes a consistent year-round consumption rate.

Well-designed longitudinal and observational studies are often used for developing dose-response measures. For example, the EPA reference dose for oral (ingestion) chronic exposure to methylmercury draws upon three studies, all of which use maternal Hg (assessed in either maternal hair or chord-blood) as an independent variable and child neurodevelopment as the dependent variable. These longitudinal,

**Assumptions Associated with Dose-Response Models**

1. Thresholds exist for certain biological outcomes (i.e. cellular necrosis at a certain dose).
2. The modeling techniques to derive the dose-response curve are valid and represent reality (including assumptions of linearity).
3. Biological variability is adequately captured by using an uncertainty factor.
4. The observational studies are sound, and accurately represented a dose-response relationship without bias.
5. The POD represents acceptable risk.
6. The end user understands all the above.

developmental studies were conducted in the Seychelles Islands, the Faroe Islands, and New Zealand (Davidson et al. 1995, 1998; Grandjean et al. 1997; Rice et al. 2000). A benchmark dose is selected from a dataset as a point of departure (POD) from an acceptable amount of health response, the point of departure is compared to other evidence and adjusted for uncertainty to arrive at the reference dose. Methods for arriving at a benchmark dose employ statistical modelling that is meant to present a relatively accurate projection of dose-response. By applying mathematical models to observational data, a dose-response curve can be estimated.

The selection of a POD (dose at which a change is "acceptable"), construction of uncertainty factors as well as their application, selection of data from suitable studies as well as their treatment, and assumptions necessary in mathematical modelling are all matters of professional judgement. And there can be different estimates derived from these models. These judgements are not typically made in consultation with the cultural values of populations such as the Anishinaabe and the indigenous communities in the Seychelles or New Zealand. The textbox above outlines the assumptions built into the reference dose model.

Dose-response modeling ensures some empirical integrity but the process makes it less accessible to researchers, affected communities, and policymakers who did not conduct the original analyses. EPA recommends that the process only be attempted by those familiar with statistical modeling, which adds further distance between the risk assessment science and translation to the public. Modelling is useful, but not intuitively understood by the layperson. The level of sophistication also shifts deliberations regarding degrees of uncertainty, and therefore concern, away from communities into the professional realm. This leaves members of vulnerable populations with the burdensome task of speculating on the costs and benefits of following professionally-derived risk communication. The relative subjectivity of "acceptable risk" is suspected but not thoroughly comprehended by the public, and the implications are further lost in translation when recommendations and advice are operationalized for public use.

*Exposure Assessment* A hazard can only impart health effects if an individual is exposed. The exposure assessment phase therefore measures or estimates the magnitude, frequency, and duration of human exposure to an agent in the environment. Future exposure resulting from changes to the environment may also be estimated. An exposure assessment requires characterization of human populations exposed to the agent along with the associated vulnerabilities and uncertainties. An organism may be exposed in different ways through Exposure Routes (inhalation, dermal, ingestion, and injection). These routes are also used to categorize dose-responses since the vector of exposure affects the relevant concentration, the rate and site of absorption, and possibly the type of health effect. In the case of MeHg, the primary route of exposure is via ingestion of fish. Almost all human MeHg exposures comes from the consumption of seafood which is also culturally and nutritionally important throughout the world. As previously discussed, fish consumption is an important aspect of Anishinaabe culture as well as many other North American Indigenous cultures. What types of fish are harvested, from where, and when all influence an exposure assessment for these groups.

In addition to exposure route, the persistence and excretion of a hazardous agent must be considered. Some chemicals are metabolized by the body, undergoing chemical change and activating a biological response in the process. Many hazardous agents, especially water-soluble chemicals, are excreted relatively quickly. Other hazardous chemicals might accumulate in certain tissues. Lipophilic chemicals such as chlorinated hydrocarbons, for example, tend to deposit and accumulate in adipose (fatty) tissue of fish and humans. Chemicals that persist and accumulate in biological media are called Persistent Bioaccumulative Toxic chemicals (PBTs). Many Persistent Organic Pollutants (POPs) are PBTs and therefore bio-magnify throughout the food chain. Biomagnification occurs when organisms at lower trophic levels (lower on the food chain) may exhibit relatively lower body burdens of PBT, however, these chemicals accumulate up trophic levels, resulting in higher body burdens in predator species.

Biomagnification is an important concept to convey with fish consumption advice. For example, salmon and walleye (both culturally-important fish) eat other fish and therefore tend to present a greater exposure risk for PBTs (such as PCBs and MeHg) to humans consuming them than fish from the same area but that are lower on the food chain. Biomagnification and bioaccumulation strongly influence exposure and therefore the level of concern. Much scientific and public attention is paid to MeHg since it is a PBT and a known hazard. This increased level of concern has led to greater attention and relatively improved understanding of these exposures worldwide. Consequently, there is more information to draw from in creating risk advisories of PBTs compared to other classes of chemicals. This relative difference in the states of knowledge for various chemicals is usually not explicitly stated nor quantified in risk advice given to the public. This is regularly seen in biomonitoring programs of fisheries and represents another important consideration when attempting to raise EHL of fish-consuming populations. This is because an important part of EHL is the ability to use sound judgement and adapt advice to one's specific context. The amount of information, and therefore certainty, regarding a risk may influence an individual's judgement to follow, or adapt, presented advice to their situation.

*Risk Characterization* Risk is a function of hazard and exposure as characterized via dose-response. During this final step, the potential for unwanted health outcomes resulting from an exposure are estimated for a given set of circumstances. Risk characterization relies on mathematical estimates of risk to communicate risk assessment findings, often conveying an unwarranted sense of precision while failing to convey the range of scientific opinion. Risk assessment findings are particularly difficult for audiences unfamiliar with the process to comprehend. Effective risk management is impeded when the target population cannot accurately understand: information about who is at risk, how they might be affected, what the severity and reversibility of adverse effects might be, how confident the risk assessors are in their predictions, and other qualitative information that is critical to decision-making.

The characterization of risk conveys the communicators' judgment as to the nature as well as presence or absence of risks. There is a great burden to the communicator to clearly and concisely present information regarding how the risk was assessed, where assumptions and uncertainties still exist, and where policy choices will need to be made. EPA's Risk Characterization Policy calls for the following principles: Transparency, Clarity, Consistency and Reasonableness (U.S.EPA 2000). The principles are outlined in the box below and they are generally representative of how the process is carried out. This chapter addresses best practices and methods for integrating cultural implications and behavioral theories, which are absent from the main principles in this box, to improving EHL through the Risk Assessment process.

**The EPA's Risk Characterization Principles**

**Transparency** – fully and explicitly disclose the risk assessment methods, default assumptions, logic, rationale, extrapolations, uncertainties, and overall strength of each step in the assessment.

**Clarity** – readily understood by readers inside and outside of the risk assessment process. Documents should be concise, free of jargon, and should use understandable tables, graphs, and equations as needed.

**Consistency** – conducted and presented in a manner consistent with EPA policy, and consistent with other risk characterizations of similar scope prepared across programs within the EPA.

**Reasonableness** – based on sound judgment, with methods and assumptions consistent with the current state-of-the-science and conveyed in a manner that is complete and balanced, informative.

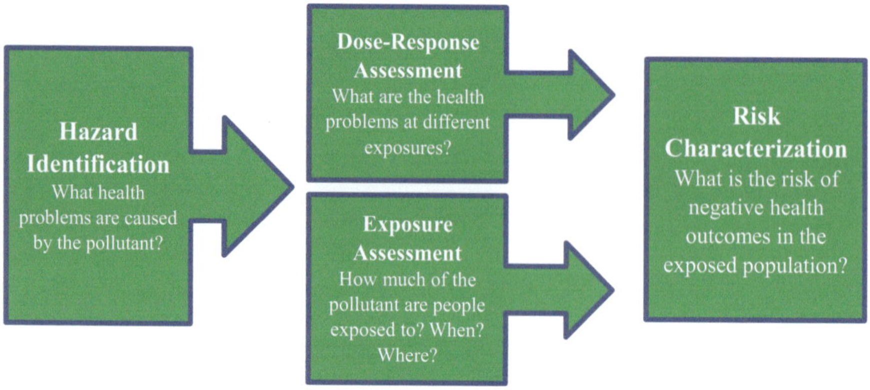


**Fig. 9.1** The steps of human health risk assessment

Risk assessment is a science-driven process leveraging quantitative data to quantify probability of outcome (risk). It is therefore not contextualized within human emotion and perception at the outset of the process. The process is stepwise and logical; thus, in practice, cultural and human contextualization is addressed downstream in the communication and management phases of Risk Characterization (Fig.9.1). The characterization process is recursive; in which concerns identified from the process lead to new research and mitigation options are considered in the refinement of the risk assessment. Risk characterization emphasizes scientific evidence. Less attention is paid to how this evidence would be received or understood by affected populations.

## Health Communication and Social Influence Theories

Once researchers recognize the need for community engagement and trust, an additional challenge of contextualizing relevant health information for stakeholders becomes clear. Communicating persuasive health messages requires more than providing relevant health information. Psychological and social factors influence whether recommendations will be accepted and followed by a given population, regardless of whether appropriate health recommendations are developed in collaboration with stakeholders. This challenge has been well-documented and describes the underlying rationale behind the field of health communication.

Much of the health communication research relevant to this discussion draws upon the Health Belief Model (HBM) (see Chaps. 1 and 3). The HBM developed from the recognition that an individual's perceived risk and their own agency, or self-efficacy, predict their health behaviors. Specifically, valid measures of perceived risk, severity, benefits, barriers, cues to action, and self-efficacy predict likelihood to enact change in health behavior. Perception of risk must be accounted for.

Many interventions are preventive in nature, which means conveying important information about a condition that people are not yet suffering from (i.e. communicating risk). This is fundamentally difficult for people to understand, as any risk communication requires convincing individuals of consequences with which they may have no direct experience. The issue is further complicated when the consumers of risk assessment advice already possess historical perceptions of risk, divorced from or falsely informed by previous health messages or publicly available media reporting (such as public media, internet, and news media outlets) that failed to consider cultural nuances. Such was the case with the Great Lakes Anishinaabe tribes; researchers identified an existing perception of risk regarding local fish contamination. In this instance, an additional challenge to accurately representing risk is modifying inaccurate, but established beliefs about the risk.

The traditional fisheries of the Anishinaabe are important sources of nutrition including fatty acids (including omega-3 and omega-6 polyunsaturated fatty acids (PUFA-3/6)), and selenium (Se). Both PUFA-3 and Se are hypothesized to counteract the negative effects of MeHg exposure in fish consumption, somewhat mitigating the danger of subsistence fishing practices. Recent trends in contaminant reduction (Carlson et al. 2010; Chang et al. 2012; Sadraddini et al. 2011; Salamova et al. 2013) and confirmation of high PUFA-3/6 ratios in Great Lakes fish has rekindled interest in traditional sources of fish nutrients (Julvez et al. 2016; Mozaffarian and Rimm 2006; Turyk et al. 2012). However, culturally-naive advisories promote a prevailing attitude that these fish should not be consumed at levels beyond occasional sport fishing. The first appearance of this notion stems from a 1992 Environmental Protection Agency (EPA) report "Tribes at Risk"(U.S.EPA 1992) which estimated that Anishinaabe subsistence practices (hunting, fishing, and gathering) placed them at greater risk due to MeHg and PCB contamination (U.S.EPA 1992). Though the intent was likely to advocate tribal health, the document signaled the creation of a stigma. When surveyed, Anishinaabe participants reported to eat 420 g/week of fish, but when measured they consumed 78.4 g/week of fish (Dellinger 2004). By comparison, the EPA and United States Department of Agriculture (USDA) both recommend a safe and healthy fish consumption rate of 240 g/week, (U.S.EPA 2004; USDA 2015) highlighting an unintended reaction to the EPA report.

The Anishinaabe tend to overestimate fish consumption, suggesting a perceived importance of the food, congruent with the culture. However, the actual rate of consumption is considerably lower (below advisory recommendations and below levels known to be therapeutic for several health endpoints). Therefore, the risk from consuming fish normally associated with subsistence may be perceived as higher than intended. This phenomenon is supported by studies of the general public, who, when targeted by advisories, tend to reduce fish intake instead of selecting safer fish (Shimshack et al. 2007; Shimshack and Ward 2010) or perceive greater risk than intended (Niederdeppe et al. 2015; Scherer et al. 2008; Shimshack and Ward 2010).

Severity of perceived consequences for contracting a given condition, measured as emotional arousal, is another dimension of the HBM. Individuals must believe they are at risk, and that the potential for harm is severe enough to warrant action.

This explains the documented efficacy of fear appeals in health communication because it taps into the need for individuals to perceive information as personally relevant or highly disconcerting if they are to attend to a message. Fear appeals and other states of emotional arousal such as guilt have been extensively discussed in the persuasion literature (O'Keefe 2002). Most notable and relevant to the present discussion is that fear appeals are most persuasive when the message receiver perceives that the modified behavior could relieve the associated anxiety, specifically regarding exposure to contaminants through fish consumption. These considerations of fear appeal also underscore the importance of the remaining factors in HBM: perceived costs and benefits of the intervention, routes for affecting the problem, and self-efficacy. The individual's perception of the proposed action's efficacy in preventing or ameliorating the condition, as well as the potential costs of action must be considered. In other words, what are the sacrifices an individual must make to enact the behavioral change? The individual must also access strategies to enact the desired change, and finally, the individual must perceive him or herself as capable of enacting that change. These factors pertain to the individual's perception and beliefs regarding the health intervention and predict behavioral change. However, in the example regarding excessive reduction in Anishinaabe fish consumption, the perceived severity was exaggerated thus contributing to an undesirable behavioral change, a reminder that additional contextual factors must be considered.

The Health Belief Model established the importance of psychological and contextual factors of influence when a health message is interpreted and applied towards lasting or temporary behavioral change. It was one of the first attempts to account for factors other than the simple threat of negative outcomes in public health intervention. Notably, the model accounts for moderating variables such as demographics (ethnicity, age, sex etc.), psycho-social variables (personality, social class, peer pressure), and structural variables (knowledge of/contact with disease). Such factors naturally vary with culture and the culturally-bound environment in which an individual is socialized. Therefore, HBM accounts for contextual factors necessary in understanding the challenges of improving EHL in cultures such as the Anishinaabe. The cultural context influences, not only the relevance of consumption habits, but also perceptions of risk. Other existing social influence theories delve deeper into the processes by which external stimuli affect human decisions. Specifically, theories concerning the adaptation of communicated persuasive messages can further inform the cultural tailoring of risk messages. Two such theories are the Elaboration Likelihood Model (ELM) (Petty and Cacioppo 1986) and Unimodel (Kruglanski and Thompson 1999).

Richard Petty and John Cacioppo (1986) proposed ELM in terms of two routes, central and peripheral, through which message receivers might be persuaded. They described the central route as when a message receiver is compelled to think logically and extensively about the message. Individuals are more likely to do so when the message has high personal relevance and the evidence is robust. The peripheral route was proposed to describe how incidental cues are more salient in the persuasive process: attractiveness, perceived motives, credibility of speaker etc. Factors proposed to influence which route is followed involve motivation, or attitudes towards the message, and ability, or capacity for depth of engaging the message. Petty and Cacioppo (1986) argue that the central route is more likely to spur lasting change, while the peripheral route might incite more temporary changes. ELM therefore stresses the importance of assessing the likelihood of the receiver to be motivated to think about the message. Therefore, the peripheral or central route of persuasion should be pursued based upon motivation to elaborate.

In the case of fish consumption and the Anishinaabe, the cultural heritage of subsistence as well as the history of legal battles to defend subsistence practices provides that motivation to elaborate upon the safe usage of fish as a subsistence resource. In contrast to this example, another study found that Hispanic males in the U.S. were more likely to disengage from a health messaging program if their belief in destiny was high (Cameron et al. 2017). With the Anishinaabe, the incidental cues of fish consumption messaging can either discourage or enhance the central route in which the logical information (i.e. information generated using the risk assessment process) is employed to improve EHL. For example, non-tribal governmental advisories are viewed with suspicion whereas advisories that were clearly generated by and for Native Americans are viewed more favorably. This is particularly true when Native American-generated fish advisories use aesthetics that match the local culture.

While ELM is arguably the most influential process model theory of persuasion, there are alternatives. For instance, Chaiken, Liberman, and Eagly (originally in 1989) described a similar, dual-process model called the Heuristic Systematic Model (HSM) (Chaiken et al. 2012). HSM stresses that information processing will take the route of least resistance, and like ELM it conceives two routes: systematic (comparable to the central route in ELM) and the heuristic route (comparable to the peripheral). Thus, HSM stresses considerations such as accessibility to heuristic cues in governing how information is processed and pro-attitudinal messages might be more persuasive. Additionally, HSM proposes that both routes can co-occur, allowing for consideration of biases, attenuation, and additivity of each route's effects. This matches observations with Anishinaabe fish consumption advice in which peripheral/heuristic route (i.e. cultural aesthetics and governmental origin of messaging) add weight to the systematic/central route (i.e. quantitative risk estimates).

Given the similarities between ELM and HSM a unifying model may be most helpful in guiding non-theoretical, applied work. The Unimodel draws upon both of the dual-process models (ELM and HSM) (Kruglanski and Thompson 1999). The Unimodel was proposed to explain further the interplay between heuristic cues and message content. Kruglanski and Thompson (1999) argued that it is not necessary to distinguish the two routes of persuasion, as the ELM and HSM do. Rather, they viewed both paths simply as different species of persuasive evidence or proof. Consequently, the cause of persuasion is not the cue or message content, but the persuasive process itself. In Unimodel, both heuristic cues and message contents follow a syllogistic pattern akin to formal logic. Thus, the message receiver deduces a conclusion from separate propositions that are asserted (in this case accepted) as true. Therefore, no functional difference in the routes of persuasion exists, but there is contextual difference as to which type of evidence is sufficient for persuasion. They implicated contextual characteristics such as length and complexity of information, the order of arguments in the message, and relevance to the message receiver as important, rather than indicating two separate routes of persuasion. The Unimodel of persuasion thus works to disentangle the process of persuasion from the validity of evidence (systemic/central persuasion route). Most importantly Kruglanski and Thompson (1999) suggest that this is how human cognition works and that distinguishing between heuristic cues and message content needlessly confuses the process of persuasion. In the Unimodel, difficult-to-process or complicated information will matter more only when the end-user's motivation and ability to interpret that information is high, while easy to process information (whether cue and heuristic or simple substantial evidence) matter more when motivation or ability are low. Both routes are valuable since individuals within a culture may share certain values and identify with similar aesthetics, but their cognitive aptitudes and other innate motivation levels will naturally vary from person to person.

Despite their differences, similar lessons can be applied from each these models of persuasion. First, context influences persuasion, which is of particular interest to those seeking to craft culturally-relevant messages to improve EHL. The most influential messages may be those which request very little change from or fit comfortably within the beliefs of message receivers. This would distinguish advice that reaffirms Anishinaabe subsistence culture from advice that implicitly ignores it. The Unimodel or HSM, in such a case, would suggest that even simple heuristic cues of credibility might be persuasive, while the ELM might argue that more familiar arguments would be more likely to be elaborated in the hearer's thoughts. Either way, there are routes of persuasion that are comparably less strenuous and more efficient than others. Therefore, a persuasive agent, such as an intervention to promote EHL, should work in consideration of the expectations of the audience, not the expectations of the message sender. This can be accomplished without compromising the underlying logic used to generate quantitative risk estimates.

## Motivated Reasoning and Bias

The lessons of social influence theory also underscore the importance of bias when a message is interpreted. When faced with new information and persuasive messages, humans employ motivated reasoning (Dieckmann et al. 2017; Kunda 1990; Scurich and Shniderman 2014) to process and filter that information. Motivated reasoning describes the unconscious tendency to assimilate information in a manner biased towards reaching an outcome influenced by *a priori* conditions such as an individual's culture (further discussed below) or innate personality traits. This form of bias affects the way in which information is processed. Information that accords with past experiences, personality traits, and cultural preferences is subconsciously weighted as credible while information that does not promote these preferences is weighed as less relevant or credible. The phenomenon is related to cognitive dissonance in which mental stress, or discomfort, occurs through the holding two or more contradicting beliefs. An aversion to this stress can manifest itself as bias. Not only does the process lead to potentially erroneous assumptions, the subconscious nature of this bias may lead an individual to believe their conclusion is rational. Motivated reasoning must be considered during the risk communication step of environmental risk assessments. Since culture informs these factors, the culture of the target population must also be carefully considered.

These biases can completely overshadow the procedural and scientific rigor described previously, especially when the benefits of the scientific process and rigorous investigation are not well-understood by the intended beneficiaries. In many cases, such as with the Anishinaabe, a history of distrust towards non-Native American sources of information strongly influences the acceptability of a message. Many quantitative predictions and recommendations regarding the environment, sociopolitical events, economics, and health are expressed in forms that also reflect substantial uncertainty (i.e. a range of predicted values). The best estimate within a range of values depends on how the range was generated. Despite a potential reluctance to admit uncertainty of one's predictions, it is important to communicate some representation of the known, unknown, or unknowable so that decision makers and/or stakeholders can appropriately weigh the pros and cons of different actions. How the uncertainty in these forecasts is communicated to the lay public may greatly affect perceptions of risk and subsequent policy decisions. The bias of motivated reasoning affects perceptions, and therefore, comprehension of quantitative data including uncertainty ranges.

**Biased Interpretations**

All populations exhibit patterns of bias. Through discussions and focus groups across 13 years, the Anishinaabe have presented two basic patterns of bias towards interpretation fish consumption advisories: (1) risk aversion, in which the perceived level of contamination is greater than exposure evaluations warrant; and (2) dismissal, in which the calculated risks are perceived as trivial relative to the cultural, economic, and nutritional values (Dellinger et al. 2017). Given previous data suggesting the Anishinaabe under-consume fish on average, the latter is a small, but important, minority.

As seen with the Human Health Risk Assessment process, conclusions drawn from science and interdisciplinary formalized processes contain uncertainties. Therefore, the inferences (i.e. specific recommendations for risk management) drawn from these conclusions require interpretations based on human judgement. Among Anishinaabe, this judgement is influenced by their beliefs and values (i.e. subsistence food and food sovereignty) which are a product of personal preferences and culture. It is a matter of opinion, not fact, that twice as many children performing below the 5th percentile (the POD for the EPA MeHg reference dose) in neurological tests is the threshold of acceptable MeHg exposure (Rice et al. 2000).

However, it is a benchmark based on methodical, educated judgement. The recommendations generated by risk assessments tend to adhere to a precautionary principle (better safe than sorry). The precautionary attitude may translate well under certain circumstances but can be distorted when extrapolated from the risk assessment and characterization process to human behavior. For example, the previous description of the EPA's dose response calculation for MeHg should demonstrate that those recommendations are quite protective, or perhaps over-protective. Nevertheless, risk-averse adults are often appalled to realize that, using the established dose response level to protect developing fetuses, certain walleye should not be consumed more than twice a month. Despite the previous example of spring spearfishing, many adults do not eat walleye twice a month, every month, when planning to become pregnant. Nevertheless, aversion to this advice, combined with the cultural significance of walleye to Great Lakes Native American communities, can lead to unintended human behaviors. These include the consequences of the previously described biases: (1) individuals may avoid more fish than is necessary (or healthy) and (2) individuals may ignore the advice that appears unreasonable and overprotective.

Within the context of social influence theories: the importance of fishing and fish consumption to the Anishinaabe may limit the persuasiveness of the often difficult-to-process advisory message content. Although the message is highly relevant, this does not necessarily increase the likelihood of elaboration in the message receiver. Rather, adequate elaboration upon message content, and thus accurate interpretation, requires relatively high levels of motivation and cognitive ability. As such, without appropriately situating the message within the existing cultural framework of the receiver, accurate interpretation of both information and relevance is unlikely. The unintended consequences may be that simple cues (e.g. fish can be dangerous) have the most impact, leading not to deeper consideration of the arguments, but either exaggerated aversion or dismissal of the message due to existing biases. For this reason, risk communication requires a deep understanding of risk assessment protocols as well as social influence theories, which include cultural implications, to adequately protect human health as well as respect values that vary between cultures and individuals.

## The Need to Understand Cultural Context

The knowledge generated from environmental health research serves to inform people from the scientific community as well as the public. The culture and terminology of the scientific community differs from that of the public. Further variation in culture is readily apparent among various communities and microcultures (i.e. urban and rural, European and American, Native American and Non-Native, etc.). It is general knowledge that different cultures interpret information differently, but it is not safe to assume that cultural boundaries will follow traditional patterns. For instance, recent intercultural communication findings have demonstrated that socio-political factors, such as urbanization and democratization, are stronger predictors of the traditional cultural dimensions of individualism and collectivism than national boundaries (Kim et al. 2016; Kyriacou 2016). Cultural contextualization of environmental risk assessment messaging therefore requires two important considerations for translation: (1) how would a layman interpret the message and (2) how would an individual from a given culture interpret this message? Further complicating the matter is the level of disagreement between various professionals who debate the degree of concern for key variables within a risk analysis.

The uncertainties, and resulting professional disagreements, regarding any environmental analysis push environmental education into a nuanced art-like realm. Indeed, those trained in art theory and methodologies, such as composition, color theory, and art history, may in some ways be better prepared to communicate across disciplines and populations. It may not be immediately obvious to scientists nor the general public that much of painting, illustration, sculpture, drawing and the other arts are formalized processes with technical training at their foundations. The end-product, which ideally invokes emotion and a desired reaction, obscures the underlying technical prowess and knowledge. Tapping into the approaches used by artists may therefore benefit those attempting to communicate scientific findings to audiences with low scientific or basic literacy or cultural constraints.

To improve Environmental Health Literacy, research must be translated into messaging that is profound, meaningful, readily understandable, and, if possible, intuitive. Intuitive translations of data may require the use of aesthetics, ergonomics, and technology that allows for precise calculation of relevant metrics without burdening the user. When observing impressionist paintings such as Monet's "Woman With a Parasol – Madame Monet and Her Son" one can nearly feel the sun shining through the image. This is possible precisely because Monet built the composition employing an understanding of what colors would hit the retina on a clear day in a lush meadow. The viewer is not required to individually process each color or understand how it was applied to interpret the message, but the composition is nonetheless technical in nature. This blend of technical knowledge and impressionism to achieve impact must be achieved in the translation of science. As argued below, the eye (or ear) of an artist can be employed to invoke valuable context to different cultures.

## Methods and Lessons from the Field: A Case Study of Anishinaabe Fish Consumption Advisories

This section focuses on an example of how messaging to improve EHL was successfully employed with Native American tribes in the Upper Great Lakes region of the United States. The Chippewa Ottawa Resource Authority (CORA) in Sault Ste. Marie, Michigan, has been monitoring contaminant concentrations in the fillet portions of lake trout (*Salvelinus namaycush*) and lake whitefish (*Coregonus clupeaformus*) since 1991 from the waters of lakes Superior, Huron, and Michigan. CORA represents the fisheries interests of five Ojibwa and Ottawa tribes (collectively called Anishinaabe) whose ancestors ceded lands through the 1836 Treaty of Washington but retained the rights to hunt and fish on those lands and waters. The contaminant monitoring program was initiated to address the emerging threat of PBT contamination in subsistence species. The CORA-regulated commercial fishery annually harvests about 2 million kg of lake whitefish and 0.5 million kg of lake trout from the three upper Great Lakes, and many Native American families derive income and subsistence from the fishery (Taylor and Ferreri 2000). Despite the cultural history (Quaife 1947) and known benefits of eating fish (Mozaffarian and Rimm 2006; Mozaffarian and Wu 2011, 2012; Turyk et al. 2012), the Anishinaabe report to consume only one third of the recommended daily fish intake (Dellinger 2004; USDA 2015). As previously described, local fish consumption advisories were developed by scientists without consideration to the cultural context of the tribes. An aversion to the culturally-insensitive state and federal advisories is often suggested as a factor influencing the low consumption rates. The associated loss of traditional food consumption then creates a nutritional health disparity for indigenous North Americans (Ho et al. 2008; Schell 2012; Whiting and Mackenzie 1998). The decreased access to fishery and other traditional foods has coincided with a general trend of decreased dietary quality, and increased obesity, and cardiovascular disease among Native Americans (GLITEC 2011; Story et al. 2003).

Improved efforts to share fish contaminant and nutrition data with the Anishinaabe communities are important priorities for tribal governments. In 2003 the Agency for Toxic Substances and Disease Registry (ATSDR) provided funding to the Inter-Tribal Council of Michigan (ITCM) who successfully generated culturally-relevant fish consumption advisories in collaboration with researchers from the University of Wisconsin-Milwaukee (UWM), Medical College of Wisconsin (MCW), and CORA (CORA 2006; Dellinger et al. 2006, 2014; Moths et al. 2013). These efforts included a brochure and a documentary titled: "Nindamwaa Giigoon: An Anishnaabe Guide to Eating Fish" (Rostenkowski et al. 2007). Both the brochure and documentary engaged the target population by intermingling risk messaging with culturally-relevant cues. Furthermore, these materials were produced in collaboration with representatives from the tribal populations. To generate risk estimates for this documentary, investigators at the Great Lakes Intertribal Fish and Wildlife Commission (GLIFWC) collaborated with their academic partners to estimate Anishnaabe self-reported consumption rates (Dellinger 2004). These rates were then compared to existing data gathered by ITFAP from the 1836 treaty waters (Dellinger et al. 2006; Dellinger 2007). The results of this risk assessment were used to generate illustrations and recommendations for "Nindamwaa Giigoon" as well as a brochure (both currently distributed by CORA). The materials were designed to efficiently conceptualize the safest fish to eat without requiring the viewer to cross-reference tables of contaminant data as is standard practice in local advisories.

The brochure was evaluated by ITCM in 2006 via focus groups during site-visits to three of the five Michigan Upper-Peninsula tribes. All of the participants (n = 100) stated that the concepts presented in the brochure were helpful and 57% reported that it changed their behavior (Dellinger et al. 2006; Dellinger 2007). In 2007, two hundred "Nindamwa Giigoon" videos were distributed at grocery stores, restaurants, and health clinics on or near reservations in the ITCM service area. Participants completed a "pre-viewing" survey (N = 78) and a "post-viewing" survey (N = 56). About half (51%) of the respondents said they would consider eating different species of fish. The majority (88%) of respondents reported that the video helped them understand which pollutants are problematic (Dellinger et al. 2008). The culturally-tailored model for developing these interventions had demonstrated its value in the focus groups and questionnaires. The video and brochure were successful; however, the team recognized a need and an opportunity to leverage new technology to ensure lasting change and to update the data as new information became available. It is important to provide the Anishinaabe public with up-to-date fish contaminant data from their traditional fisheries as these data are collected by CORA on an annual basis. In response to these opportunities, the team developed a mobile app, "Gigiigoo'inaan", under National Institute for Environmental Health Sciences (NIEHS) funding to integrate risk and benefit concepts using Anishinaabe data (the data collected by CORA) (Dellinger et al. 2015–2017). The app reports risk and benefit data from the tribal fisheries using culturally-tailored imagery and messaging in a personalized, interactive format (Fig. 9.2).

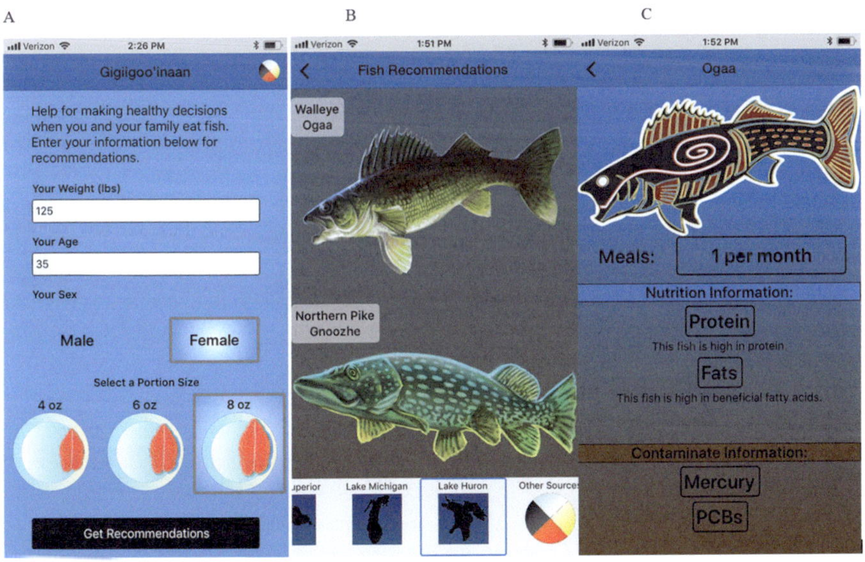


**Fig. 9.2** Screen shots from Gigiigoo'inaan: main data entry page (**a**), main results page (**b**), and advisory details page (**c**). Orginal art by: Matthew Dellinger

## Development of Gigiigoo'inaan

Gigiigoo'inaan means "our fish" in Ojibwe and thus the name is a cultural cue that the software is meant for Anishinaabe tribal members. Gigiigoo'inaan was programmed by the UWM Mobile Innovation Lab to the specifications of the research partners at MCW, ITCMI, and CORA. The software was initially built using a popular scripting language (PHP 5.6.6) that is used to develop websites, native apps (downloaded to a device), and web apps (run through web browser). The software works on both Android™ and IOS™ allowing for on-demand calculation of safe consumption rates for thirteen species of fish (lake trout, whitefish, walleye, yellow perch, smelt, herrring, bass, burbot, northern pike, farmed atlantic salmon, tuna, cod, and tilapia) based on user-input data. The user-input includes age, weight and sex as the modifying factors for individualized recommendations. The introductory page states that the app helps users to choose which fish are safest to eat. On the first results page (after personalization data is entered) fish are ranked in order of the most to least recommended meals for consumption. The input data is applied to reference dose to generate risk quantifications, in this case meal frequencies that fall below specified risk levels. These frequencies are used by the software to generate and present personalized recommendations.

When selected from the initial output page a given fish displays specific consumption advice using safe consumption ranges based on the user input. These fish can also be browsed by water body (Lakes: Huron, Michigan, Superior, and Other Sources) to represent the options within the fisheries of the Anishinaabe tribes. Distinct from conventional advisories, the software references the user data (bodyweight and selection of illustrated 4 oz., 6 oz., and 8 oz. serving sizes) to calculate a ratio of fish grams/meal/bodyweight which can be linked to pre-determined safe consumption ranges per fish using the appropriate guidelines determined by CORA. Using the software will allow users to generate personalized recommendations without the need to cross-reference multiple tables, check caveats, or memorize acronyms. The contaminant and nutrition data are therefore presented in a user-friendly format, reducing the burden of interpretation and increasing likelihood of elaboration. The risk assessment methodology (i.e. selection of dose-response metrics) to generate the recommendations were selected in dialogue with ITCM and CORA tribal members. The recommendations are then generated using a database of fish contaminants primarily collected by CORA thereby increasing the credibility to the tribal end-users, who tend to trust their own governmental agencies over the federal, local, and state agencies.

As covered earlier in this chapter, improving EHL requires a navigation of overlapping domains to convey a message that could be warped by human perceptions. Therefore, it is necessary to tailor all risk assessment recommendations to the cultural context of the target population whilst also introducing aesthetic approaches that are familiar and resonate with that audience. To ensure the cultural appropriateness of the illustrations used in the app, it features custom-made digital images made in the Woodland Style founded by the Anishinaabe artist Norval Morrisseau

(Fig. 9.2c). These familiar images allay concerns that the advisory was developed outside the culture by unsympathetic scientists. By including Woodland styled images of fish in Gigiigoo'inaan, the creators were signaling to the Anishinaabe that they are either familiar with, or part of, Northern Great Lakes Native American culture.

Gigiigoo'inaan modernizes culturally-tailored advisories by integrating a community-engaged design approach with software development and risk assessment. Consistent with the original lessons of HBM, work on the video and brochure were based on important perceptions of preventive messaging that must be considered for them to be adopted. This was especially exemplified by the reluctance to adopt advice that appeared stigmatizing to Anishinaabe culture. The impact of the tribally-informed advisories is partially explained by examining the social influence theories that describe the development of opinions and attitudes, whether dual-process (as in ELM and HSM) or single process (as in the Unimodel) (Kruglanski and Thompson 1999). Focus group responses confirmed that culturally congruent messaging with the Anishinaabe leverages elements from these models: information is provided in a contextually relevant format, increasing likelihood of message consideration, and the message provides sufficient levels of technical detail to allow for lasting behavioral change. The software, Gigiigoo'inaan, was therefore designed to influence these perceptions by providing cues (such as culturally-tailored artwork [Fig. 9.2c]) but also by presenting recommendations based on community-driven data collection founded on a rigorous body of science. Scientific authority and cultural sensitivity are therefore employed simultaneously, in an accessible and non-burdensome message medium.

Gigiigoo'inaan recommendations can be generated as needed, on any mobile device or computer and be used in settings such as supermarkets, fishing trips, and at restaurants. It also includes specific advice on the balance between nutrition and contaminant risk, providing more motivation to heed the advice. Aesthetic and cultural cues situate the information (as in the peripheral route of dual process models), and the potential for engagement and consideration with not only the behavioral change, but also the scientific data, is maximized (as in the central route). In summary; the software is dynamic enough to pursue behavior change using mechanisms posited by multiple social influence theories.

The Gigiigoo'inaan research team was appropriately comprised of academic and Anishinaabe collaborators, which offered necessary cultural insights to the project development. However, the systematic integration of community feedback was required so that the local culture could influence the research. Further credibility of the advice presented by Gigioo'inaan was due to the development process which followed a model of community feedback and input from tribal scientists and policy makers (as illustrated in Fig. 9.3). Much like the planning and research phase of a risk assessment, the project was founded on previous experience and other published works on fish consumption and Anishinaabe environmental health concerns including the research team's 13 years of experience working with tribal organizations on health research and environmental issues. Once the concept of the mobile app was developed in discussion with colleagues who are both health professionals

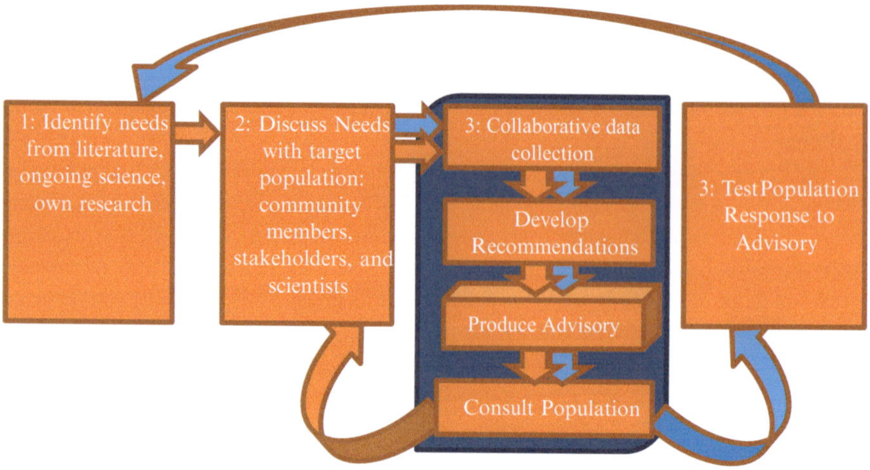


**Fig. 9.3** Conceptual model outlining community-engaged risk communication process, the goal of which is to improve environmental health literacy. The community and academic partners are consulted throughout. The messaging is developed iteratively based on these inputs and updated using data that was preferably gathered with participation and/or cooperation from members of the target population (è.g. contaminant data from fish caught by local fishermen). Orange arrows represent the first pass, which has already occurred, blue arrows represent future iterations of the process

and tribal member, the project commenced in cycles. Programming cycles proceeded quarterly in which CORA and ITCMI Anishinaabe scientists consulted on the design and format providing trouble-shooting, and recommendations related to style and usability, among other recommendations. It is anticipated that annually the software will be piloted to multiple small groups of Anishinaabe fish consumers for their feedback. The feedback mechanisms include focus group discussion, anonymous response data from a questionnaire, and, more recently, a questionnaire recording specific fish choices the participants make while using the app. Employment of Anishinaabe facilitators ensures that focus groups (termed talking circles by the Anishinaabe) provide minimally-inhibited dialogue between the community members and the research team. Finally, the fish contaminant data used to generate the recommendations are updated annually based on CORA's spring/summer fish sampling.

The personalization of the recommendations serves the purpose of allowing users to generate advice that matches their fishing culture. In the past, generalized advisories provided simplified consumption scenarios and were repeatedly disregarded as either overprotective, insensitive to the fishing culture, and/or outdated. Since Gigiigoo'inaan takes into account the users self-reported weight and sensitivity category (reproductive female, child, or all others) many users found they were able to remain within the protective guidelines of ATSDR and EPA whilst eating more fish than they previously perceived to be safe. Furthermore, the Anishinaabe people preferred the advice to come from the fish collected by their tribal scientists en lieu of state-run programs (Dellinger 2007; Dellinger et al. 2017).

## Integrating Feedback and Synthesizing Best Practices

Gigiigoo'inaan's development was iterative. A pilot version was created to meet challenges identified in the other CORA advisories. Upon testing, further challenges were identified. This first iteration tested by Anishinaabe end users attempted to provide quantitative data reflecting the benefits of eating fish relative to the risks. It contained a novel risk/benefit estimation using PUFA-3 and MeHg data presented by Ginsberg et al. (Ginsberg and Toal 2009; Ginsberg et al. 2015). The estimates consider adult cardiovascular disease risk and infant neurodevelopment risk. Like the methods that federal agencies use to derive reference doses for toxic agents, the Ginsberg et al. (2009, 2015) method uses published dose-response data to estimate predicted changes in risks to identical health endpoints (i.e. increased risk of heart disease mortality and diminished neurological performance associated with MeHg compared to the decreased risk from the associated PUFA-3). Although the method is not meant as a commentary on the biological mechanisms of MeHg toxicity, it is a useful indicator of health tradeoffs for consuming fish which are the almost exclusive exposure route for both MeHg and PUFA-3. The Ginsberg et al. (2009, 2015) calculations were therefore used to rank fish by benefit in the primary output page of Gigiigoo'inaan using updated data in the hopes of presenting a positive message at the outset of the advisory.

Focus group discussions during the iterative design of Gigiigoo'inaan, summarized by the Anishinaabe facilitators, revealed important concerns and themes for refining the advisory. When asked about fish consumption, two concerns were highlighted: (1) that fish consumption was a healthy and culturally important practice and (2) exposing themselves and their families to chemical contaminants that might pose health risks. Some resented any restrictive recommendations saying, "The app is too negative, it's going to scare people away from eating fish." Another participant said, "Warnings can turn people off from consuming fish." Nevertheless, in a concurrent questionnaire of 101 Anishinaabe app users, most (87.1%,) of participants stated they would feel more confident feeding fish to their families if they had regular access to Gigiigoo'inaan. As indicated at the introduction of this chapter, stigmatizing fish consumption via risk assessment was another prominent concern in discussion. These sentiments highlight the need to communicate the benefits of local fish species along with the risks. To accommodate this the partnering fish monitoring program at CORA assessed the nutritional value of their sampled fish. These data included PUFA-3 and selenium (an important micronutrient in fish). The benefit data was placed prominently in the details page for each fish along with personalized meal recommendations (Fig. 9.2c).

Many Anishinaabe app-testers appreciated the culturally-tailored components of the Gigiigoo'inaan including the use of Ojibwe fish names and woodland art to represent the fish (Fig. 9.2b). During the questionnaires, most (80.4%) of 101 the participants agreed that the app was useful, culturally appropriate (87.6%) and helped them identify the best fish to eat (87.1%). The focus group discussions, as well as the ARS data, reiterated previous findings (Dellinger et al. 2006, 2008) that the Anishinaabe are culturally motivated to consume fish. Recommendations along these lines included integrating more cultural features into the app such as use of Anishinaabe Great Lake names, fish preparation and cooking instructions, cultural stories, and access to details on treaty rights. Some requested further traditional teachings to complement the cultural artwork. Despite the positive portrayal of local fishing culture, many respondents also expressed concerns about safely eating fish and feeding their families, particularly for vulnerable groups such as women and children. Users therefore appreciated the ability to change serving sizes and weights to calculate safe consumption for their whole family. Suggestions for future software versions include more data summaries of PBTs in lakes and fish, further access to the data and calculations, and explanations of the recommendations. Respondents expressed an interest to better understand the meaning of contaminant levels. One participant requested, "…how does mercury and contaminants affect you – should I really be worried? How will it really affect me?" These results suggest that Gigiigoo'inaan users were motivated to elaborate upon the message provided by quantitative risk estimates and other scientific data, consistent with the central route of persuasion in ELM. A further advantage of using interactive software and an iterative design process, is the ability to refine and add these features in response to feedback as the tool is used.

Another challenge with developing the fish advisories was the inconsistency between noting the benefits of fish consumption vs. the risk from PCB contaminants. The iterative design and community feedback approach demonstrated its flexibility when testing an app feature to present risk/benefit quantifications to the user. CORA data has consistently demonstrated that tribal harvests in the area offer considerable health benefits from PUFA-3, especially when compared to popular alternatives such as tuna, tilapia, swordfish, and processed fish sticks (Dellinger and Ripley 2016; Dellinger et al. 2017). This presentation of fish healthfulness was met with mixed reactions both from the focus groups and from the research team during development. Although the desire for positive messaging has been unanimous from Anishinaabe focus groups throughout development, it was clear that the primary metric of interest was: "how much can I safely eat?" Thus, many were confused by the discordance between the app reporting a highly beneficial fish (i.e. lake trout) and the advice that it should be eaten less often than a fish with lower PUFA-3 s such as whitefish (which is also lower in PCBs). It also presents a conundrum in which advisories lead to the primacy of consumption limits based on risk assessment. Suggesting consumption limits is generally viewed negatively by the Anishinaabe as illustrated from the focus group reports and surveys.

Presenting benefits as Hg vs PUFA-3 can mislead the consumer since fish with relatively low PUFA-3 and MeHg (such as perch) could seem "less beneficial" which would be an overgeneralization. For example, the PUFA-3 benefit coefficient for neurodevelopment in Ginsberg et al. (2015) is calibrated to match observational studies that demonstrate health benefits from consuming fish meals. The positive health associations found in the epidemiological literature do not necessarily reflect an exclusive PUFA-3 effect. Consumption of species like perch and walleye may impart more benefits that the model accounts for. It is therefore crucial that advisories seeking to convey a fair representation of fish consumption benefits include other, thoughtfully placed, information. This may include PUFA-3/6 ratios, selenium, and traditional/cultural knowledge regarding the cultural and historical value of each fish. The following version of the app therefore included more species of local fish and ranked them by their highest recommended consumption rate based on PBTs and reference doses. Also, a more sensitive calculation for upper-limit PBT advice to accommodate women who may become pregnant and small children was added. The approval rating of the second version subsequently increased from 75% to 87.6% between these iterations. Advice was also carefully crafted to clearly emphasize advice that promotes consumption of fish to gain the nutritional benefits as well as exercise cultural practices whilst minimizing PBT exposure. At the most recent evaluation, ideas for future iterations included: dietary tracking and budgeting, further personalization of recommendations based on fish length, as well as fish cooking recipes and traditional knowledge from Anishinaabe culture. As before, these features would be discussed with Anishinaabe stakeholders and added to Gigiigoo'inaan via the process illustrated in Fig. 9.3.

Other Great Lakes tribal consortia work with academic partnerships to culturally-tailor the relevant risk assessment science to their service areas. For example, fish-consuming tribes who spear walleye in the spring months may respond better to MeHg advice based on specific lakes with an emphasis on smaller sizes and caution for young children and women rather than generalized species advice for year-round consumption. The Great Lakes Indian Fish and Wildlife Commission (GLIFWC) oversee the fishing activities of the tribes to the west of CORA. GLIFWC has an extensive program to inform Anishinaabe tribal members from northern Wisconsin, Michigan, and Minnesota who harvest and consume walleye about the health risks of consuming these fish, and to encourage harvest and consumption practices that reduce exposure to MeHg. GLIFWC has performed probabilistic exposure analysis of MeHg among tribal members who consume walleye to fine-tune their advisories. These advisories are color-coded maps of lakes that the Anishinaabe harvest walleye from (DeWeese et al. 2009). The potential for median MeHg exposure to exceed the mercury reference dose among children and women of child-bearing age who consume only larger fish from red-coded lakes was deemed a concern. GLIFWC therefore developed walleye consumption advice for these individuals: to not eat walleye larger than 51 cm (20 inches), to not eat walleye from lakes color-coded red, and to consume smaller walleye from lakes other than those color-coded red. The maps also encourage tribal members other than children and women of child-bearing age to eat fewer meals if they choose fish greater than 51 cm from any lake (Foran et al. 2010). The GLIFWC walleye maps exemplify the need for contextualization as cultural norms shift, even within a common overarching culture such as the Anishinaabe. Anishinaabe who live further from the Great Lakes (Northern WI in the case of GLIFWC) place more emphasis on walleye consumption during their spring spear fishing season of inland lakes than do the CORA tribes who are more focused on consuming fish from the Great Lakes. The maps guiding consumption of walleye from certain areas and of smaller sizes was therefore warranted, whereas the CORA Anishinaabe tribes required a tool to discern between Great Lakes fish species.

## Conclusions

Environmental health science provides valuable information that is relevant to public health policies, interventions, and recommendations. EHL is a crucial part of the process to operationalize scientific findings that can improve health outcomes. Figure 9.4 displays a Venn diagram in which HHRA, policy, and science intersect with one another to create overlapping domains. These domains are not labeled in the figure since they could be described by many names depending on the context. One might say, however, that environmental health literacy is required to understand the overlap in these domains and subsequently translate that knowledge into health-altering behavior. The histories of the Health Belief Model, subsequent behavioral theories, and risk assessment processes demonstrate benchmarks of progress towards effective communication between science and the lay public. This communication is essential to building improved EHL. Environmental hazards to health range from the relatively straight-forward (Minamata disease) to the relatively abstract (low-to moderate neurological effects of MeHg). This chapter emphasized the importance of using an iterative feedback process to culturally contextualize risk messaging when translating these abstractions into health-promoting behavior. Furthermore, since most EHL initiatives are preventive rather than therapeutic,

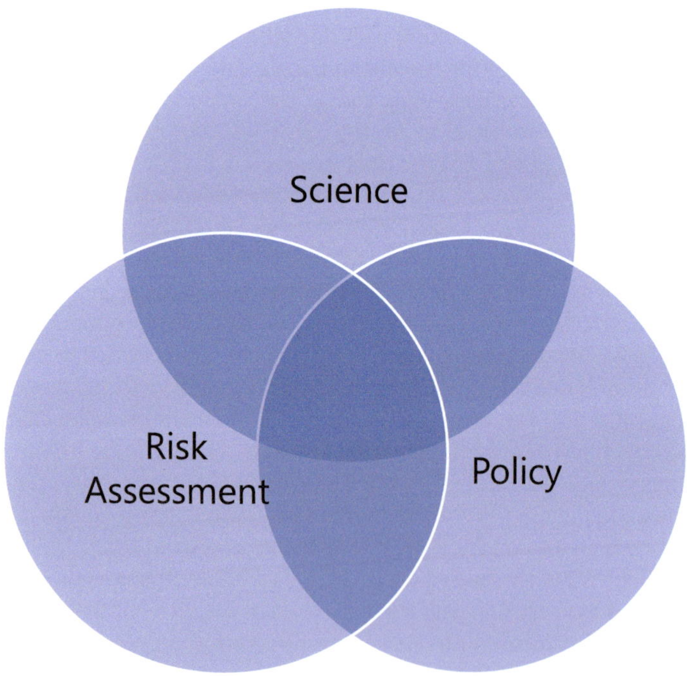


**Fig. 9.4** A Venn diagram describing the relationship between human health risk assessment, science, and policy

showcasing the benefits of effective communication will remain a challenge. Cultural context can be a barrier to EHL if not properly understood and integrated into the process. By employing the strategies described in this chapter, culture switches from a barrier to an empowering asset.

By invoking multiple disciplines in behavioral health, health communication, and risk assessment, researchers can tailor the fruits of their work to benefit diverse populations. As the process moves from risk assessment into risk characterization, assumptions and quantifications of uncertainty must be shared with and explained to relevant stakeholders as much as possible. Involving risk assessors who are integrated with the target culture and/or are part of that culture will increase source credibility and acceptance of messaging. Consistent with HBM and theories of social influence and persuasion, cultural and aesthetic cues should not be overlooked as efficient methods for motivating the audience to synthesize information and elaborate on the logic used to guide recommendations. Though these lessons have been applied to fish consumption advice in Anishinaabe populations, they are broadly applicable to other scenarios.

The challenges associated with EHL are not insurmountable. However, they represent significant obstacles to the desired outcomes and central goal of research and the academy in general. The pursuit of knowledge and spirit of inquiry is not merely for the benefit of science, but is also intended to benefit society. This would require that specialists consider the intended audience in communicating their findings. Yet, misinterpretations of scientific findings abound in public discourse. This may occur due to a limited understanding of scientific terminology and methodology, the innate limitations of interpreting aggregated findings, and the human factors described in this chapter (and Chap. 1). Improvement of EHL requires a recursive, iterative process by which end-user feedback influences design and risk assessment assumptions. Cultural compatibility is ensured by thoughtfully placed aesthetic cues as well as designing advice that is personalized and responsive. Care must be taken to avoid assumptions or advice metrics that may set a stigmatizing tone to a given culture. When this occurs, a cultural group will lose trust in the advice and develop cognitive biases towards the science producing the advice. These biases can become historically entrenched and result in unintended, counterproductive behavior patterns. This chapter has presented some examples in which these pitfalls were addressed, and best practices were followed. The generation of fish advice in partnership with Anishinaabe tribes, with personalized, easy to calculate recommendation metrics, and culturally-familiar aesthetic cues has generated an atmosphere of enthusiasm for EHL advice as well as improved trust in the process. These successes were achieved through an understanding of social influence theory as well as risk assessment processes. One size does not fit all at the risk characterization stage. However, social influence theory and risk assessment are universally transferrable across cultures if properly interwoven with community engaged research practices.

## References


Burger, J., & Gochfeld, M. (2011). Conceptual environmental justice model for evaluating chemical pathways of exposure in low-income, minority, native American, and other unique exposure populations. *American Journal of Public Health, 101*(Suppl 1), S64–S73. https://doi.org/10.2105/ajph.2010.300077

Cameron, L. D., Durazo, A., Ramirez, A. S., Corona, R., Ultreras, M., & Piva, S. (2017). Cultural and linguistic adaptation of a healthy diet text message intervention for Hispanic adults living in the United States. *Journal of Health Communication, 22*(3), 262–273. https://doi.org/10.1080/10810730.2016.1276985

Carlson, D. L., De Vault, D. S., & Swackhamer, D. L. (2010). On the rate of decline of persistent organic contaminants in lake trout (Salvelinus namaycush) from the Great Lakes, 1970–2003. *Environmental Science and Technology, 44*(6), 2004–2010. https://doi.org/10.1021/es903191u.

Chaffee, S. H. (1991). *Explication*. Newbury Park: Sage Publications Inc.

Chaiken, S., Liberman, A., & Eagly, A. (2012). *Psychology of attitudes*. London: SAGE Publications Ltd. Retrieved from http://sk.sagepub.com/navigator/psychology-of-attitudes. https://doi.org/10.4135/9781446263037

Chang, F., Pagano, J. J., Crimmins, B. S., Milligan, M. S., Xia, X., Hopke, P. K., & Holsen, T. M. (2012). Temporal trends of polychlorinated biphenyls and organochlorine pesticides in Great Lakes fish, 1999–2009. *Science of the Total Environment, 439*, 284–290. https://doi.org/10.1016/j.scitotenv.2012.09.019

CORA. (2006). Eat fish but choose wisely - a family guid to eating fish. *Chippewa-Ottawa Resource Authority.* Retrieved from http://www.1836cora.org/Publications.php

Davidson, P. W., Myers, G. J., Cox, C., Shamlaye, C. F., Marsh, D. O., Tanner, M. A., et al. (1995). Longitudinal neurodevelopmental study of Seychellois children following in utero exposure to methylmercury from maternal fish ingestion: outcomes at 19 and 29 months. *Neurotoxicology, 16*(4), 677–688.

Davidson, P. W., Myers, G. J., Cox, C., et al. (1998). Effects of prenatal and postnatal methylmercury exposure from fish consumption on neurodevelopment: Outcomes at 66 months of age in the Seychelles child development study. *JAMA, 280*(8), 701–707. https://doi.org/10.1001/jama.280.8.701

Dellinger, J. A. (2004). Exposure assessment and initial intervention regarding fish consumption of tribal members of the upper Great Lakes region in the United States. *Environmental Research, 95*(3), 325–340. https://doi.org/10.1016/j.envres.2003.07.012

Dellinger, M. (2007). *Using translational research to create culturally relevant fish consumption advisories.* (Master of Science), University of Wisconsin-Milwaukee, UW Libraries.

Dellinger, M. J., & Ripley, M. P. (2016). Mercury risks versus nutritional benefits of tribal commercial fish harvests in the Upper Laurentian Great Lakes. *Human and Ecological Risk Assessment: An International Journal, 22*(4), 1036–1049. https://doi.org/10.1080/10807039.2015.1133240

Dellinger, M., Dellinger, J., Haverkate, R., Ripley, M., & Lemoine, N. (2006). *From the lab to the community: Helping tribal members choose the safest fish to eat.* Paper presented at the Indian Health Service Research, Albuquerque, NM.

Dellinger, J., Haverkate, R., Hickey, A., Dellinger, M., Pritchard, D., Farmer, A., et al. (2008). *'Eat more fish but choose wisely' risk reduction strategy for Michigan Anishnaabe tribes: Preliminary evaluation of DVD.* Paper presented at the Native American Health Research, Poertland, Oregon.

Dellinger, J. A., Dellinger, M. J., & Yauck, J. (2012). Chapter 14: Mercury Exposure in vulnerable populations: Guidelines for fish consumption. In M. S. Bank (Ed.), *Mercury in the environment: pattern & process* (pp. 289–300). Berkeley: University of California Press.

Dellinger, J. A., Moths, M. D., Dellinger, M., & Ripley, M. P. (2014). Contaminant trends in freshwater fish from the Great Lakes: A 20 year analysis. *Human and Ecological Risk Assessment, 20*(2), 461–478.

Dellinger, M. J., Pingatore, N., Ripley, M. P., & Cassidy, L. (2015–2017). *A fish consumption advisory to promote Anishinabe environmental health literacy* Medical College of Wisconsin: National Institute for environmental health sciences. 1R21ES025788–01.

Dellinger, M. J., Olson, J., Clark, R., Pingatore, N., & Ripley, M. P. (2017). Development and pilot testing of a model to translate risk assessment data for Great Lakes Native American communities using mobile technology. *Human and Ecological Risk Assessment: An International Journal*, 1–14. https://doi.org/10.1080/10807039.2017.1377596

DeWeese, A. D., Kmiecik, N. E., Chiriboga, E. D., & Foran, J. A. (2009). Efficacy of risk-based, culturally sensitive Ogaa (walleye) consumption advice for Anishinaabe tribal members in the Great Lakes region. *Risk Analysis: An Official Publication of the Society for Risk Analysis, 29*(5), 729–742. https://doi.org/10.1111/j.1539-6924.2008.01198.x

Dieckmann, N. F., Gregory, R., Peters, E., & Hartman, R. (2017). Seeing what you want to see: How imprecise uncertainty ranges enhance motivated reasoning. *Risk Analysis, 37*(3), 471–486. https://doi.org/10.1111/risa.12639

Foran, J. A., Deweese, A. D., Hudson, M. J., & Kmiecik, N. E. (2010). Evaluation of mercury exposure reduction through a fish consumption advisory program for Anishinaabe tribal members in Northern Wisconsin, Michigan, and Minnesota. *Journal of Environmental and Public Health, 2010*, 802584. https://doi.org/10.1155/2010/802584

Garrison, N. A. (2012). Genomic justice for native Americans: Impact of the havasupai case on genetic research. *Science, Technology, & Human Values, 38*(2), 201–223. https://doi.org/10.1177/0162243912470009

Ginsberg, G. L., & Toal, B. F. (2009). Quantitative approach for incorporating methylmercury risks and omega-3 fatty acid benefits in developing species-specific fish consumption advice. Environ Health Perspect, 117(2), 267–275. https://doi.org/10.1289/ehp.11368

Ginsberg, G. L., Toal, B. F., & McCann, P. J. (2015). Updated risk/benefit analysis of fish consumption effects on neurodevelopment: Implications for setting advisories. *Human and Ecological Risk Assessment: An International Journal, 21*(7), 1810–1839. https://doi.org/10.1080/10807039.2014.992851

GLIFWC. (2003). *Ojibwe health study report to university of Wisconsin–Milwaukee: Project components 1–4. Project report 99–2*. Retrieved from Odanah, WI, USA.

GLITEC. (2011). *Community health data profile: Michigan, Minnesota, and Wisconsin Tribal Communities, 2010*. Retrieved from Lac du Flabeau, WI.

Grandjean, P., Weihe, P., White, R. F., Debes, F., Araki, S., Yokoyama, K., et al. (1997). Cognitive deficit in 7-year-old children with prenatal exposure to methylmercury. *Neurotoxicology and Teratology, 19*(6), 417–428 doi: S0892-0362(97)00097-4 [pii].

Harada, Y. (1967). Congenital Minamata disease. In T. Tsubaki & K. Irukayama (Eds.), *Methylmercury poisoning in Minamata and Niigata Japan*. New York: Elsevier Scientific Pub. Co.

Ho, L., Gittelsohn, J., Sharma, S., Cao, X., Treuth, M., Rimal, R., et al. (2008). Food-related behavior, physical activity, and dietary intake in first nations - a population at high risk for diabetes. *Ethnicity & Health, 13*(4), 335–349. https://doi.org/10.1080/13557850701882936.

Hoover, E., Cook, K., Plain, R., Sanchez, K., Waghiyi, V., Miller, P., et al. (2012). Indigenous peoples of North America: Environmental exposures and reproductive justice. *Environmental Health Perspectives, 120*(12), 1645–1649. https://doi.org/10.1289/ehp.1205422

Johns, T., Powell, B., Maundu, P., & Eyzaguirre, P. B. (2013). Agricultural biodiversity as a link between traditional food systems and contemporary development, social integrity and ecological health. *Journal of the Science of Food and Agriculture, 93*(14), 3433–3442. https://doi.org/10.1002/jsfa.6351

Julvez, J., Mendez, M., Fernandez-Barres, S., Romaguera, D., Vioque, J., Llop, S., et al. (2016). Maternal consumption of seafood in pregnancy and child neuropsychological development: A longitudinal study based on a population with high consumption levels. *American Journal of Epidemiology, 183*, 169–182. https://doi.org/10.1093/aje/kwv195

Kim, S., Dellinger, J., Ahn, S., Cramer, E. M., Nicolini, K., Marks, L., & Wagner, J. (2016). *Diminishing cultural differences: Transitioning from cross- to inter-cultural communication.* Paper presented at the 102nd annual convention of the National Communication Association, Philadelphia, PA.

Kruglanski, A. W., & Thompson, E. P. (1999). Persuasion by a single route: A view from the Unimodel. *Psychological Inquiry, 10*(2), 83–109. https://doi.org/10.1207/S15327965PL100201

Kunda, Z. (1990). The case for motivated reasoning. *Psychological Bulletin, 108*(3), 480–498.

Kyriacou, A. P. (2016). Individualism–collectivism, governance and economic development. *European Journal of Political Economy, 42*, 91–104. https://doi.org/10.1016/j.ejpoleco.2015.11.005

Moths, M. D., Dellinger, J. A., Holub, B., Ripley, M. P., McGraw, J. E., & Kinnunen, R. E. (2013). Omega-3 fatty acids in fish from the Laurentian Great Lakes tribal fisheries. *Human and Ecological Risk Assessment, 19*(6), 1628–1643.

Mozaffarian, D., & Rimm, E. B. (2006). Fish intake, contaminants, and human health - evaluating the risks and the benefits. *Jama-Journal of the American Medical Association, 296*(15), 1885–1899. https://doi.org/10.1001/jama.296.15.1885

Mozaffarian, D., & Wu, J. H. (2011). Omega-3 fatty acids and cardiovascular disease: Effects on risk factors, molecular pathways, and clinical events. *Journal of the American College of Cardiology, 58*(20), 2047–2067. https://doi.org/10.1016/j.jacc.2011.06.063

Mozaffarian, D., & Wu, J. H. (2012). (n-3) fatty acids and cardiovascular health: Are effects of EPA and DHA shared or complementary? *The Journal of Nutrition, 142*(3), 614s–625s. https://doi.org/10.3945/jn.111.149633

Niederdeppe, J., Connelly, N. A., Labuer, T. B., & Knuth, B. A. (2015). Using theory to identify beliefs associated with intentions to follow fish consumption advisories among anglers living in the Great Lakes region. *Risk Analysis*. https://doi.org/10.1111/risa.12408

O'Keefe, D. J. (2002). *The Persuasion Handbook: Developments in theory and practice*. Thousand Oaks: SAGE Publications, Inc. Retrieved from http://sk.sagepub.com/reference/hdbk_persuasion. https://doi.org/10.4135/9781412976046

Petty, R. E., & Cacioppo, J. T. (1986). The elaboration likelihood model of persuasion. *Advances in Experimental Social Psychology, 19*, 123–205.

Quaife, M. M. (1947). *The western country in the 17th century: The memoirs of Lamothe Cadillac and Pierre Liette*. Chicago: The Lakeside Press.

Rice, G., Swartout, J., Mahaffey, K., & Schoeny, R. (2000). Derivatino of U.S. EPA's oral reference dose (RfD) for methylmercury. *Drug and Chemical Toxicology, 23*(1), 41–54. https://doi.org/10.1081/DCT-100100101

Rostenkowski, A., Dellinger, M., & Gallum, T. (Writers). (2007). *Nindamwaa Giigoon: An Anishnaabe guide to eating fish [DVD]*. In D. H. Petering (Producer). Chippewa Ottawa Resource Authority: Agency for Toxic Substances and Disease Registry.

Sadraddini, S., Azim, M. E., Shimoda, Y., Bhavsar, S. P., Drouillard, K. G., Backus, S. M., & Arhonditsis, G. B. (2011). A Bayesian assessment of the PCB temporal trends in Lake Erie fish communities. *Journal of Great Lakes Research, 37*(3), 507–520. https://doi.org/10.1016/j.jglr.2011.06.005.

Salamova, A., Pagano, J. J., Holsen, T. M., & Hites, R. A. (2013). Post-1990 temporal trends of PCBs and organochlorine pesticides in the atmosphere and in fish from lakes Erie, Michigan, and superior. *Environmental Science & Technology, 47*(16), 9109–9114. https://doi.org/10.1021/es401895g

Schell, L. M. (2012). Transitioning from traditional: Pollution, diet and the development of children. *Collegium Antropologicum, 36*(4), 1129–1134.

Scherer, A. C., Tsuchiya, A., Younglove, L. R., Burbacher, T. M., & Faustman, E. M. (2008). Comparative analysis of state fish consumption advisories targeting sensitive populations. *Environmental Health Perspectives, 116*(12), 1598–1606. https://doi.org/10.1289/ehp.11372

Scurich, N., & Shniderman, A. (2014). The selective allure of neuroscientific explanations. *PLoS One, 9*(9), e107529. https://doi.org/10.1371/journal.pone.0107529

Sheehy, T., Kolahdooz, F., Schaefer, S. E., Douglas, D. N., Corriveau, A., & Sharma, S. (2015). Traditional food patterns are associated with better diet quality and improved dietary adequacy in aboriginal peoples in the northwest territories, Canada. *Journal of Human Nutrition and Dietetics, 28*(3), 262–271. https://doi.org/10.1111/jhn.12243.

Shimshack, J. P., & Ward, M. B. (2010). Mercury advisories and household health trade-offs. *Journal of Health Economics, 29*(5), 674–685. https://doi.org/10.1016/j.jhealeco.2010.05.001

Shimshack, J. P., Ward, M. B., & Beatty, T. K. M. (2007). Mercury advisories: Information, education, and fish consumption. *Journal of Environmental Economics and Management, 53*(2), 158–179.

Story, M., Stevens, J., Himes, J., Stone, E., Rock, B. H., Ethelbah, B., & Davis, S. (2003). Obesity in American-Indian children: Prevalence, consequences, and prevention. *Preventive Medicine, 37*(6 Pt 2), S3–S12.

Taylor, W., & Ferreri, C. P. (2000). *Great Lakes fisheries policy and management: A binational perspective*. East Lansing: Michigan State University Press.

Tobias, J. K., & Richmond, C. A. (2014). "That land means everything to us as Anishinaabe....": Environmental dispossession and resilience on the North Shore of Lake superior. *Health & Place, 29*, 26–33. https://doi.org/10.1016/j.healthplace.2014.05.008

Turyk, M. E., Bhavsar, S. P., Bowerman, W., Boysen, E., Clark, M., Diamond, M., et al. (2012). Risks and benefits of consumption of Great Lakes fish. *Environmental Health Perspectives, 120*(1), 11–18. https://doi.org/10.1289/ehp.1003396

U.S.EPA. (1992). *Tribes at risk: The Wisconsin tribes comparative risk project*. Retrieved from Washington DC, USA.

U.S.EPA. (2000). *Science policy council handbook: Risk characterization*. Washington, DC: Office of Research and Development.

U.S.EPA. (2004). What you need to know about mercury in fish and shellfish. *U.S. Environmental Protection Agency*. Retrieved from http://water.epa.gov/scitech/swguidance/fishshellfish/outreach/advice_index.cfm

USDA. (2015, December, 2015). 2015 – 2020 Dietary guidelines for Americans. *U.S. Government Printing Office*. 8th. Retrieved from http://health.gov/dietaryguidelines/2015/guidelines/

Whiting, S. J., & Mackenzie, M. L. (1998). Assessing the changing diet of indigenous peoples. *Nutrition Reviews, 56*(8), 248–250.

# Chapter 10
# The Use of Digital Communication Channels to Enhance Environmental Health Literacy

**Gary L. Kreps, Kevin Wright, and Amelia Burke-Garcia**

**Abstract** There is tremendous potential to utilize new and emerging digital channels of communication to help raise public environmental health literacy. Modern society is in the midst of a communication revolution that is rapidly expanding public use of digital communication systems and tools for many goals and purposes. Yet, there is a plethora of complex, and sometimes contradictory, information concerning environmental health risks that consumers struggle to understand. Consumers additionally struggle with determining what are the most important environmental health risks to be aware of concerning health issues such as cancers and heart disease, and what can be done to avoid these health risks? This chapter examines the growth and influence of digital communication as channels for helping consumers access, evaluate, and utilize relevant health information concerning environmental health risks. Different digital communication channels and platforms are identified as potentially strong venues for disseminating important environmental health information. Guidelines for designing and implementing environmental health education messages are examined to promote the most effective use of digital communication channels to reach and influence intended audiences. Strategic design principles, such as enhancing system interactivity, interoperability, ease of use, immediacy, adaptability, accessibility, and cultural sensitivity are examined and applied to the dissemination of environmental health information. Strategies for working in close collaboration with representatives from key populations to promote user-centered design of health information systems is highlighted. Relevant theory and research concerning the applications of digital information technologies for health information dissemination are reviewed and applied to promoting environmental health literacy.

**Keywords** Health information technology · ehealth · Digital information systems · mhealth · Digital divide · Usability · Interactivity · Tailored messaging · Feedback systems · User centered design

---

G. L. Kreps (✉) · K. Wright
George Mason University, Gaithersburg, MD, USA
e-mail: gkreps@gmu.edu

A. Burke-Garcia
George Mason University, Westat, MD, USA



S. Finn, L. R. O'Fallon (eds.), *Environmental Health Literacy*,
https://doi.org/10.1007/978-3-319-94108-0_10

## Background

Digital communication channels such as the internet, social media, and mobile applications are becoming increasingly powerful venues for disseminating relevant health information, including information about environmental health issues, to health care consumers, health care providers, and policy makers (Lefebvre and Bornkessel 2013). This approach is a direct result of the rapidly expanding technologically-based information revolution that has been transforming communication in modern society. Growth in use of digital health communication channels have dramatically changed the nature of health communication and health education (Lefebvre and Bornkessel 2013; Kreps 2015). The Pew Research Center's Internet and Technology Program reported in 2017 that 83% of a representative sample of American adults responded that they are either very or somewhat interested in seeking health information news online, with health information identified as the leading topic of interest for American Internet users (Horrigan 2017).

The Internet is a powerful digital information platform that provides unparalleled opportunities to reach a broad range of individuals with vital health information – and to tailor messages to audiences' literacy levels, cultural orientations, and communication channel preferences (Wen et al. 2010). When it comes to those that participate in the health care system, e.g. consumers, providers, and caregivers, use of the Internet and other digital communication channels to gather relevant health information and to communicate about important health issues is increasing For example, patients regularly search websites for health information when preparing to interact with their health care providers about their health concerns and to help them make key health care decisions, leading to dramatic changes in the delivery of care and on consumer-provider health communication (Bowes et al. 2012). Yet, it is not easy for many consumers to fully understand health risk information they may encounter online, such as warnings about potential chemical, food, air, sun, or water related environmental health risks, especially when the information provided about these risks and is complex and sometimes contradictory (Finn and O'Fallon 2017). Care must be taken to present environmental health risk information in ways that different audiences can understand (Burke-Garcia et al. 2017; Cleary et al. 2017).

Social media platforms have become powerful sources that typically deliver relevant health information and support within established and trusted social networks (Burke-Garcia et al. 2017). There are currently 2.5 billion social media users globally (Colucci-Gray et al. 2006) and 70% of Americans now use social media to connect with one another, engage with news content, and share information (Nutbeam 2008; Fox and Duggan 2013). Compared to other generations, millennials are the most active users of social media, with 88% of Americans ages 18–29 using Facebook and 75% of them having created at least one social media account (Pew Research Center 2010). Moreover, while fewer older adults use social media than younger individuals, the number of older people using social media are still relatively high with 62% of adults ages 65 and older using Facebook, and the number of these older social media users is growing (Greenwood et al. 2016). In fact, the Pew Research Center reports that as of 2016, use of Facebook is on the rise, in part due to the growing number of older adults who are joining the site. Further, these numbers of older social media users are substantially increased from just 6 years ago when it was reported that only 6% of those aged 65 and older use social media (Pew Research Center 2010). Based on the increasing use of social media, as well as both the high level of trust among members of these social networks and their abilities to identify and interact effectively with one another, there are great opportunities to promote the dissemination of health information, and enhance EHL with the use of social media channels (Kreps et al. 2016).

Mobile health information channels for digital communication (mHealth), such as smart phones, tablets, and wearable devices, are providing the public with easy access to relevant health information wherever they are and whenever they need health information (Kreps 2017a, b). Evidence suggests that the strategic design and implementation of powerful digital communication channels (such as social media networks, blogs, websites, and health system portals) has the potential to help raise public understanding about important health issues and enhance environmental health literacy (EHL) (Kreps 2017a, b). It is critically important that health information is presented to different audiences in ways that individuals can easily access, understand, and utilize the information for guiding relevant health decisions that they need to make to promote their health and the health of others (Amann et al. 2015; Alpert et al. 2017a, b; Kreps 2017a, b).

Evidence suggests that use of multiple channels of digital communication may be the best way to disseminate relevant health information (Buhl et al. 2016). Presenting overlapping health information over multiple digital channels can enhance exposure and understanding through communication redundancy and reinforcement (Kreps and Neuhauser 2013). Multi-channel digital health communication has tremendous potential for use to help raise levels of environmental health literacy. However, care must be taken in the design of coordinated multi-channel digital health communication programs, so they communicate relevant environmental health information effectively to key audiences. For example, coordinated and overlapping presentations of relevant health risk information can be effectively disseminated via social media platforms, trusted health websites, health system portals, online discussion groups, and even via blogs. Strategic selection of different combinations of media channels can be designed to reach different audiences, such as at-risk youth, parents, patients, health care providers, or policy makers, who are likely to utilize different digital media channels.

## Theoretical Frameworks Concerning the Use of Digital Communication

Surveillance research has shown that increasing numbers of health care participants (including health care consumers, providers, and caregivers) regularly use digital communication channels to gather relevant health information and to communicate about important health issues (Kreps et al. 2007; Fox and Duggan 2013; Finn and

O'Fallon 2017; Horrigan 2017). This research has confirmed that health care professionals and informal caregivers use digital health information channels to update their knowledge about the latest health care findings to inform treatment decision-making and to guide their delivery of care (Bennett et al. 2004; Fox and Brenner 2012). The expanding use of digital communication platforms is creating new opportunities for diverse audiences to access relevant health information, learn about health conditions, monitor health, direct health care, share health information, discuss and compare problems and solutions with peers, and make informed and collaborative health care decisions (Lefebvre and Bornkessel 2013). While more traditional older health information websites typically only enable users to access static information, more advanced digital health information systems, such as interactive websites, social media, and mhealth (mobile health) programs, now are enabling users to exchange messages, ask questions, and receive adaptive responses to address specific health information needs, bridging both mass and interpersonal communication (Kreps 2015; Kreps 2017a, b).

Walther has described digital communication channels as having the unique capacity to combine mass and interpersonal forms of communication through media convergence by disseminating messages to broad audiences, while also engaging users in interpersonal interactions (Walther 1996; Walther 2007). Combining mass and interpersonal communication is especially valuable for examining complex health issues because it enables in-depth analysis of health information through interaction and collaborative health decision-making, which are essential communication processes for achieving health education, health promotion, and health behavior change goals (Chaffee 1986; Reardon and Rogers 1988; Buhl et al. 2016; Strekalova 2017). This combination of mass and interpersonal communication modes is especially important for enhancing understanding of health information and enhancing EHL through message repetition, redundancy, and feedback loops (Kreps 2009; Kreps 2014a, b; Kreps 2017a, b). Interpersonal feedback loops, something that Karl Weick refers to as communication cycles, is particularly valuable for reducing uncertainty (equivocality), and enhancing understanding about and informed responses to complex health situations (Kreps 2009). The widespread use of digital health information channels has dramatically changed the nature of health communication, often converging mass and interpersonal communication in three potentially distinctive and important ways (Kreps 2017a, b).

Type 1 media convergence occurs in a sequence of impersonal to interpersonal interactions, when individuals first obtain existing online health or medical information, with the original intention, subsequent decision, or prompting, to follow up on the information in a discussion with a health professional in person or with peers in online or offline settings (McMullan 2006; Savage et al. 2014; Kreps 2017a, b).

Type 2 media convergence involves fostering interpersonal health communication discussions on digital discussion systems, such as online social support groups (Rubinelli et al. 2013; Kreps 2014a, b; Kreps 2017a, b). Both types of convergence influence the ways health care providers and consumers interact about health care and health promotion (Suziedelyte 2012; Kreps 2017a, b)

Both Type 1 and Type 2 media convergence are relevant to the use of digital health channels for disseminating health and risk information for raising EHL. For example, with Type 1 convergence when consumers might access environmental health information through digital channels, such as a health website, the information they gather often elicits follow-up communication with health care providers, other experts, or even with family members and friends. These interactions with others are most likely to help enhance awareness and understanding about environmental health issue. According to Weick's Model of Organizing, engaging in discussions with others is an important communicative strategy for reducing uncertainty and increasing understanding, especially when the interactions are with knowledgeable others (Hu et al. 2012; Kreps 2017a, b). Consumers regularly search websites for health information when preparing to interact one on one with their health care providers about health concerns to help guide health care decision-making, leading to dramatic changes in consumer-provider health communication and the delivery of care (Finn and O'Fallon 2017). Bowes, Stevenson, Ahluwalia, and Murray found that patients who search for health information on the Internet expect their doctors to discuss the information with them, explain it, contextualize it, offer a professional opinion about the information in making treatment decisions (Bowes et al. 2012). This dialogue is a crucial communication process for helping to promote EHL for all participating interactants, whether the interactions are with health care providers, patients, family members, scientists, or with members of social networks.

Type 2 convergence with digital media has even greater potential than Type 1 convergence for promoting EHL because consumers are provided with opportunities to engage in relevant interactions concerning complex health issues right online (Kreps 2017a, b). Modern digital health information systems provide ample opportunities to disseminate interactive environmental health information. For example, health care system participants regularly use digital channels (such as email, computer portals, blogs, social media platforms, online discussion groups, and even texting) to communicate with one another (Bennett et al. 2004). Interaction on online support groups provides needed information and social support to users, and encourages increased consumer-provider health communication (Hu et al. 2012). Research in health communication, social support, and computer-mediated communication has shown distinctive advantages of online discussions that depart from the qualities of traditional face to face encounters. Online support groups and social media communities are frequently used to connect members of social networks as powerful sources of information, advice, and encouragement for users who identify closely with one another due to their shared experiences (Wen et al. 2011). Burke-Garcia and colleagues found that participants in online mommy blogs, a form of social media, provided one another with influential health information about parenting, health promotion, and prevention behaviors (Burke-Garcia et al. 2017). A study of interactions on the online patient community PatientsLikeMe.com found that patients shared relevant health information with one another to increase their knowledge about their conditions, leading to enhanced patient empowerment, literacy, and even participation in and compliance with health care recommendations (Petersson and Erlingsdottir 2015).

Affective communication on digital health information systems can also help manage emotional health concerns related to communication about frightening health risks, such as cancers (Weick 1979; Kreps and Neuhauser 2013); Walther (1996, 2007) explains that interactive digital media not only bridge mass and interpersonal communication, but also influence communication intimacy, from impersonal to hyperpersonal (Walther 1996; Walther 2007). The facilitation of hyperpersonal interactions can enhance communicating complex and emotionally intense health information between health care providers and consumers, aiding health decision-making. Serious health challenges often demand the exchange of complex health information during emotionally intense situations, necessitating attention to effective content and relational communication. Users of a clinic-based patient health portal wanted the system to provide them with opportunities to communicate online with other patients who shared similar health challenges (Alpert et al. 2017b). Participation in online support groups were found to provide important therapeutic outcomes for military spouses who were experiencing stress concerning the status of their deployed partners (Rea et al. 2015). In general, online support groups are also good channels for communicating about emotional health concerns, promoting social support and coping, which can be most valuable when communicating about scary carcinogenic environmental cancer risks (Wen et al. 2011; Hammond 2015).

In addition, the ability to preserve and review health messages on digital discussion systems can enhance user comprehension and retention of complex environmental health information. For example, access to and review of preventive health information via health portals, for instance, has been shown to enhance health education and encourage adherence with medical recommendations (Bolanos et al. 2012; Alpert et al. 2017b). Social media also provide important review and interaction capacities that can help enhance EHL (Report 2016). A review of the literature reveals that the use of different social media platforms (including Facebook and other interactive online programs) has been found to encourage participants to engage in follow-up protocols, and generate active dialogue in communities nationwide (Rizvi et al. 2011; Report 2016). This suggests that social media platforms are potentially influential communication media for not only disseminating environmental health information, but also for increasing understanding of health information, and for applying the information to engaging in relevant health behaviors.

Concerns have been raised about the relative quality of health information acquired by consumers via digital communication channels, as well as about consumers' abilities to understand complex health information, so it is important to make sure that environmental health websites contain the latest and best health information (Kreps et al. 2007; Alpert et al. 2017a). A challenge with the rapid growth of the internet is the numerous health-related websites that may not provide the latest and most accurate environmental health information to the public. Some of these websites may be offered by commercial concerns that are more interested in enticing consumers into buying their products than in providing useful health information. Consumers must carefully evaluate the sources of online health information, as well as the accuracy and timeliness of the information, to make sure the information provided is valid and reliable. It is important for consumers to become sophisticated users of online information, developing digital communication literacy (which is a crucial part of overall health literacy in the modern world), to know what online health information is most accurate and appropriate for guiding their health decisions. Health communicators can enhance EHL by educating consumers about evaluating the quality of health information disseminated via different media, as well as by insuring that the environmental health risk information provided on digital communication channels is presented in ways that is easily understandable and relevant to different audiences in accordance with their language proficiencies, cultural backgrounds, educations, and levels of health literacy (Fox and Brenner 2012). The rapid growth and plethora of information on the digital space presents communicators with a key challenge – building the skills to determine information quality.

## Research and Outreach Examples for Promoting EHL

There are many digital communication strategies that have demonstrated potential for effectively disseminating relevant environmental health information that can enhance consumers' EHL. Mommy blogs have shown great promise for communicating environmental health information to consumers. These mommy blogs are well utilized by many new mothers who trust the bloggers to provide them with relevant and helpful health information. A recent intervention study utilizing mommy blogs to disseminate environmental risk information to mothers was designed to test the utility of this digital platform for environmental health education (Kreps et al. 2016). This project builds upon an earlier preliminary study concerning HPV vaccine promotion that utilized mommy blogs successfully to educate mothers about vaccinating their sons and daughters against the HPV virus (Burke-Garcia et al. 2017). By analyzing blog reader comments, findings from this pilot project suggest that these mommy bloggers can drive health behavior consideration amongst their readers (Burke-Garcia et al. 2017). Moreover, this study revealed through interviews with the bloggers that as a result of engaging with them, they also had offline conversations with their children and extended family about HPV. This study illustrated that online social networks, such as mommy blogs, are frequently used and influential channels for disseminating health information between mothers because of the close identification, shared perspectives on health issues, and established communication patterns between mommy bloggers and their audiences of mothers. Moreover, this study suggests that the influence of these types of outreach strategies can have effects beyond the digital environment. However, this preliminary study also suggested that importance of providing the mommy bloggers with up-to-date, accurate, and evidence based health information to share on their blogs (Burke-Garcia et al. 2017). This research also suggests that mommy blogs have great potential as influential communication channels for disseminating family-relevant environmental health risk information to mothers who are key influencers in seeking, distributing, and applying health information to promote the health of their family members (Kreps et al. 2016).

There is also great promise for the use of mobile digital communication systems (mHealth) to provide relevant environmental health information to mobile communication device (such as smartphones) users wherever they are and whenever they need health information (Fox and Brenner 2012). Preliminary work is taking place in this arena. Recent work by a team in UCLA has shown the value of an integrated system to help children with asthma. A specialized smart watch based wireless system was developed that incorporated environmental sensing and ecological momentary assessment to identify emergent environmental asthma risks and asthma symptoms, enabling early identification of potential asthma attacks, as well as quick risk responses and interventions (Buonocore et al. 2017). Similar mHealth applications are being designed for use in communities where there are known environmental health risk hot spots, such as areas where there are industrial waste contaminant hot spots, to automatically alert residents with specialized applications installed on their smartphones whenever they approach an area with potential environmental health risks. The smartphone applications could be programmed to provide the residents with warning information and advice concerning the contamination risks. Other mHealth applications can be designed for use on smartphones to provide relevant warning and response health information to residents who live in urban areas, such as Beijing, China, where there are frequent days when the air pollution exceeds healthy levels. For example, a mobile phone text message system has been designed to warn pregnant women about avoiding air pollution risks to help them avoid these environmental risks (Jasemzadeh et al. 2018). These mobile applications could warn residents when they are approaching areas with dangerous air pollution levels, provide them with timely information about strategies for avoiding this environmental health risk, and help them make informed health behavior decisions for promoting health.

In addition, there are wonderful opportunities to utilize multiple reinforcing digital communication channels in tandem with other channels of communication to effectively disseminate relevant environmental health information. Buhl et al. (2016) effectively presented related health information on multiple communication channels in a demonstration research project to help enhance consumers participation in an exercise program, demonstrating that increased audience exposure to health-related messages across different communication channels helped to increase long-term participation in this health promotion program (Fitzpatrick-Lewis et al. 2010; Buhl et al. 2016). Environmental health information dissemination programs can also effectively utilize multiple complementary digital communication channels, such as to coordinated use of key websites, portals, and social media platforms to provide consumers with information about relevant environmental health risks and advice about risk prevention. The redundancy built in to these multi-channel digital health communication programs, that could include posts on social media platforms such as Instagram, followed up by similar text messages, email notes, and/or blog posts, can increase consumer exposure to important health messages, reinforce understanding about environmental health risks, and encourage adoption of relevant prevention and health promotion behaviors, while also enhancing EHL (Fox 2011). For example, in the HPV vaccine intervention study using mommy bloggers, the researchers found that blog posts about vaccination often initiated posts and discussions on other social media platforms, such as Facebook and Twitter, that amplified the dissemination of health information provided initially on the blogs (Burke-Garcia et al. 2017). Moreover, given such findings, the opportunity of accessing this kind of information via mobile communication systems may help make critical health information available when caregivers are trying to make healthcare decisions for others in real time. While there are some examples of the potential, one must consider how these systems are designed to achieve their intended goals and to raise EHL.

## Methods for Enhancing Use of Digital Channels for Promoting EHL

There are important strategic design principles that can enable digital health communication channels to be used effectively in raising EHL (Kreps and Neuhauser 2015; Kreps 2017a, b). With the growth of new and exciting HIT opportunities comes the daunting responsibility to design digital health tools that communicate effectively with a diverse array of health care consumers, providers, and policy makers (Kreps and Neuhauser 2015). These tools must be designed to effectively communicate the right information needed by different audiences at the right time, in the right place, and in the best ways to guide health care and health promotion. Care must be taken, however, to make sure that digital health tools are optimally interactive, interoperable, easy to use, engaging, adaptable, and accessible for diverse audiences (Kreps and Neuhauser 2013; Alpert et al. 2017b). Kreps (2017a, b) identifies seven major strategic communication factors to take into account when designing digital health information systems that has great relevance for communicating environmental health information (Kreps 2017a, b).

### *Tailor Digital Information Systems and Messages to Users*

Digital health communication system users typically want answers to very specific questions about environmental risks to their health or to the health of significant others, making it vitally important to design systems that can provide them with the right information that they need. Too often, however, digital health communication systems provide frustratingly broad overviews of a range of related health issues when users are looking for specific answers to their health problems (Alpert et al. 2017b). Tailored communication systems can be programmed to collect information on key user factors (from questions posed to users or through analysis of existing data), and use these data to develop a repertoire of targeted message responses that match users' unique health information needs (op den Akker et al. 2015).Smart, adaptive information systems that employ artificial intelligence can even go one step further by recalling queries/searches users that has been collected over time to predict users' unique needs and provide them with the most relevant and appropriate tailored health messages (Kreps and Neuhauser 2013; Sadasivam et al. 2016).

Designing tailored health communication systems is becoming increasingly sophisticated for predicting user information needs and providing pertinent answers to health questions (op den Akker et al. 2015; Sadasivam et al. 2016). Smart adaptive tailored health information systems have great potential to provide salient health information for improving EHL and for enhancing health outcomes (Piette et al. 2015). For example, Dr. Alan Greene, a pediatrician and clinical faculty member at Stanford University School of Medicine established in 1995 what has become a very popular pediatric consumer health information website (see http://www.drgreene.com) that provides parents (and others) with tailored information about caring for children. Initially, Dr. Greene provided individual answers to questions users posed about child care, but over time he collected, archived, and updated the responses provided to questions, so that now, with use of a message tailoring system, users can automatically be provided with archived answers to the specific questions they ask. This tailored health information system can provide system users with pertinent health information to answer their specific questions to help them make evidence-based health decisions. Similar tailored environmental health information systems can be designed to provide pertinent environmental health risk information to users online.

## *Make Digital Systems Easy and Convenient*

Digital health information systems are often difficult for many health care consumers to use due to overly complex design features and limited user computer system competencies (Kellerman and Jones 2013). Research suggests that patients are often frustrated with cumbersome usability factors with many basic digital health systems, including difficulties logging on, navigating, and accessing specific health information, and wanting mobile access to health information to increase convenience and timeliness of answering their health questions (Alpert et al. 2017b). While health information systems generally work well for computer experts who design them, they do not always work well for typical users, even those users with advanced levels of education. Consumers who may need relevant information about environmental health risks may not necessarily be sophisticated digital health system users, so it will be important to design digital tools for these consumers that are easy for them to access and navigate. Care must be taken to design environmental health information systems to meet unique user computer skills and information needs, taking a user-centric approach to system design (Alpert et al. 2017b). This guideline for designing health information systems also suggests the need to identify and analyze the digital communication competencies of audiences that need environmental health risk information, conduct usability tests of digital health information systems with these users to identify any challenges to access and use, and then, apply the usability data gathered to refine information systems so they are easy and convenient for these audience members to use (Fitzpatrick 2013).

## *Provide Relevant and Clear Information on Digital Systems*

Many consumers have difficulties understanding the health information provided on digital health information systems due to health literacy limitations (Fox and Brenner 2012; Amann et al. 2015). Evidence suggests that too many health information systems use overly complex medical language, complicated statistics, and do not include needed explanations and examples to enhance user understanding (Alpert et al. 2017a). Moreover, few systems enable users to ask questions and provide answers which can clarify their understanding of health information presented (Rubin 2012). Patients also want digital systems to provide them with specific, relevant health information (Alpert et al. 2017b). Digital systems need to incorporate message systems that are targeted to users' literacy levels and information needs, incorporating language that is easy for them to understand and apply, and providing vivid examples (visual, auditory, and language-based) that users can identify with (Kreps and Neuhauser 2010).

Research shows that using illustrations with consumers who possess limited levels of health literacy can increase understanding and recall of health information (Meppelink and Bol 2015). This suggests the need for careful audience analysis and user-centered design in developing multi-modal and user-responsive digital messaging systems to deliver relevant environmental health information (Kreps 2014a, b). Health information systems can be designed to meet the level of EHL of different audiences, so that experts (health care providers, scientists, and others) can receive detailed technical environmental health risk information, including direct access to relevant research reports, while laypersons can be provided with simpler messages, with examples and illustrations, to help them understand complex environmental health risk information.

## *Design Systems that Provide Feedback and Promote Interaction*

Too many digital health information systems provide static one-way communication with users, without incorporating and promoting active feedback and interaction (Kreps 2014a, b) Users are not able to seek clarification about complex health information presented or to dig-in to gather additional specific and in-depth health information. By increasing the interactivity of health information systems with chat boxes, discussion boards, and links to connect with others, these digital systems can enhance user understanding, recall, system use, and their application of information provided (Rubinelli et al. 2013; Brady et al. 2015; Rider et al. 2016). These feedback and interaction systems can significantly enhance the dissemination of environmental health information.

## *Design Systems that Are Responsive to User Needs and Feelings*

Health concerns can be a very personal and emotional issue for many people, especially when dealing with serious health threats. Confronting serious health challenges can be both physically and emotionally debilitating. For example, communities that are exposed to dangerous contaminants and environmental risks due to problems such as industrial waste or climate hazards can feel trapped, helpless, and frustrated by these dangers and adverse outcomes that may affect individual, family, and community health. Effective health information systems should be designed to help consumers address both their physiological and psychological information and support needs (Kim et al. 2015). However, there is a frequent bias in the ways digital health information systems present information that focuses much more on content issues related to health in comparison to emotional issues, failing to adequately meet the full range of consumers' health information needs (Kreps and Neuhauser 2013). Affective computing focuses on programming information systems to adapt and respond to users' emotional concerns (Calvo et al. 2014). Intelligent interactive health information systems can be designed to interact with consumers sensitively and empathically about important health issues, providing social support, counseling, and even health behavior change interventions (Lisetti et al. 2013; Ring et al. 2013; Calvo et al. 2014). Affective information systems are designed to respond to users' expressed emotional states so they are responsive to users' emotional needs and feelings.

## *Make Digital Systems Interesting and Immediate*

Many digital health information systems are dull and boring for users, perhaps due to their focus on delivering detailed health information without providing sufficient compelling contextual information (Kreps 2014a, b). Research suggests that digital health system users want health information presented vividly and personally, with relevant examples and stories (Alpert et al. 2017b). There is a need to design health information systems that communicate with immediacy, to fully engage and activate users (Bolanos et al. 2012; Kreps and Neuhauser 2013). Immediate digital health information systems are interesting and exciting, using sound, color, movement, drama, personalization, humor, responsiveness, and intensity (Kreps and Neuhauser 2013). Bickmore, Schulman, and Yin demonstrate that use of relational interactive virtual human agents can help promote engagement and immediacy (Bickmore et al. 2010). A large body of educational communication research illustrates that increases in teacher immediacy enhances learning, liking, cooperation, perceptions of credibility and commitment, suggesting that increasing the immediacy of digital health information systems can also accrue these positive influences with users (Kreps and Neuhauser 2013).

### *Connect Users to Trusted Social Networks*

The social networks individuals belong to have powerful influences on the information they have access to, the health behaviors they engage in, and the health decisions they make (Kim and Kreps 2014; Kim et al. 2015; Valente 2015). Social network members can provide one another with especially trusted and influential health information (Rogers 2003; Valente 2015). Digital health information systems can effectively connect users to trusted social networks for disseminating heath information and support (Lefebvre and Bornkessel 2013; Burke-Garcia et al. 2017). Digital health information systems can be fruitfully designed to connect individuals with shared interests and experiences to exchange relevant information and provide needed support. Strategic use of interactive social media to access health information and communicate with relevant others about environmental health concerns can provide important opportunities to influence users' health behaviors and improve health outcomes (Moorhead et al. 2013; Burke-Garcia et al. 2017).

## Future Directions

There are several promising directions for effectively utilizing digital health communication systems to disseminate relevant environmental health information to key audiences (such as the general public, patients, health care providers, policy makers, parents, teachers, scientists), and to enhance EHL among members of these audiences. As described earlier in this chapter, strategic design of digital communication programs should help enhance system interactivity, interoperability, ease of use, immediacy, adaptability, accessibility, and cultural sensitivity to promote effective dissemination of environmental health information. To achieve these strategic design goals, it will be important to work in close collaboration with representatives from the key audiences. There are unique preferences between different audience segments for the use of specific digital channels, the credibility of different health information sources, and the design of specific messages that will capture their attention and influence them (Kayser et al. 2015).

Good evaluation research is needed to guide the development and refinement of digital communication tools and programs. Audience analysis research should be conducted to identify the best ways to communicate with members of different segmented populations (Kreps 2014a, b). For example, audience analysis research can provide data establishing the health literacy levels of members of key audiences, especially their EHL levels, for guiding the design of relevant and understandable health information dissemination programs. Needs analysis research can help identify the most salient environmental health issues for different populations. Formative evaluation research can help to design digital health education programs and messages. Formative evaluation research often provides good opportunities to connect with and establish community-based participative partnerships with members of key audiences, to work with them to design, implement, evaluate, and sustain health education programs through a process of user-centered design (Neuhauser and Kreps 2014). There is also tremendous demand to conduct in-depth and rigorous evaluation research to assess the influences of digital programs on the diverse audiences that use them and on relevant health outcomes (Fitzpatrick 2013; Kreps 2014a, b).

Health care delivery systems have the opportunity to build interactive social media into existing health information systems, such as hospital websites and patient portals to increase interactivity and the use of feedback loops to make sense of complex health information topics (Gallant et al. 2011). Information about environmental health risks is often highly technical, especially when describing chemical risks and biological responses, necessitating opportunities for laypersons to ask questions about these risks that will enhance understanding. How do these risks relate to me? How can I avoid these risks? How can I find out if these risks have already influenced my health? What can I do to minimize negative effects from potential exposures to environmental health risks? Interactive health information systems can help answer these relevant questions and promote informed health decision-making.

Health promotion specialists may not always recognize that effective communication is a complex and fragile human process that demands strategic design, careful monitoring, and responsive adaptation. Promoting EHL will demand new partnerships between environmental health scientists, health communication experts, digital information system designers, public health promotion practitioners, health system administrators, health care providers, consumer advocates, policy makers, and many others, that will draw upon unique skills, knowledge, and resources (Krist et al. 2016; Finn and O'Fallon 2017). Often, it appears that digital health system designers are more enamored with the technical elegance and innovation of new information technologies than with the utility of these tools for health care consumers and providers (Kreps and Neuhauser 2013). What they should be asking is whether the technologies are easy for these audiences to understand and use? (Amann et al. 2015; Meppelink and Bol 2015).

New ehealth tools must be designed to fit comfortably within the policies, practices, and technical infrastructure that are built into existing health and social systems. These new ehealth tools should be affordable and accessible for all intended audiences (Kreps et al. 2007). The messages delivered on ehealth programs need to be designed so that diverse populations of users can understand and apply the health information provided. The best digital health information systems are adaptive, interactive, and self-correcting. These ehealth programs provide interesting, relevant, and engaging information for users (Kreps and Neuhauser 2013). Efforts must be taken to make digital information systems sustainable and adaptive to meet health information needs over time. Future efforts to develop digital tools for raising EHL have to be designed to complement other health communication channels, be easy for health care professionals to use, and communicate effectively with diverse audiences of consumers (Lefebvre and Bornkessel 2013; García-Betances et al. 2015; Kayser et al. 2015; Krist et al. 2016). With these recommendations in mind, there is rich potential to utilize digital health information systems to enhance the abilities of all audience members to identify and utilize the best available evidence-based health information to adapt to environmental health risks.

## References

Alpert, J. M., Desens, L., Krist, A. H., Aycock, R. A., & Kreps, G. L. (2017a). Measuring health literacy levels of a patient portal using the CDC's clear communication index. *Health Promotion Practice, 18*(1), 140–149.

Alpert, J. M., Krist, A. H., Aycock, R. A., & Kreps, G. L. (2017b). Designing user-centric patient portals: Clinician and patients' uses and gratifications. *Telemedicine and e-Health, 23*(3), 248–253.

Amann, J., Rubinelli, S., & Kreps, G. (2015). Revisiting the concept of health literacy. The patient as information seeker and provider. *European Health Psychologist, 17*(6), 286–290.

Bennett, N. L., Casebeer, L. L., Kristofco, R. E., & Strasser, S. M. (2004). Physicians' Internet information-seeking behaviors. *Journal of Continuing Education in the Health Professions, 24*(1), 31–38.

Bickmore, T., Schulman, D., & Yin, L. (2010). Maintaining engagement in long-term interventions with relational agents. *Applied Artificial Intelligence, 24*(6), 648–666.

Bolanos, F., Herbeck, D., Christou, D., Lovinger, K., Pham, A., Raihan, A., Rodriguez, L., Sheaff, P., & Brecht, M.-L. (2012). Using Facebook to maximize follow-up response rates in a longitudinal study of adults who use methamphetamine. *Substance Abuse: Research and Treatment, 6*, 1–11.

Bowes, P., Stevenson, F., Ahluwalia, S., & Murray, E. (2012). 'I need her to be a doctor': Patients' experiences of presenting health information from the internet in GP consultations. *British Journal of General Practice, 62*(604), e732–e738.

Brady, S. S., Sieving, R. E., Terveen, L. G., Rosser, B. S., Kodet, A. J., & Rothberg, V. D. (2015). An interactive website to reduce sexual risk behavior: Process evaluation of TeensTalkHealth. *JMIR Research Protocols, 4*(3), e106.

Buhl, M., J. Famulare, C. Glazier, J. Harris, A. McDowell, G. Waldrip, L. E. Barnes and M. S. Gerber (2016). *Optimizing multi-channel health information delivery for behavioral change*. 2016 IEEE Systems and Information Engineering Design Symposium (SIEDS).

Buonocore, C. M., R. A. Rocchio, A. Roman, C. E. King and M. Sarrafzadeh (2017). *Wireless sensor-dependent ecological momentary assessment for pediatric asthma mHealth applications*. 2017 IEEE/ACM International Conference on Connected Health: Applications, Systems and Engineering Technologies (CHASE).

Burke-Garcia, A., C. N. Berry, G. L. Kreps and K. B. Wright (2017). *The power & perspective of mommy bloggers: Formative research with social media opinion leaders about HPV vaccination*. Proceedings of the Annual Hawaii International Conference on System Sciences.

Calvo, R. A., D'Mello, S., Gratch, J., & Kappas, A. (2014). *The Oxford handbook of affective computing*. Oxford: Oxford Library of Psychology.

Chaffee, S. H. (1986). Mass media and interpersonal channels: Competitive, convergent, or complementary. In G. Gumpert & R. Cathcart (Eds.), *Inter/media: Interpersonal communication in a media world* (Vol. 57, pp. 62–80). New York: Oxford University Press.

Cleary, E. G., Patton, A. P., Wu, H. C., Xie, A., Stubblefield, J., Mass, W., Grinstein, G., Koch-Weser, S., Brugge, D., & Wong, C. (2017). Making air pollution visible: A tool for promoting environmental health literacy. *JMIR Public Health Surveill, 3*(2), e16.

Colucci-Gray, L., Camino, E., Barbiero, G., & Gray, D. (2006). From scientific literacy to sustainability literacy: An ecological framework for education. *Science Education, 90*(2), 227–252.

Finn, S., & O'Fallon, L. (2017). The emergence of environmental health literacy-From its roots to its future potential. *Environmental Health Perspectives, 125*(4), 495–501.

Fitzpatrick, S. (2013). *Emerging usability evaluation tools and techniques in web-based health communication*. National Conference on Health Communication, Marketing, and Media.

Fitzpatrick-Lewis, D., Yost, J., Ciliska, D., & Krishnaratne, S. (2010). Communication about environmental health risks: A systematic review. *Environmental Health, 9*(1), 67.

Fox, S. (2011). *The social life of health information*. From http://www.pewinternet.org/2011/05/12/the-social-life-of-health-information-2011/

Fox, S. and J. Brenner. (2012). *Family caregivers online*. From http://www.pewinternet.org/2012/07/12/family-caregivers-online/

Fox, S. and M. Duggan. (2013). *Health online 2013*. From http://www.pewinternet.org/2013/01/15/health-online-2013/

Gallant, L. M., Irizarry, C., Boone, G., & Kreps, G. L. (2011). Promoting participatory medicine with social media: New media applications on hospital websites that enhance health education and e-patients' voices. *Journal of Participatory Medicine, 3*, e49.

García-Betances, R. I., Fico, G., Salvi, D., Ottaviano, M., & Arredondo, M. T. (2015). On the convergence of affective and persuasive technologies in computer-mediated health-care systems. *Human Technology, 11*(1), 71–93.

Greenwood, S., A. Perrin and M. Duggan. (2016). *Social media update 2016*. From http://www.pewinternet.org/2016/11/11/social-media-update-2016/

Hammond, H. (2015). Social interest, empathy, and online support groups. *The Journal of Individual Psychology, 71*(2), 174–184.

Horrigan, J. B. (2017). *How people approach facts and information*. From http://www.pewinternet.org/2017/09/11/the-elements-of-the-information-engagement-typology/

Hu, X., Bell, R. A., Kravitz, R. L., & Orrange, S. (2012). The prepared patient: Information seeking of online support group members before their medical appointments. *Journal of Health Communication, 17*(8), 960–978.

Jasemzadeh, M., Khafaie, M. A., Jaafarzadeh, N., & Araban, M. (2018). Effectiveness of a theory-based mobile phone text message intervention for improving protective behaviors of pregnant women against air pollution: A randomized controlled trial. *Environmental Science and Pollution Research, 25*(7), 6648–6655.

Kayser, L., Kushniruk, A., Osborne, R. H., Norgaard, O., & Turner, P. (2015). Enhancing the effectiveness of consumer-focused health information technology systems through eHealth literacy: A framework for understanding users' needs. *JMIR Human Factors, 2*(1), e9.

Kellerman, A. L., & Jones, S. S. (2013). What it will take to achieve the as-yet-unfulfilled promises of health information technology. *Health Affairs, 32*(1), 63–68.

Kim, W., & Kreps, G. L. (2014). The association between online health information-seeking behavior and social support in social networks among Korean Americans. *Health Behavior and Policy Review, 1*(5), 381–394.

Kim, W., Kreps, G. L., & Shin, C.-N. (2015). The role of social support and social networks in health information–seeking behavior among Korean Americans: A qualitative study. *International Journal for Equity in Health, 14*(1), 40.

Kreps, G. L. (2009). Applying Weick's model of organizing to health care and health promotion: Highlighting the central role of health communication. *Patient Education and Counseling, 74*(3), 347–355.

Kreps, G. L. (2014a). Achieving the promise of digital health information systems. *Journal of Public Health Research, 3*(3), 128–129.

Kreps, G. L. (2014b). Evaluating health communication programs to enhance health care and health promotion. *Journal of Health Communication, 19*(12), 1449–1459.

Kreps, G. L. (2015). Communication technology and health: The advent of ehealth applications. In J. A. D. Lorenzo Cantoni (Ed.), *Communication and technology* (Vol. 5, pp. 485–495). Berlin: De Gruyter Mouton Publications.

Kreps, G. L. (2017a). Online information and communication systems to enhance health outcomes through communication convergence. *Human Communication Research, 43*(4), 518–530.

Kreps, G. L. (2017b). The relevance of health literacy to mHealth. *Information Services & Use, 37*(2), 123–130.

Kreps, G. L., & Neuhauser, L. (2010). Editors' introduction, ehealth and health promotion. *Journal of Computer-Mediated Communication, 15*(4), 527–529.

Kreps, G. L., & Neuhauser, L. (2013). Artificial intelligence and immediacy: Designing health communication to personally engage consumers and providers. *Patient Education and Counseling, 92*(2), 205–210.

Kreps, G. L., & Neuhauser, L. (2015). Designing health information programs to promote the health and well-being of vulnerable populations: The benefits of evidence-based strategic health communication. In C. A. Smith & A. Keselman (Eds.), *Meeting health information needs outside of healthcare* (pp. 3–17). Chandos Publishing.

Kreps, G. L., Gustafson, D., Salovey, P., Perocchia, R. S., Wilbright, W., Bright, M. A., & Muha, C. (2007). The NCI digital divide pilot projects: Implications for cancer education. *Journal of Cancer Education, 22*(1 Suppl), S56–S60.

Kreps, G., Wright, K., & Burke-Garcia, A. (2016). *Using targeted social media to communicate culturally sensitive and interactive environmental risk information about breast cancer to mothers. Breast Cancer and the Environment Research Program (BCERP) Annual Grantees Meeting*. Durham: National Institute of Environmental Health Sciences.

Krist, A. H., Nease, D. E., Kreps, G. L., Overholser, L., & McKenzie, M. (2016). Engaging patients in primary and specialty care. In B. W. Hesse, D. K. Ahern, & E. Beckjord (Eds.), *Oncology informatics* (pp. 55–79). Boston: Academic Press.

Lefebvre, R. C., & Bornkessel, A. S. (2013). Digital social networks and health. *Circulation, 127*(17), 1829–1836.

Lisetti, C., Amini, R., Yasavur, U., & Rishe, N. (2013). I can help you change! An empathic virtual agent delivers behavior change health interventions. *ACM Transactions on Management Information Systems, 4*(4), 1–28.

McMullan, M. (2006). Patients using the internet to obtain health information: How this affects the patient–health professional relationship. *Patient Education and Counseling, 63*(1), 24–28.

Meppelink, C. S., & Bol, N. (2015). Exploring the role of health literacy on attention to and recall of text-illustrated health information: An eye-tracking study. *Computers in Human Behavior, 48*, 87–93.

Moorhead, S. A., Hazlett, D. E., Harrison, L., Carroll, J. K., Irwin, A., & Hoving, C. (2013). A new dimension of health care: Systematic review of the uses, benefits, and limitations of social media for health communication. *Journal of Medical Internet Research, 15*(4), e85.

Neuhauser, L., & Kreps, G. L. (2014). Integrating design science theory and methods to improve the development and evaluation of health communication programs. *Journal of Health Communication, 19*(12), 1460–1471.

Nutbeam, D. (2008). The evolving concept of health literacy. *Social Science & Medicine, 67*(12), 2072–2078.

op den Akker, H., Cabrita, M., op den Akker, R., Jones, V. M., & Hermens, H. J. (2015). Tailored motivational message generation: A model and practical framework for real-time physical activity coaching. *Journal of Biomedical Informatics, 55*, 104–115.

Petersson, L. and G. Erlingsdottir (2015). *Will PatientsLikeMe. com affect the doctor–patient relation and the work environment of doctors*. 19th Triennial Congress of the International Ergonomics Association (IEA 2015) in Melbourne.

Pew Research Center. (2010). *Millennials: Confident. Connected. Open to change* (pp. 1–149). Washington, DC: Pew Research Center.

Piette, J. D., Farris, K. B., Newman, S., An, L., Sussman, J., & Singh, S. (2015). The potential impact of intelligent systems for mobile health self-management support: Monte Carlo simulations of text message support for medication adherence. *Annals of Behavioral Medicine, 49*(1), 84–94.

Rea, J., Behnke, A., Huff, N., & Allen, K. (2015). The role of online communication in the lives of military spouses. *Contemporary Family Therapy, 37*(3), 329–339.

Reardon, K. K., & Rogers, E. M. (1988). Interpersonal versus mass media communication: A false dichotomy. *Human Communication Research, 15*(2), 284–303.

Report, e. (2016). *US ad spending: eMarketer's updated estimates and forecast for 2015–2020*. New York: eMarketer.

Rider, B. B., Lier, S. C., Johnson, T. K., & Hu, D. J. (2016). Interactive web-based learning: Translating health policy into improved diabetes care. *American Journal of Preventive Medicine, 50*(1), 122–128.

Ring, L., B. Barry, K. Totzke and T. Bickmore (2013). *Addressing loneliness and isolation in older adults: Proactive affective agents provide better support*. 2013 Humaine Association Conference on Affective Computing and Intelligent Interaction.

Rizvi, S. L., Dimeff, L. A., Skutch, J., Carroll, D., & Linehan, M. M. (2011). A pilot study of the DBT coach: An interactive mobile phone application for individuals with borderline personality disorder and substance use disorder. *Behavior Therapy, 42*(4), 589–600.

Rogers, E. M. (2003). *Diffusion of innovations*. New York: Simon and Schuster.

Rubin, D. L. (2012). Listenability as a tool for advancing health literacy. *Journal of Health Communication, 17*(sup3), 176–190.

Rubinelli, S., Collm, A., Glässel, A., Diesner, F., Kinast, J., Stucki, G., & Brach, M. (2013). Designing interactivity on consumer health websites: PARAFORUM for spinal cord injury. *Patient Education and Counseling, 93*(3), 459–463.

Sadasivam, R. S., Cutrona, S. L., Kinney, R. L., Marlin, B. M., Mazor, K. M., Lemon, S. C., & Houston, T. K. (2016). Collective-intelligence recommender systems: Advancing computer tailoring for health behavior change into the 21st century. *Journal of Medical Internet Research, 18*(3), e42.

Savage, S. V., Kwan, S., & Bergstrand, K. (2014). Virtual health: The impact of health-related websites on patient-doctor interactions. In J. J. Kronenfeld (Ed.), *Technology, communication, disparities and government options in health and health care services* (pp. 93–116).

Strekalova, Y. A. (2017). Health risk information engagement and amplification on social media: News about an emerging pandemic on Facebook. *Health Education & Behavior, 44*(2), 332–339.

Suziedelyte, A. (2012). How does searching for health information on the internet affect individuals' demand for health care services? *Social Science & Medicine, 75*(10), 1828–1835.

Valente, T. W. (2015). Social networks and health behavior. In *Health behavior: Theory, research and practice* (Vol. 5, pp. 205–222).

Walther, J. B. (1996). Computer-mediated communication: Impersonal, interpersonal, and hyperpersonal interaction. *Communication Research, 23*(1), 3–43.

Walther, J. B. (2007). Selective self-presentation in computer-mediated communication: Hyperpersonal dimensions of technology, language, and cognition. *Computers in Human Behavior, 23*(5), 2538–2557.

Weick, K. E. (1979). *The social psychology of organizing*. Reading: Addison-Wesley Pub. Co.

Wen, K. Y., Kreps, G., Zhu, F., & Miller, S. (2010). Consumers' perceptions about and use of the internet for personal health records and health information exchange: Analysis of the 2007 Health Information National Trends Survey. *Journal of Medical Internet Research, 12*(4), e73.

Wen, K. Y., McTavish, F., Kreps, G., Wise, M., & Gustafson, D. (2011). From diagnosis to death: A case study of coping with breast cancer as seen through online discussion group messages. *Journal of Computer-Mediated Communication, 16*(2), 331–361.

# Chapter 11
# Using Augusto Boal's Theatre of the Oppressed in a Community-Based Participatory Research Approach to Environmental Health Literacy

**John Sullivan**

**Abstract** In 2002 the Community Outreach & Engagement Core (COEC) program of the Sealy Center for Environmental Health & Medicine and the NIEHS-funded Center in Environmental Toxicology (CET) at the University of Texas Medical Branch at Galveston TX (UTMB) implemented a unique outreach and education methodology based on the pedagogical principles of Paulo Freire and Augusto Boal's community-oriented applied theatre system, the Theatre of the Oppressed (TO). This project – dubbed *Community Environmental Forum Theatre* – grew out of the UTMB COEC's adoption of a Community Based Participatory (CBPR) approach to environmental public health research, with a complementary emphasis on enhancing Environmental Health Literacy (EHL) within project partner communities. The dialogic engagement embodied in Freire's "democratizing pedagogy" and its mirror image in Boal's inclusive and multi-perspectival body of dramatic technique have proven effective in promoting EHL via sustainable engagement with community partners in research.

**Keywords** CBPR · Environmental public health · Environmental health literacy · Augusto Boal · Paulo freire · Theatre of the oppressed · Image theatre · Community environmental forum theatre · Local knowledge · Risk communication

---

J. Sullivan (✉)
Sealy Center for Environmental Health & Medicine at UTMB, University of Texas Medical Branch, Galveston, TX, USA



S. Finn, L. R. O'Fallon (eds.), *Environmental Health Literacy*,
https://doi.org/10.1007/978-3-319-94108-0_11

## Introduction

It is essential to acknowledge that the CET's "adoption" of TO as a linchpin in their community-based environmental public health outreach process should not be interpreted as a rejection of the many carefully formulated and validated processes in current use, or of methodologies and procedures suggested in the ASDHHS / Clinical & Translational Science Awards Consortium's "Principles of Community Engagement" (McCloskey et al. 2011). After a series of "trial balloon" preliminary projects, the TO arts-based approach was deemed appropriate and effective in promoting multi-generational inclusion at the grassroots in community environmental public health research. It also had larger goals that facilitated EHL:

- Integration of sound science into community dialogues to prioritize environmental health needs
- Development of effective educational formats for disseminating basic toxicology and risk characterization concepts
- Engagement with an environmental justice community often skeptical of scientific methodology and motives
- Reconciliation of rationalist and democratic approaches to cumulative risk and environmental justice in community environmental public health projects

This chapter endeavors to explain the philosophical and operational relationships among Community Based Participatory Research (CBPR), Freire's Pedagogy and Boal's Theatre of the Oppressed. Beginning with a brief overview of the historical and artistic development of Theatre of the Oppressed, and a closer examination of congruency among TO core principles and Freire's Pedagogy, the focus shifts to a discussion of TO as a process complementary with the values and practice principle of CBPR, and a vehicle for enhancing EHL. This discussion is followed by a closer look at how TO outcomes are incorporated into more effective community-advocacy, on one hand, and more inclusive and local knowledge-based environmental public health research, on the other.

The chapter also highlights examples of discrete projects focused on Community Environmental Forum Theatre, either as a community performance, a series of project-focused performances targeting a specific community, or a workshop using TO structures and techniques. This includes a discussion of working principles for organizing, implementing and evaluating TO projects from the perspective of a community artist informed by a CBPR/environmental public health perspective. This chapter concludes by addressing both the challenges and possible pitfalls of using TO in a research-oriented, rather than strictly community-oriented, context. It also examines TO's benefits for conducting research and for understanding the social dynamics of CBPR coalitions.

## A Very Short History of Augusto Boal's Theatre of the Oppressed

TO is a participatory theater that fosters democratic and cooperative forms of interaction among participants. Theater is emphasized not as a spectacle but rather as a language designed to: (1) analyze and discuss problems of oppression and power; and (2) explore group solutions to these problems through an inclusive artistic process (Picher, TOP-Lab, NY, NY). This unique blend of theater games, group image-making structures, and improvisational performance formats has spread from its origins in Augusto Boal's Arena Theatre based in Sao Paulo, Brazil to hundreds of countries on every continent. The International Center for Theatre of the Oppressed estimates that 20,000 theater artists, worldwide, identify themselves as TO practitioners, and formal Centers for the Theatre of the Oppressed thrive in major cities across Europe and North America, and in Rio de Janeiro, Brazil (Boal, rev. 2002).

Augusto Boal's Theatre of the Oppressed originated in 1971 as both a book – Theatre of the Oppressed – and an embodied artistic response to problems in Brazil's economy and social culture. TO has proven to be a dynamic, democratic and openly political form of applied theater with potential for a rich diversity of uses. Since 1971, TO principles have been used to develop several variants: Invisible Theatre and Forum performances on a wide-range of social issues such as environmental justice, racism, poverty, homelessness, violence against women, and religious and ethnic conflicts; Rainbow of Desire workshops that focus on the deeper, more personal implications of larger social oppressions; and Legislative Theatre that seeks to leverage a more direct influence on public policy (Boal 1998).

TO was originally conceived as a topical Newspaper Theatre that dramatized the implications of current events in Sao Paulo and Brazil. TO slowly morphed into the present day interactive Forum influenced by Boal's outreach to urban *favelas* and agricultural workers in the Brazilian back-country, and literacy campaigns structured around the methods and philosophy of Brazilian educator, Paulo Freire. Boal's exile to Buenos Aires, Argentina brought TO into yet another repressive political context, and Invisible Theatre was born to inject politically charged provocations directly into the social life of the city. This period also saw the growth of TO's Image Theater techniques as Boal sought to establish the basis for a dialogue among indigenous people and the descendants of European conquerors (Boal 1985).

By 1986, the Brazilian government had undergone a democratic transformation and Boal returned home to establish the Center for the Theatre of the Oppressed in Rio de Janeiro. In 1992, he was elected as a *vereador* – city council-person – in Rio and immediately hired members of his theater group to conduct a revolutionary experiment in democratic practice. He called this system Legislative Theatre, and he and his political/theatrical staff used the Forum to assess the needs and ideas of his constituents, and to formulate specific bills that addressed these priorities. After his term of office expired, Boal returned full-time to his life in theater, and established various TO projects in schools, impoverished *favela* districts within Rio de Janeiro and Sao Paulo, and prisons throughout Brazil (Boal 1998).

## Core Principles of TO Relevant to Environmental Health Research

TO developed in response to lack of freedoms and opportunities in Brazil, and Latin American society. Its central purpose is to empower people to dramatically analyze their situation while building a sense of community and creating an action agenda for changing the situations represented on stage when the audience and actors return to their lives in the world. While many streams of thought converge in TO, many of these core concepts are part of the lineage of CBPR and other forms of action-oriented research.

Core principles of TO are described in detail elsewhere (Boal 1985); this brief introduction highlights the key points that are relevant to CBPR and efforts to raise EHL.

### *Commitment, Truth and Authentic Dialogue Are More Important Than Technique and Production Values*

Though many permanent TO ensembles are highly accomplished as performers in a number of dramatic genres, some of the most exciting and socially rewarding TO projects involve ordinary citizens with no training in theater or any other art form. These groups often produce exciting performances fueled by the urgency of their own real needs and focused on issues and ideas crucial to their communities. This is similar to the engagement of community members in CBPR research who are motivated to learn the skills needed to address pressing environmental concerns within the context of their local sociopolitical and economic conditions. Similar to nonprofessional actors learning TO performance skills, community engagement in CBPR confers socially and educationally rewarding benefits that can empower individuals, raise EHL and lead to action (Boal 1992).

### *Boal, Freire and CBPR Share a Method Based on Thought, Action, Dialogue and Group Dynamics*

Both Freire and Boal use a popular education model that involves perspectives from multiple points-of-view to reconfigure existing cognitive maps. This transformative method, termed *praxis,* focuses on the actions of autonomous individuals – or co-learners – in a mutually supportive relationship of dialogue, reflection and action. This iterative process develops critical awareness, or *conscientização* by inspiring learners to deconstruct (*decodify*) oppressive elements that distort and limit their horizons, and to reconfigure (*recodify*) their lives as a process that fulfills their true potentials. Boal's process structures these relationships dramatically, through cooperative games, image-making and performance (Freire 1971). It mirrors the CBPR context by analyzing environmental health disparities and environmental justice concepts such as disproportionate exposure, multiple stressor effects, cumulative risk and the underlying effects of social determinants of health, all in a site-specific context that creates a "group portrait" of health indicators and social dynamics (Freire 2000).

## *TO Stimulates and Activates a Process That Results in Embodied Truth*

The great German playwright/director, Bertolt Brecht, developed a playwriting/performance style he called Epic Theatre to goad his audiences into analyzing the social circumstances that molded the lives of his characters (Mitter 1992; Schutzman 1994;). Boal carries this experiment to the next level with his idea that audience members could shed their passive role as spectators and break the fourth wall that separates them from the performers, leading to *"an activating and dynamizing catharsis"* that energizes the audience and compels them to in some way continue their efforts to address local issues (Boal 1998, rev. 2002). This process is comparable to translational CBPR that seeks to move environmental health research findings into public health action, policy change and improved health care to recognize, treat or prevent environmentally induced disease (Wallerstein & Duran).

## CBPR and TO Are Complementary Processes

Community Based Participatory Research (CBPR) is an overarching approach to population health knowledge creation based on a reconfiguration of the power-sharing dynamic among communities and researchers. CBPR empowers citizens to step out of their traditional position as objects of research by highly trained specialists and take a role as active subjects in various aspects of the research cycle (Katz 2004). CBPR, Freire and TO (as used in the research context) all share a developmental lineage. The common practice values of CBPR and TO directly challenge the idea that citizens without scientific training have no place in the domain of science – what Marvin Legator calls the *myth of exclusivity* (Legator and Strawn 1993). CBPR values such as trust, mutual respect, self-determination, inclusion, multidirectionality of communication, and solidarity are mirrored in both Freire's popular education process of developing critical awareness (*conscientização*) and Boal's Theatre of the Oppressed – particularly the inclusive studio process of image making (Siqueira and Sullivan 2009).

The practice principles of CBPR – most notably the concepts of co-learning, striking an equitable balance between knowledge making and action, the concept of negotiating reciprocal partnerships, building on the strengths and resources of project partners, and co-ownership of the outcomes of the process – resonate with the *procedural democracy* of Freirean praxis and TO dramaturgy (Wallerstein & Duran, Sullivan). As applied to EHL, TO has also proved useful in developing communication models and culturally literate risk communication messages within short-term and long-form CBPR contexts (Sullivan and Lloyd 2006; Ward et al. 2008; Rustveld et al. 2013).

TO clarifies and enhances the power and granular richness of community research by adding local knowledge. TO enables researchers to gather an assortment of anecdotal data about community beliefs, attitudes and life-ways that shed light on practices and exposure pathways crucial to environmental public health research (Horster 1992; Corburn 2005; Sullivan and Parras 2008).

When scientific investigators use TO to incorporate community input, it reinforces collaborative knowledge creation, also a key process in the CBPR approach. This approach carries the potential to expand the epistemological parameters of research with contextual knowledge that includes experience beyond the ken of current models (Corburn 2005; Fischer 2009).

Both CBPR and TO employ similar strategies that utilize existing grassroots networks to boost the scope, reach and cost effectiveness of outreach and dissemination efforts (Sullivan). Since the goal of CBPR is almost always to influence change in community health, norms, systems, programs, and policies, the use of local knowledge data developed through TO segues cleanly into more effective advocacy on the part of community partners as a project outcome (Sullivan 2010). Finally, TO's emphasis on equity through inclusion of a wide spectrum of voices promotes greater awareness of distributive justice implications of the research process and findings, by asking questions such as "who are the people at risk, and what do they need, not merely what level of risk is acceptable" (Corburn 2005).

TO has also proven invaluable to promoting dialogue among community members and researchers on differences in world-views, priorities and assumptions – bridging what Frank Fischer terms the *local knowledge/expertise divide*. (Horster 1992; Fischer 2009). Using various techniques of Boal's Image Theatre, community Forum workshop participants build images showing the importance of their community history, their primary health and quality of life concerns, important narratives of struggle and community success, local knowledge of pollution sources and exposure pathways, marginalization and lack of access to needed resources. This information can help to combat preconceptions of their community by researchers that often reinforce negative social stereotypes (Wing 2005). It can also inform research design to ensure that the research addresses local concerns and provides the tools and skills necessary to raise EHL.

Research participants similarly use TO to make images of concepts, attitudes and values comprising the "typical" scientific point of view. Through TO, non-scientists have the opportunity to think about the fundamental scientific principles that underlie the research process and form the basis for the more specific understanding that EHL conveys. This includes examining scientific beliefs about commitment to "absolute" objectivity, the view of health effects as mechanistic process outcomes, and the need to simplify complex site-specific relationships for "cleaner" analytic outcomes. Participants have the opportunity to question why, in their experience, local knowledge is always trumped by sophisticated analytic and scientific expertise (Wing 2005). Through this process, TO can help to develop a more honest view of their own perspectives, as well as the scientific point-of-view, and devise ways of working that maximize the strengths and assets of each. This process does not automatically result in integration of the scientific and public perspectives on research, but by acknowledging and attempting to understand the essence of these images, the research process going forward is potentially less fraught, and more grounded in the distinct perspectives of all the stakeholders involved (Boal 1998).

## How Programmatic Uses of TO in a CBPR Context Create an Outreach Modality and a Tightly Interwoven Pedagogy for Environmental Health Literacy

The use of TO as a major aspect of the community outreach toolkit, shaped the process that evolved into Community Environmental Forum Theatre at the Sealy Center for Environmental Health and Medicine and NIEHS P30 Center in Environmental Toxicology at the University of Texas Medical Branch (Galveston TX). The research team's initial outreach focused on fence-line communities proximate to refining industry facilities already in process of engaging around issues of air quality, petrochemical regulations, and health impacts. In this way, it offered communities throughout the Texas/Louisiana Gulf Coast petrochemical belt an opportunity to use the tools of Image Theater and improvisation to develop a wide range of useful toxicological concepts, a working knowledge of risk, and analytic tools to understand how toxic exposures, risk burdens, and socio-economic factors contribute to environmental (in)justices that severely impact their families. This *"Tox, Risk & Community Stress"* curriculum incorporates basic toxicology, community ethnography, social epidemiology and environmental justice concepts. This coordinated program is based on an overarching concept of Cumulative Risk described in a National Environmental Justice Action Committee (NEJAC) document, *"Ensuring Risk Reduction in Communities with Multiple Stressors: Environmental Justice and Cumulative Risk/Impacts"* (NEJAC-EPA 2004; Sullivan and Parras 2008).

The **Tox** section employs participatory image structures and Sociometry exercises to illustrate toxicology concepts germane to community needs for information. The "tox" component encompasses exposure concepts such as exposure pathways, fate and transport, and dose response parameters – magnitude, duration, frequency, and timing. The curriculum also covers susceptibility factors, vulnerable populations, bioaccumulation, bio-magnification, and bioavailability, among other topics, and provides a brief introduction to community-based, popular epidemiology. (Legator and Strawn 1993; Sexton et al. 2003; ATSDR 2005).

The **Risk** component stems from an image-based ethnography process in which participants assemble site-specific snapshots of exposure pathways, risk perceptions, risk and action priorities, personal experience with EJ and environmental health issues, and community power dynamics to create image maps of community assets, and stressors. Sociometry exercises allow participants to determine intragroup safety and toxic abatement priorities, as well as reveal the spectrum of group experience regarding the personal effects of toxic exposures. (ATSDR 2005).

Physical, social and economic **Community Stress** factors complicate the understanding of risks posed by toxic emissions and related health outcomes. To understand the role these factors play in risk, the facilitator and participants create short scenes and improvisational exercises exploring the real world human outcomes of cumulative community stress burdens from chronic toxic exposures, environmentally-induced health problems, lack of access to needed health care and nutritious food, and other well documented socio-economic indicators of health, opportunity and social justice disparities. Stressors originate from various sectors of our social-economic reality and Forum scenes illustrate how these burdens converge upon and oppress the most vulnerable communities. The concepts of multiple stressor effects, cumulative risk, and the influence of these factors on a community's ability to identify, mitigate and ultimately recover from chronic toxic assaults closely guides planning scenes for the public Forum. (ATSDR 2005; NEJAC-EPA 2004).

## Elements of TO: A Working Process for Communication and Education

Theatre of the Oppressed, as it is used in creating public Forums, is composed of two equally vital parts: a transformational workshop process that broadens the personal and social perspectives of each actor/participant and a performance event that incorporates the audience as ***spect-actors*** in an interactive dialogue for and with an entire community.

**Facilitation** TO workshops must be guided by an experienced facilitator or facilitation team. The TO facilitator guides the group – dialogically – toward consensus on ground rules and parameters, creates a master plan for each workshop, keeps the process sequentially on-track, and maintains a safe emotional container for processing of ideas and feelings during this transformational experience. Ultimately, the facilitator is responsible for teaching performance skills, enabling self-discovery, creating the conditions for a smooth flow of events, directing focus, bolstering confidence, maintaining safety, peacemaking, ensuring privacy and confidentiality (as necessary) and gently molding the fruits of group research into an attractive Forum Theatre experience (Rohd 1998).

**TO Workshop Process** The TO process – in an EHL/CBPR context – usually begins with an evening or day-long workshop followed by a day-long studio process (i.e., a performance/intervention). Two components of the workshop session are described below as examples of the process and content explored by participants. This includes development of an idea bank (for further development during the performance/intervention process) and performing Sociometry exercises. Sociometry, in this context, is a process of defining, mapping, measuring and evaluating choices and group dynamics. The workshop process includes adapting a series of Sociometric exercises that use "reconfigurative movements through space to probe and query group attitudes, preferences and experiences with respect to choices, preferences, and priorities" (Sternberg and Garcia 2000; Hale 2005; Marineau (2007); Sullivan 2010).

These Sociometry configurations include the *sliding scale*, *face-off* and the *exploding atom*. Sliding scale involves participants repositioning their bodies to indicate where they stand on a scale of 1–10 vis-à-vis the issue statement. The *face-off* graphs a simple Yes/No reaction to the statement; Yes now steps 8–10 steps away from a line and turns to face the No's. The *exploding atom* first configures as a tight circle; participants respond to the issue statement by a) staying close to the center to show their strong connection to the statement, or b) moving away from the center to show their degree of disconnect or marginalization from the statement. The finished configuration resembles a solar system or exploded diagram of an atom (Sullivan 2010).

The workshop process also includes development of an idea bank. The Box below shows examples of the types of questions posed and activities conducted during this part of a typical TO workshop session. These ideas are developed to reflect the specific parameters of individual research projects however, each variation is intended to elucidate the viewpoint of the *spect-actors* about their local environmental issues.

**Image Ethnography Idea Bank**

- Worst case scenario/worst environmental fear
- How do people around you (your neighbors) view their risk?
- How do people around you react when you give them a wake-up call about the environment?
- What does the local EJ power dynamic look like? (this will need prior discussion)
- Make images of examples of multiple stressors experienced; lack of health care access, crumbling infrastructure, mental health issues & effects on families, lack of jobs, deteriorating educational prospects, abandonment by government & private sector, industrial safety uncertainties + environmental health problems & ecosystem degradation
- Make a picture of how these stressors compound to create cumulative burdens for the community
- Start with cumulative burden image (Real)/make an image of what you'd like the community to look & feel like (Ideal) (a peel away process)
- How do we see our community? (perceptions of assets, history, cultural treasures)
- How do others see our community? (snapshot of a community's place in the social dynamic)

These ideas help the participants arrive at a deeper understanding of how toxic exposures have affected their community. It also provides feedback to scientists on the level of the group's environmental health literacy related to the linkage between toxic exposures and environmental health outcomes. Understanding the level of EHL of the various research participants, as well as the sociopolitical context in the affected community, ultimately helps scientists communicate more effectively about the results of research.

**TO Studio Process** The following overview presents a very basic synopsis of the Theatre of the Oppressed *studio process* from the opening "check-in" with participants, through the analytic image-making structures, story creation and scene-building activities, rehearsal, and culminating in closure dialogue and closing structures. This process is essentially the same regardless of the type of community working through the process, and the focus issues that provide the content of the final Forum performance. Early sessions focus more closely on the interactive games that develop positive chemistry and improvisational facility within the group, while later sessions will use these games as warm-ups for efforts to develop images-based fragments as fully plotted stories with a central conflict and well-defined choice points for the actors and audience (Rohd 1998; Boal 2000; rev. 2002).

*Opening Structure* A "check-in" for the group. This will involve a simple game or series of game-like structures that allow each participant some personal/public space to express hopes, fears, misgivings, affirmations, personal goals – literally, how they feel about the issue and the project at that moment.

*Activation: Sensory Retuning, Community Building, Trust Exercises* Exercises and game structures that promote cooperation, fluid movement, and to familiarize participants with using their bodies as expressive tools. This exercise encourages participants to connect with each other and the group on verbal and non-verbal levels. Participants, quite literally, retune their senses as a preparation for the subsequent image work and performance (Boal 2000).

*Image Theatre Structures* An exploration of Boal's sequence of body sculpture images to express action, concepts, feelings and essential qualities. These sculptural images will become the primary language of the intra-group dialogue on the individual and community impacts of the focus issues, and each person's lived reality in relation to these ideas. The primary process in building scenes for performance is the creation of a *Core-Conflict image* that encapsulates the essence of the complex struggle imbedded in the real-world situation mirrored in the scene. The Community Environmental Forum Theatre program has also added image structures like *pairs* and *tableaus in motion* developed in Playback Theatre (Salas 1996; Boal, rev. 2002).

*Improvisational Structures* A series of improvisational exercises are used throughout each session to learn the mechanics of building an effective, unscripted story, improvisationally. Eventually, these improvisations become animating factors for the most powerful Core-Conflict image – as determined by group consensus – and these formerly static images begin to walk and talk like a play.

*Issue Deepening* A series of Image exercises, readings, personal stories and shared feelings and impressions stemming from a relationship to the focus issues. This Freirean process of collectively deconstructing the causal elements and personal/community impacts of the central conflict encourages all participants to share both subjective (feelings, ambivalences, frustrations) and objective material (from formal and casual sources) that add nuance and verisimilitude to the scene. In an environmental health context, the facilitator would introduce toxicological materials, project-specific risk implications, and concepts of community vulnerability, and multiple stressor impacts at this juncture. This segment of the activity sequence provides nuanced content for potential scenes and sets the tone for upcoming scene-construction and rehearsal (Rohd 1998; Boal, rev. 2002).

*Story-Boarding and Scene Construction* A section of the workshop that translates the group's Core-Conflict images and more extended stories into representation scenes. The Story Board develops from a Core Conflict that represents the emotional (and sometimes chronological) center of the scene. The scene is then *regressed backward in time-place* so that all the actors directly involved in the scene can retrace what their character might be doing/where it could be coming from before the conflict occurs. Working again from the Core–Conflict, the scene is now *progressed forward* and the actor/characters leave the image for their next destination, immediately after the conflict. In an environmental health context, this would help participants explore how the conflict developed (and where), their role in the conflict, and what they did/where they went after the conflict. The outcome of this process is the unembellished core of the community performance. At this point, the group will usually develop 2–3 additional scenes – all related to the central issues that constitute the focus issues for the Forum under construction. Each scene is formed around the actions of a clearly defined ***protagonist*** (a character whose efforts to resolve a difficult situation are blocked), an ***antagonist*** (the primary character who blocks the efforts of the protagonist), ***effective allies*** (characters who could possibly help the protagonist if they chose to act on her/his behalf) and ***accomplice/collaborators*** (characters who perceive they benefit from the original oppression and tacitly support the ***antagonist***) (Schutzman 1994; Boal, rev. 2002).

*Group Processing of the Workshop or Performance Session* The group is encouraged to individually release any tensions, voice any needs, ask additional questions, and provide any comments or information that might deepen or clarify the experience and ensure that each person's safety and personal autonomy issues have been addressed. This part of the process may also occur at other points throughout the session to reflect on feelings and ideas prompted by specific images or exercises. This process applies equally to scenes performed for a community audience that did not participate in the workshop.

The Forum facilitator assists this *spect-actor* intervention process, first activating the audience with warm-up games and exercises, leading the audience – after they've watched each scenario once – in voting for which scene they would like to run through the Forum process, and finally de-briefing the *spect-actor*, the other

Forum actors, and the audience after each intervention. During this de-briefing, the facilitator asks five sets of questions in this sequence:

1. (***for the audience***) What did you see? How was this action different from what the protagonist originally did?
2. (***for the spect-actor***) How did you feel during your intervention? This question opens the door for community members to express their frustration with the antagonist or apathetic potential allies who sat by passively. This moment also allows them to tell their neighbors how empowered they felt while engaged in the struggle for justice, regardless of the outcomes.
3. (***a two part question, again, for the spect-actor***) Why did you choose this specific intervention and how would you evaluate the results?
4. (***for the other actors, especially the antagonist***) How did the spect-actor's intervention force you to change your own actions, or reconsider your beliefs while remaining in character?
5. (***for the audience***) What do you think of what happened? Do you have any suggestions for actions that might make this intervention more effective? This question gives audience members the opportunity to offer their own evaluation of how effective they think each intervention would be in the real world (Sullivan and Parras 2008; Sullivan 2010).

These direct conversations with the audience after each intervention invariably introduce fresh perspectives and often uncover formerly submerged facts about the situation that were not mentioned during the workshop. For example, audience members employed by polluting industries often mention facts associated with accidents, upsets or standard practices with toxic stock chemicals that are not common knowledge in the community. Spectators with connections to the public health, social services and political sectors often make similar revelations about official "rationing" of local resources. These facts on the ground accumulate and add to the context and knowledge base for future evidence-based community advocacy, and enhance the scope and range of the research data-base.

## Flipping the Script on Environmental Injustice: How TO & Forum Theatre Offers a Rehearsal for Action in the Real World

Scenes and images created by community members in the workshop process are performed for the community as interactive Forum theater. In Boal's dramatic system, these "open" forum scenes – with no closure and no satisfying solutions to the environmental health and justice problems represented in the theatre piece – activate the audience to step onto the stage and into the drama with new actions they propose to advance the cause of health and environmental justice (Boal 1985; rev. 2002).

The Forum process gives potential allies the chance to develop an effective new rhetoric for presenting information and arguing their case (McCarthy and Galvao 2002). Rehearsing assertive action in the Forum on behalf of their neighborhood's health and the community's environment transforms participants, who formerly accepted toxic assault and chronic social-political neglect as a fact of life in environmentally degraded neighborhoods. Allies/spect-actors also have the opportunity to rehearse networking skills, to practice risk communication with their neighbors, and to question central assumptions about the sociopolitical causes of their exposure to toxic substances. An "ideal Forum" leaves actors, spect-actors and audience energized, motivated, actively vigilant and more aware of the environmental injustices they have seen portrayed on stage. Freire calls this a state of "patient impatience," but Boal is less ambiguous. There is no place for passivity at a Forum; Theatre of the Oppressed primes spectators to apply their powers of critical awareness to deconstruct what they see in order to "prepare for action in and on the world" (Freire 2004; Boal 2005).

## TO/CBPR Case Study Profiles: A Wide Spectrum of Scope, Range and Goals

The NIEHS CET at UTMB (Galveston) directly co-sponsored a succession of TO projects with "fence-line" communities under the rubric of Community Environmental Forum Theatre. The term fence-line, in this context, refers to a community closely proximate to an industrial facility producing air or water-borne emissions. The following section features four CET projects conducted in Port Arthur TX, Buffalo, NY, and in Houston TX. They were chosen to illustrate how TO has raised the EHL of residents participating in research projects with widely varying community settings, issue foci, purposes, durations, and outcomes. (See Appendix 1 for a full listing of TO projects with an even wider range of purposes). Each of these examples reflects very local conditions, local concerns, and specific project goals, however in each case, the TO approach provided the potential to raise awareness of environmental influences on health and increase the EHL of participants.

## Community In–Power & Development Association (CIDA) (Port Arthur TX)

Collaborating Partners: CIDA, UTMB CET (2004).

The purpose of this exercise was to increase membership recruitment for community organizations, develop greater environmental health literacy with respect to oil refining, chemical manufacturing and hazardous waste storage sources of air pollution and the subsequent enhancing of evidence-based advocacy skills. An additional purpose was to build a platform for subsequent community research with the CET CBPR team.

The specific issue was the poor air quality in an African-American neighborhood in west Port Arthur neighborhoods closely proximate to an extremely large oil refining operation. The cumulative risks that increase vulnerability for this population include the overall local density of point-source facilities and waste storage, regulatory compliance history issues, the community members' overall lack of effectiveness with advocacy, and their lack of credibility with decision-makers. The focus of EHL efforts was to increase understanding of chemical exposures, daily and occasional exposure pathways, differences between higher level 8 hour occupational exposures and 24/7 chronic low level exposures to residents, possible disease outcomes associated with probable exposures such as asthma and respiratory distress, and what to do when an emissions event occurs. The extent of emissions released into the local Port Arthur air shed and rough numbers of active cancer and asthma cases in the west side community were documented by CIDA before the Forum project (Kelley 2004; Sullivan 2008).

The outcomes of this Community Forum included clarification of short-term goals and objectives for the community partner (CIDA) agreement to work with the CET on community-based research to address air emission impacts on community health, indoor air quality, and asthma incidence in the effected neighborhoods. Additional outcomes included the community partner collaborating on various NIEHS grants, working with the Sealy Center on healthy homes themes, providing technical assistance on toxic waste storage and incineration issues, co-participation in a neighborhood DIY mapping experience with the Pacific Institute, consultation on the US EPA EJ Showcase project (EPA Region 6, 2010–2012), and a healthy neighborhoods collaboration between CIDA and UTMB Preventive Medicine & Community Health. All of these subsequent activities provided enhancements to community levels of EHL stemming from the baseline that developed during the initial Forum experience (Kelley 2004; Sullivan and Lloyd 2006) (Figs. 11.1–11.3).

## 858 E Ferry Street Superfund Site (Buffalo NY)

Collaborating Partners: Toxic Waste & Lupus Coalition, Ujima Theatre.

The purpose of the Capstone Project was to review the historical process associated with abatement at a toxic Superfund site, re-emphasize important environmental health concepts around toxic compounds found on-site, honor prime movers among local activists on the project, and to establish community-wide goals for moving forward with sustainable community development of the site.

The research that inspired the TO activities focused on heavy metal and PCB residues left on-site from previous industrial smelting and casting operations. The specific issues in this community were the impacts of the community's exposures to

**Fig. 11.1** "Fire in the Hole": CIDA Forum actors create an image of their worst-case catastrophic environmental scenario – an explosion and fire at a nearby petro-chemical facility. (Port Arthur TX 2004, photo by Bryan Parras)

**Fig. 11.2** Image Theatre – "Circle of Protection": a fence-line family builds a "circle of protection" when one of their young children is diagnosed with Acute Lymphocytic Leukemia. (Port Arthur TX 2004, photo by Bryan Parras)

a variety of heavy metals (such as lead and chromium) and PCB's, in particular, the specific health impacts on children, and an investigation of a localized Lupus disease cluster and its possible association with on and off-site toxic compounds. The ultimate aim of the TO process was to provide insights that would support the development of a plan for land use to benefit the entire community after abatement actions are completed (Williams et al. 2009).

The focus of EHL efforts in this project was related to understanding the risks from exposure to lead and other heavy metals and mapping the potential exposure pathways such as wind-blown metals that contaminated soil in nearby neighborhoods. The TO exercises and subsequent raising of EHL were related to understanding the connections among exposures and auto-immune diseases, and the environmental and sociopolitical factors needed for sustainable urban land use (Sullivan 2008; Williams et al. 2009).

**Fig. 11.3** Frozen Scene: "What Are You Going to Do About This?!" Multi-gen Community Coalition Confronts Industry Liaison Personnel About the Most Recent Emission Event that Spilled into Their Neighborhood. (Port Arthur TX, photo by Bryan Parras)

The TO experience focused on a neighborhood proximate to 855 E. Ferry Street, the Superfund site. The goal of the workshop and performance were to solidify dialogic connections among neighborhood organizations, faith-based communities, and city/state government representatives to address ongoing environmental justice issues. After the Forum performance – with help from audience spect-actors – community stakeholders collaboratively developed a list of possible uses for the former industrial site for submission to city planners. The Forum also enabled an important intra-community dialogue on alternative interpretations of the project process, thus empowering community members to voice concerns and to determine how research is conducted in their community (Williams et al. 2009; Sullivan 2008) (Figs. 11.4–11.8).

## Project COAL (Communities Organized Against Asthma & Lead) (Houston TX)

Collaborating Partners: *De Madres a Madres* (a social services provider), City of Houston Bureau of Children and Community Environmental Health, Harris County Casa de Amigos (public access clinic), TEJAS (Texas Environmental Justice Advocacy Services), Texas Southern University Environmental Justice Law Clinic, Nuestra Palabra: Latino Writers Having Their Say (KPFT FM 90.1), MECA – Houston (Multi-Cultural Education & Counseling Through the Arts), Latino Learning Center – Houston, Communities in the Schools program (HISD), and the

**Fig. 11.4** Community-building/sensory retuning trust exercise: "Floating in the circle" promotes trust and sensory awareness. (Buffalo NY, photo by Bryan Parras)

UTMB Sealy Center for Environmental Health and Medicine and NIEHS P-30 Center in Environmental Toxicology.

The purpose for this TO project reflects the overall focus of the community partner, *De Madres a Madres* on the importance of educating families – particular young mothers – near north side Latino neighborhoods, and city-county agencies on the developmental threat from lead in Houston's older housing. Another key element of this project was related to asthma; the project added an assessment of environmental lead concentrations and asthma triggers in pre-1978 Houston housing stock. The TO experience led to the development of risk messages on lead and asthma and the dissemination of these messages and results of environmental health assessments to affected community members. The TO Forum also provided a platform for evaluating the content and cultural fluency of the project's risk messaging through *spect-actor* interactions during performances. Additional foci were community access to nutritious food, safety needs, and how the relationship of immigration status to social service access in the neighborhoods served by the community hub organization (Sullivan 2005; Sullivan et al. 2006).

The EHL focus was the introduction to the concept of environmental health risk, with a specific focus on childhood lead exposure risk. Performance troupe participants learned about asthma pathogenesis, the influence of triggers on asthma, and exposure pathways for lead and asthma triggers in the context of their cultural beliefs and folk medicine approaches related to lead and asthma. A bi-lingual, native Spanish speaker from de Madres a Madres staff was trained to facilitate the performance events and this proved a major key to the project's success. EHL facts from Project's COAL focus areas were carefully folded into the dialogue for each show (NIEHS EJ 2015; Sullivan et al. 2008).

**Figs. 11.5–11.6** Image theatre: Community actors make images showing community/researcher tension dynamic. (5) Researcher reaches out to neighborhood activist but, (6) the activist ultimately rejects the process in favor of a more familiar and action-oriented grassroots approach. (Buffalo NY, photo by Bryan Parras)

The performances circulated through the neighborhoods – with Holy Name School auditorium and the HISD Community in the Schools program as magnets – in two waves. Wave 1 coincided with the recruitment and initial data- gathering phase of the research cycle. Forums focused on inviting spect-actors from the community performances to collaborate with the actors in developing a respectful, culturally literate bilingual risk message that synced with the community's lived realities. Wave 2 deployed as results from homes surveys, lead and air sampling, and ELISA assays were being analyzed and interpreted by researchers and de Madres a Madres staff. These performances featured close-focus vignettes that illustrated common misperceptions (mostly underestimates) of environmental health threats in the neighborhood. Spect-actors were invited to replace the Forum protagonist and use facts gleaned from the project – and skillful, improvised persuasion – to convince

**Figs. 11.7–11.8** Image theatre: (7) "Help Us Mr. Scientist": Science arrives in the neighborhood with a plan and resources; (8) Community problems are more complex than the tools available – the scientist is surprised, everyone asks "What Comes Next?" (Buffalo NY, photo by Bryan Parras)

the actors playing members of the community resistant to the risk message. The intent was to perform a model of informal, peer-to-peer risk communication and to encourage this ongoing conversation in all the target neighborhoods. In all, *El Teatro lucha* did 28 performances and 5 workshops across the near north side (Sullivan 2005).

Outcomes of this project include closer cooperation among all agencies involved with childhood lead poisoning prevention in Houston, and an increase in the environmental health literacy and research skills of community members and the neighborhood organization. Personnel from *De Madres a Madres* were trained as Citizen Scientists to collect indoor air samples targeting residential carbon monoxide, formaldehyde, and mold spores (i.e., asthma triggers). They also collected paint chip and dust samples, and outdoor soil samples along the drip lines of residential structures for laboratory analysis to quantify residential lead in the neighborhoods. In addition to these benefits for community members, the scientists involved in the research became more cognizant of how cultural beliefs – such as the practice of traditional folk medicine – influence risk perception (Sullivan 2005; Sullivan et al. 2006; Ward et al. 2006).

The increase in community EHL pertaining to lead exposure and asthma risks led to the development of an autonomous, bilingual TO troupe – *El Teatro lucha por la salud del barrio* – that continues to provide or consult on performances and TO facilitation services throughout the Houston Metropolitan area (Adams and Goldbard 2001; Goldbard 2006).

At the mid-point of Project COAL (August 2005), participants from the collaborating community-based organizations, and the City of Houston, the UTMB Centers research teams, representatives from the HISD Communities in the Schools, and members of the TO facilitation team convened a TO hybrid workshop to troubleshoot and repair channels of communication within the project. The primary purposes of this gathering were to ensure that families living in homes identified as moderate/high lead risk were connected with City of Houston programs that provide residential lead abatement, and that risk messages pertaining to childhood lead exposure were consistent and effective (Sullivan et al. 2006).

The workshop adapted a series of Sociometric exercises to answer questions about the Project COAL's role in connecting lead exposed families with necessary services. Sociometry structures used in this mini-project focused on communication glitches, redundancies, and confusion associated with families that "fell through the cracks" and weren't able to access programmatic efforts by the City of Houston to abate existing lead-based hazards (Sullivan et al. 2006). The Sociometry session concluded with a TO Image Theatre-based structure called "Choosing Effective Allies/Talking to Power." *De Madres a Madres* used this process – based on an embodied structure called the "Social Atom" – to identify leverage points and leaders in the Houston public health sector that could directly assist them with next steps in their lead-abatement focused community health campaign (Schutzman 1994; Sullivan and Parras 2008; Sullivan 2010).

This additional TO session also helped to identify blind alleys in communication channels among the collaborating partners and led to modifications in the City of Houston and Harris County lead abatement referral process (Figs. 11.9–11.12).

Fig. 11.9 Image theatre: El Teatro lucha por la salud del barrio sets up fluid (moving) image at the top of the Forum. (Houston TX, photo by Karla Held)

Fig. 11.10 Forum theatre: in freeze frame shot from forum, "Dad Suits Up" for safe lead abatement in the family home. (Houston TX, photo by Rene Magdeleno)

**Fig. 11.11** Image theatre workshop process: Teatro member throws up his arms in frustration when landlord tells him that doing safe lead abatement on his family's rental will hurt the rental company's bottom line. (Project COAL Forum Performance – Houston TX, photo by Maria Murillo)

**Fig. 11.12** Forum performance: "I don't want to hear about it!!" Project COAL Teatro makes image showing "conventional" neighborhood reaction to hearing about environmental issues where they live. (Project Coal Forum Performance – Hoston TX, photo by Rene Magdeleno)

## An Environmental Health Literacy Community Arts Framework for Evaluating Impacts of TO Forum

### *Projects in Public Health*

Many environmental public health research projects include an evaluation phase based on specific metrics of success established at project start. This type of evaluation is used to assess that project goals are met as well as provide an opportunity for a longer-term assessment of a project's broader impact on a scientific field or on public policy. Evaluation is particularly important in CBPR projects to ensure that research goals are achieved and scientific questions answered, as well as progress made on the community desired outcomes. Since CBPR principles and the democratic approaches of Freire and Boal influenced the development of TO at EHS research centers, the TO facilitation team likewise developed a dual lens approach to evaluating the outcomes of the Forum. Their metrics of success were based on community research goals and targeted environmental health literacy themes determined by the project focus (Lens #1), and a set of inherently more subjective Community Arts Process goals that applied more directly to the arts process, and the self-evaluation of the TO facilitator (Lens #2). The items in *Lens #1* are the most relevant to this chapter as they apply to the overarching scientific design of the project. Applying this lens enabled the investigators to assess, among other things, the value of TO for raising EHL in specific situations (Lacy 1994; Young 2012).

### *Lens #1 - Scientific Perspective*

- Did the Forum provide useful community input for design of research and evidence-based interventions (*integrating local knowledge, expanding scope of available data*)?
- Did the Forum promote an atmosphere of trust and dispel any unwarranted community skepticism toward scientific collaboration (*gauging* the effectiveness of *TO as a bridging process*)?
- Did the workshop process effectively communicate toxicological concepts, ideas of risk, and assessments of reasonable scientific risk, and was this translational efficacy tracked and measured (*documented enhancement of/growth in environmental health literacy*)?
- Did this Forum provide a platform for future collaboration (*research momentum and sustainability: a difficult to measure phenomenon that may mature over time*)?

## Using TO in a Public Health Context Does Pose Challenges

It would be disingenuous to claim that Theatre of the Oppressed is the best outreach and/or educational modality for every community-based research context. Or that managing and facilitating TO projects ever becomes a by-rote and uneventful process. Over the course of the Community Environmental Forum Theatre project, CET TO facilitation team members have been required to use their improvisational mind set to remove or avoid a variety of conceptual and physical obstacles, and to solve many (mostly logistical) problems to keep their projects moving forward. The following list represents some of the more significant challenges encountered in doing this work:

- The facilitator needs a broad understanding of the health context for the project: environmental exposures and relationship to disease outcomes, concept of exposure pathways for mapping with Image Theatre, some concept of regulatory policy and its actual history of application in the project community, some knowledge of barriers to health care and other social services access in the community, some knowledge of the power dynamic in the community (relationship of town or city to neighborhood, race dynamics/socio-economic factors). This understanding of both site-specific science and the lived reality of the community in relation to both the scientific focus and the daily experience of cumulative risk harkens back to cultural fluency and allowing the necessary time for sufficient on-site research prior to the first TO workshop with community actors.
- Ideally, a facilitator should be personally/physically familiar with the community, and vice-versa. This may entail taking a "Toxic Tour" facilitated by community members, attending community meetings focused on secondary issues that may or may not be directly relevant to the project, and having conversations with public health practitioners whose practice area includes the focus community.
- Facilitator for these projects should be relatively adept in using the TO process in a variety of social, cultural, and linguistic contexts. This may indicate a need for specific training in cultural fluency or choosing a facilitator who can speak the local language, which may influence the timeline for the project.
- Improving public health and sustaining EHL and the increased awareness developed during TO requires multiple stages of engagement beyond the initial Forum. Ideally, the process would segue into a more systematic study of specific exposures/health impacts pertinent to the community partner. This reflects TO's function as a bridge; but the bridge needs to have a real destination. What Boal calls the *catharsis of activation* that follows a well-managed Forum predisposes Forum participants to press for more energetic advocacy, action-oriented research focused on their identified needs, and ultimately, an activist posture on the issue (Boal, 1998). If these potentials find no useful outlet, both community hope for needed change and enthusiasm for succeeding research and continued struggle may ebb. The outcome of a Forum with no segue into research to action could

reinforce the status quo and produce cynicism toward any further collaboration on research (Camacho 1998).

- Communities with environmental justice issues tend to experience marginalization and sometimes outright oppression within their local/regional power dynamic. The use of close contact modes of outreach like TO encourage personal expression and valorizes the opinions expressed by all stakeholders. These issues may be sources of fruitful dialogue during the project. But they may also be sources of contention (Sullivan 2008). TO allows these issues to be respectfully and authentically addressed regardless of the complications (Sullivan 2004).

Other challenges relate to the logistics and evaluation of TO performances in the context of EHS research and its role in research, outreach, education and development of EHL. This includes:

- The time, effort and physical energy necessary for organizing, managing, facilitating and processing a TO Forum project is significant. The time necessary for full community participation can be extensive (up to 15 contact hours) and finding a common schedule may be difficult. This also applies to research team members who choose to fully participate in the workshop/performance process. Stipends for community participants are absolutely necessary as compensation for the "off-the-clock" personal time they invest in completing a Forum project.
- Because there are no existing *"sabermetrics"* to quantify many of this modality's most significant impacts, and some of these are firmly rooted in the subjective/affective realm, the benefits and challenges of TO must be carefully pitched to both communities and public health scientists. Video documentation of the TO process is growing (much of material is archived or available for purchase at *www.ptoweb.org*). Video and anecdotal reactions from prior participants are always the most convincing forms of testimonial evidence.

Given these challenges, TO can not be construed as a trans-contextual panacea for outreach, education and maintaining communication channels within a CBPR context. Sometimes it is an appropriate modality and all partners readily agree to go that route. And at other times, a proposal to use TO falls flat for a variety of reasons – the most common, simple unfamiliarity with the process. This modality can be a powerful outreach component in community-based research, but it should never be the only tool available (Lacy 1994; McCloskey et al. 2011; Young 2012; CfCH&D 2017).

## TO Is an Undeniable Challenge, but Its Rewards Are Worth the Challenges

In conclusion, TO has proven to be effective in enhancing environmental health literacy within the context of public health oriented community research. This modality helps build capacity for citizens and community-based organizations to collaborate on increasingly more complex public health research. Because of its improvisational and non-directive nature, TO opens new possibilities, provides unique and often unexpected perspectives on issues that potentiate the development of fresh solutions to previously confounding public health problems.

The philosophical core, sequential structure and democratic knowledge-making mechanics of the Forum process reflect the values and working principles of Community Based Participatory Research (Wallerstein and Duran 2003; Fischer 2000). This makes TO an excellent platform for deepening the inter-group dynamics that make CBPR such a rich and ultimately satisfying approach to public health research, and accelerates the time frame for doing such research on a truly granular level.

The TO process forefronts a community's local knowledge of important public health data, such as potential exposure pathways and a wide range of important community stressors that might otherwise be difficult – and more time-consuming – to identify. The addition of local knowledge to the research mix – especially in the area of environmental health sciences – adds power to the research (Corburn 2005) and strengthens the inclusive, reciprocal, multi-directional democratic process codified in CBPR (Fischer 2009; Wallerstein and Duran 2003). It also personalizes data and provides a more intimate context for the negative outcomes of the public health issues under scrutiny. Using TO promotes greater awareness of the distributive justice implications of research by drawing precise focus on "who are the people at risk, and what do they need, not merely what level of risk is acceptable" (Corburn 2005).

The most deep-seated impact of embedding TO in community research has been its facility to promote sharing of unique lived realities – what German social philosophers Husserl and Habermas would term *life-ways*, largely impenetrable from a detached perspective – and how this sharing sharply increases empathy and trust within the entire group of community members, community advocates, environmental health activists, and research team members who personally participate (Horster 1992; Siqueira 2009).

Direct participation in the TO process promotes development of affective skills, awareness of the social/political dynamic based on concrete experience and a bias toward evidence-based action. This applies to all participants, community organization members, local political/economic leaders, public health practitioners, and community research teams, alike. In the words of TO's founder, Augusto Boal: "Through the ongoing dialogue of Theatre of the Oppressed, we penetrate the mirror of our given reality [in this case: toxic exposures, lacking access to health care, safety, equity or procedural justice] and transform the image we see in that mirror. Then we bring that transformed image back into our lives" (Boal 2005).

## Appendix 1: TO-Based Projects

**Exposure Concerns and TO Process Outcomes in an environmental public health context**

| Name of Event | Location | Date(s) | Exposure Concern | Outcome |
|---|---|---|---|---|
| 2nd Ward/East Side Project with Nuestra Palabra | Houston TX | 2002 | Water quality | Advocacy for bayou clean up |
| Mothers for Clean Air/5th Ward Branch (Bruce Elementary School) | Houston TX | 2004 | Air, Soil, Water quality; proximity to Superfund site; fence-line safety issues | Clarity on exposures to air pollution, metals |
| Corpus Christi Citizens for Environmental Justice | Corpus Christi TX | 2004 | Air quality: fence-line safety issues | Cancer, asthma, and exposure pathways |
| Community In-Power & Development Association | Port Arthur TX | 2004 | Air quality: fence-line safety issues | Clarify exposure pathways, healthy homes |
| Mothers for Clean Air Leadership Development Project, Sims Bayou Nature Center | Houston TX | 2004–2006 | Air, Soil, Water quality | Techniques for peer-peer risk communication |
| Seabrook/Shore Acres: Clean Air Clearlake | Clear Lake X | 2005 | Air quality | Clarify exposure pathways, cancer risk |
| Health Care Access Project (St. Vincent's House/IMH | Anchorage Galveston TX | 2005 | Exposures in homeless population, health care access | Enhanced advocacy for access |
| AK Environmental Forum Charlton Park | Anchorage AK | 2007 | Water, Soil quality (metals) | Cancers in Native Alaskans |
| "Behind the Fence": Toxic Waste & Lupus Coalition | Buffalo NY | 2007–2008 | Air, Soil, Water quality: metals, PCBs | Lupus clusters, residential lead levels |
| Charlton Park: Mothers for Clean Air | Houston TX | 2007 | Air quality (diesel fumes) | Asthma, Enhanced advocacy |
| Moving Forward: Ports/Goods Movement Conference | Carson City CA | 2009–2010 | Air quality: Multi point source | Exposure pathways; proximity to roads, respiratory health |
| Houston Inter-Faith Workers Justice Center | Houston TX | 2010 | Air quality: fumes from barges & welding | Correct uses of safety gear, clarification of workers rights |
| International Peoples Health University/Rebel Diaz Collective | Bronx, New York City | 2011 | Air quality | Identifying point sources/exposure pathways |
| Community/Campus Partnerships for Health Conference | Houston TX | 2012 | Health care access for homeless & undocumented | Clarify avenues of access/advocacy for policy |

(continued)

| Name of Event | Location | Date(s) | Exposure Concern | Outcome |
| --- | --- | --- | --- | --- |
| Southwest Workers Union | San Antonio, TX | 2013 | Air, Water quality | Clarity on pollution sources from fracking |
| Global Health Education/Health Literacy Conference | Galveston TX | 2014–2017 | Vector -borne tropical diseases: dengue, chikungunya | International transmission process |

## References


Adams, D., & Goldbard, A. (2001). *Creative community: The art of cultural development.* New York City: The Rockefeller Foundation.

Agency for Toxic Substances & Disease Registry. (2005, revised). *Public health assessment: Guidance manual.* Atlanta: U.S. Department of Health & Human Services; Public Health Service.

Boal, A. (1985). *Theatre of the oppressed* (pp. 117–190). New York: Theatre Communications Group.

Boal, A. (1992, Revised 2002). *Games for actors and non-actors.* London: Routledge.

Boal, A. (1998). *Legislative theatre: In which the citizen makes the law through the legislator.* London: Routledge.

Boal, A. (2000). The Structure of the actor's work. In L. Goodman & J. de Gay (Eds.), *The Routledge reader in politics and performance.* London: Routledge.

Boal, A. (2005, June 3). Radio interview; with J. Gonzales. *Democracy Now,* Los Angeles, CA.

Camacho, D. (1998). The environmental justice movement: A political framework. In D. Comacho (Ed.), *Environmental injustices, political struggles: race, class and the environment* (pp. 11–30). Durham: Duke University Press.

Center for Community Health & Development, University of Kansas Community Tool Box. (2017). See: http://ctb.ku.edu/en/table-contents/overview/other-models-promoting-community-health-and-development/preceder-proceder/main

Corburn, J. (2005). *Street science: Community knowledge and environmental health justice.* Cambridge, MA: MIT Press.

Fischer, F. (2000). *Citizens, experts and the environment; the politics of local knowledge.* Durham: Duke University Press.

Fischer, F. (2009). *Democracy & expertise: reorienting policy inquiries.* Oxford/New York City: Oxford university press.

Freire, P. (1971, Revised 2000) *Pedagogy of the oppressed.* New York: Continuum International Publishing Group.

Freire, P. (2004). *Pedagogy of indignation.* Boulder/London: Paradigm Publishers.

Goldbard, A. (2006). *New creative community: The art of cultural development* (pp. 75–76, p. 221). Oakland: New Village Press.

Hale, A. (2005). *Sociometry: Expanding the meaning of the term.* www.sociometry.net/modules.php?NAME=CONTENT&pa=print_page&pid=20

Horster, D. (1992). *Habermas: An Introduction.* London: Pennbridge Books.

Katz, D. L. (2004). Representing your community in community-based participatory research: Difference made & measured. *Preventing Chronic Disease.* [serial online]; http://www.cdc.gov/pcd/issues/2004jan03_0024.htm

Kelley, H. (2004). *The code of silence. Environmental health findings of the community in-power & development association.* Austin: Seed Coalition.

Lacy, S. (1994). Debated territory: Toward a critical language for public art. In S. Lacy (Ed.), *Mapping the terrain: New genre public art*. Seattle: Bay Press.
Legator, M., & Strawn, S. (1993). *Chemical alert: A community action handbook*. Austin: University of Texas Press.
Marineau, R. F. (2007). The birth and development of sociometry: The work and legacy of jacob Moreno (1889–1974). *Social Psychology Quarterly, 70*(4), 322–325.
McCarthy, J., & Galvao, K. (2002). *ARTPAD: A resource for theatre & participatory development*. Manchester: University of Manchester Press.
McCloskey, D. J., Aguilar-Gaxiola, S., & Michener, J. L. (2011). *Principles of community engagement* (2nd ed., pp. 43–55, pp. 107–149). Developed by: US Department of Health & Human Services, National Institutes of Health, Center for Disease control, Agency for Toxic Substances & Disease Registry, Clinical & Translational Science Awards. NIH Publication: #11-7782. U. S. Government Printing Office: Washington, DC.
Mitter, S. (1992). *Systems of rehearsal: Stanislavsky, Brecht, Grotowski, and brook*. New York City: Routledge.
National Environmental Justice Advisory Council. (2004). *Ensuring risk reduction in communities with multiple stressors: Environmental Justice & Cumulative Risks/impacts*. Washington, DC: U.S. EPA.
National Institute for Environmental Health Science. (2015). Communities Organized against Asthma & Lead. In *Advancing environmental justice,* 2015*: Contributions of the National Institute of Environmental Health Sciences division of extramural research and training to environmental justice*: *1998–2012, 40–41*.
Rohd, M. (1998). *Theatre for community conflict and dialogue: The hope is vital training manual*. Portsmouth: Heinemann Drama.
Rustveld, L., Valverde, I., Chenier, R. S., McLaughlin, R. J., Walters, V. S., Sullivan, J., & Jibaja-Weiss, M. (2013). A novel colorectal and cervical cancer education program: Findings from the community network for cancer prevention forum theater program. *Journal of Cancer Education, 28*(4), 684–689. https://doi.org/10.1007/s13187-013-0530-9.
Salas, J. (1996). *Improvising real life*. New Paltz: Tusitala Publishing.
Schutzman, M. (1994). Brechtian shamanism: The political therapy of Augusto Boal. In J. Cohen-Cruz & M. Schutzman (Eds.), *Playing Boal: theatre, therapy, activism*. London: Routledge.
Sexton, K., Needham, L., & Pirkle, J. (2003). Human bio-monitoring of environmental chemicals. *American Scientist, 92*, 38–45.
Siqueira, C. E., & Sullivan, J. (2009). Popular arts & education in community-based participatory research(cbpr): On the subtle craft of developing and enhancing channels for clear conversations among cbpr partners. *New Solutions: A Journal of Environmental and Occupational Health Policy, 19*(4), 399–406.
Sternberg, P., & Garcia, A. (2000). *Sociodrama: Who's in your shoes?* (2nd ed.). Westport: Preager.
Sullivan, J. (2004). Community environmental forum theater: A dramatic model for dialogue and participatory research among citizens & scientists. *Stage of the Art; Winter, 2004*, 17–23.
Sullivan, J. (2005). *El Teatro Lucha de Salud del barrio: Theater and environmental health in Texas*. Community arts network/art in the public interest. http://wayback.archive-it.org/2077/20100906202911/http://www.communityarts.net/readingroom/archivefiles/2005/10/acrobats_of_the.php
Sullivan, J. (2008). *Behind the Fence: Forum theatre on lupus, lead poisoning & environmental justice*. Community Arts Network/Art in the Public Interest. http://wayback.archive-it.org/2077/20100906202832/http://www.communityarts.net/readingroom/archivefiles/2008/02/behind_the_fenc.php
Sullivan, J. (2010). Engaging Neighbors, Transforming toxic realities: Community environmental forum theatre using Augusto Boal's theatre of the oppressed to promote environmental justice. In McLean & Kelly (Ed.), *Creative arts in interdisciplinary practice: Inquiries for hope and change* (pp. 61–93). Detselig: Calgary.

Sullivan, J., & Lloyd, R. S. (2006, December). The forum theatre of Augusto Boal: A dramatic model for dialogue and community-based environmental science. *Local Environment: The International Journal of Justice and Sustainability, 11*(6), 627–646.

Sullivan, J., & Parras, J. (2008). Environmental justice and Augusto Boal's theatre of the oppressed: A unique community tool for outreach, communication, education and advocacy. *Theory in Action, 1*(2), 20–39.

Sullivan, J., Petronella (Croisant), S., Brooks, E., Rudkin, L., & Ward, J. (2006). Using participatory action research methods to troubleshoot communications issues in an NIEHS environmental justice partnerships project. *Texas Public Health Association Journal, 59*(2), 46–49.

Sullivan, J., Petronella (Croisant), S., Brooks, E., Maria, M., Primeau, L., & Ward, J. (2008, March). Theatre of the oppressed & environmental justice communities: A transformational therapy for the body politic. *Journal of Health Psychology, 13*(2), 166–179.

Wallerstein, N., & Duran, B. (2003). The conceptual, historical and practice roots of community based participatory research and related participatory traditions. In M. Minkler & N. Wallerstein (Eds.), *Community –Based participatory research for health* (pp. 27–52). San Francisco: Jossey-Bass.

Ward, J. B., Jr., Castillo, S., Romo, I., Murillo, M., Sullivan, J., Petronella (Croisant), S., Rudkin, L., Hicks, D., Ramanujam, V. M. S., & Brooks, E. G. (2006, November). *The value of a community-academic partnership in assisting families with exposures to lead and asthma triggers* (pp. 4–8). Boston: American Public Health Association Meeting.

Williams, E. M., Anderson, J., Lee, R., White, J., & Hahn-Baker, D. (2009). Behind the fence forum theatre: An arts performance partnership to address lupus and environmental justice. *New Solutions: a Journal of Environmental & Occupational Health Policy, 19*(4), 467–479.

Wing, S. (2005). Environmental justice, science and public health; in environmental health perspectives. *Essays on the Future of Environmental Health Research, 113*(Supplement #2), 54–63.

Young, P. D. (2012, June 1). The Suzanne Lacy network. In *Art in America magazine* (Vol. 1). http://www.artinamericamagazine.com/news-features/magazine/the-suzanne-lacy-network/

# Chapter 12
# Tactile Approaches to Help Learners Visualize Key Processes in Environmental Health Sciences

**Kathleen M. Vandiver**

**Abstract** This chapter describes how hands-on models, or manipulatives, can be employed to improve the environmental health literacy of a variety of people, from science teachers and students in classrooms, to global audiences in large festival gatherings. Environmental health concepts can be quite abstract. For example, the effect of wood smoke on human lungs. People are concerned about the exposure to toxic molecules from the smoke, but find an explanation of the chemical process by which wood smoke harms human health too difficult to fully understand. Hands-on activities and models are a visual and tactile way of communicating essential molecular environmental health concepts in an inviting way without requiring a technical background.

The MIT Edgerton Center Molecule Set (hereafter referred to as the Molecule Set) is one example of an engaging model set that employs a simple design of differently colored LEGO® bricks to represent atoms. The set was designed to teach chemical principles to middle school students, and has evolved to include new topics with a more environmental health emphasis such as climate change and air pollution. The success of the Molecule Set and corresponding lessons stems from a unique collaboration between MIT scientists and the MIT Edgerton Center. This chapter highlights the Molecule Set and other relevant examples where hands-on models have been used to communicate abstract science concepts and improve environmental health literacy.

**Keywords** STEM education · Hands-on models · Carbon dioxide · Climate change · Ocean acidification · Air pollution · Environmental health literacy · Mercury contamination · Manipulatives · Chemical models

K. M. Vandiver (✉)
Center for Environmental Health Sciences, Massachusetts Institute of Technology, Cambridge, MA, USA
e-mail: kathymv@mit.edu



S. Finn, L. R. O'Fallon (eds.), *Environmental Health Literacy*,
https://doi.org/10.1007/978-3-319-94108-0_12

## Introduction

General observations over the course of history record the usefulness of physical objects to solve problems and share knowledge, as well as the value of experiential learning for vocational training. The abacus used in China and the *quipu* (colorful knotted strings) used by the Incas are examples of such tactile approaches for mathematical calculations and record keeping. In addition, humankind has long relied on apprenticeships where a youth learns by experience, guided by a master of the craft over time. Maria Montessori recognized the importance of learner-centered experiential learning and how it facilitated a greater understanding of abstract concepts. The Montessori approach that utilizes manipulatives for teaching math concepts has been shown to be effective for younger students (Laski et al. 2015). These fundamental learning principles can also be applied to teaching molecular biology or atmospheric chemistry with hands-on manipulatives.

This approach is based on the principle that people learn by doing (Ingmire 2015). In these experiences, the learner's mind is highly engaged with planning what the hands should do. Thus, the mental learning goals can be achieved by manipulating physical objects. Achievement is best when the tasks are mentally consolidated by repetition. Additionally, the challenge of performing any manipulative task can be motivating and affect the learning outcome in a positive way as compared to a passive listening experience. Motivation, the desire to learn more, is the key to processing information (Ormrod 2008). Several researchers have compared knowledge attainment among different pedagogical delivery methods in studies comparing hands-on only lesson delivery with verbal only lesson delivery (Asokanthan 1997; Lipson 2007; Stice 1987). The latter showed that as much as 90% of the knowledge was retained when a concrete experience was given, compared to 20% retention when an abstract concept was provided alone.

## A Modelling Set with one Purpose

The set was designed with one purpose – to communicate abstract concepts in concrete ways.

Each lesson is grounded in the use of LEGO bricks and instructional mats. Building tasks using the bricks are guided by pictures on placemats to provide immediate, positive feedback. In the last step of building their model, students place their structures on top of full-scale pictures to check for correctness; this step provides immediate, positive feedback for the tactile experiences. While the target age group in school settings is grade six through grade ten, because the materials summarize key concepts quickly and well, community colleges and colleges have used them successfully for non-scientific audiences. Thus, such kit-based, hands-on approach can help students of different ages and from different backgrounds better understand how environmental exposures affect their health.

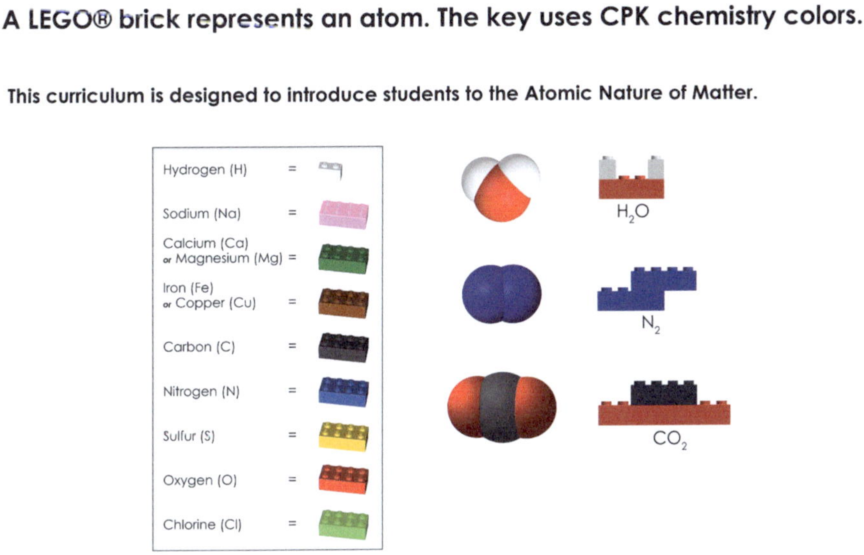


**Fig. 12.1** Colors, shapes, and connectivity are important in this model

The sets come with a well-designed teacher guide to help facilitate the classroom activities. The guides enable the teachers to be properly prepared for the classroom activities. Additionally, this guided inquiry approach allows the student to explore and discover the concepts for themselves and may stimulate the students' desire to learn more.

LEGO bricks embody a favorite classroom teaching analogy – elements as the building blocks of nature – but more importantly, the bricks help to convey several key, abstract chemical concepts. Colors represent the atoms of different elements, which helps students become familiar with standardized chemical colors, such as black for carbon, red for oxygen, and white for hydrogen. Brick sizes also have meaning and help students understand the different atom sizes. The bricks connect to each other, just like atoms in nature cling together to form molecules.[1] Students can create molecules by pressing bricks together, rather than using small sticks or springs as in other molecular modeling systems. The simplicity of the bricks keeps the students and teacher focused on the primary instructional goal – using the models to illustrate the essential introductory concepts and definitions for: atoms, molecules, elements, compounds, mixtures, and chemical reactions.

The instructional placemats emphasize that molecules take specific, functional shapes. Students use the placemats to confirm they have assembled the bricks in the correct shape. Although exact chemical bond angles cannot be duplicated with the bricks, each molecular model has a designated, stylized space-filled shape that is consistent across lessons. For example, the $CO_2$ model has one carbon attached to two separate oxygens (See Fig. 12.1.)

[1] The word "molecule" comes from the Latin words meaning "little lump." ("Mole" = mass or lump of matter and "cule" = tiny or little.)

## Filling Educational Gaps

The atomic nature of matter is one of the major cross-cutting concepts listed in the Next Generation Science Standards (NGSS). Furthermore, the NGSS identifies working with models as one of the eight key science and engineering practices for students. Thus the curriculum avoids the pitfall of many well-designed stand-alone environmental health science programs that are provided to school districts by university staff that cannot be adapted because there isn't enough time in the public-school year to schedule these lessons.

Since an understanding of modern biology has become increasingly focused on molecular content, the Molecule Set helps to address student misconceptions about science and lack of basic chemistry training. The efficacy of this lesson series in helping students better understand abstract concepts has been carefully evaluated and subsequently promoted in the National Science Teachers Association publication, *Toward High School Biology* (Hermann-Abell et al. 2016). The teacher guide and the student workbook employ several modeling methods for atoms and the workbooks include lesson plans that utilize LEGO bricks as atoms, which focus on core chemistry concepts and National Science Standards.

These lessons assure School districts that classes will be well aligned in the context of modeling both disciplinary core ideas, and connecting crosscutting concepts about the atomic nature of matter. Specifically, the Molecule Set embodies how models, including physical representations of atoms, mathematical descriptions, and graphing techniques can be used to explain scientific phenomena across many scientific fields. The inclusion of environmental health content in the Molecule Set curriculum helps make abstract molecular concepts understandable by providing an example of real-world problems.

As previously stated, models make abstract concepts more concrete. This fact is well demonstrated when students learn about subscripts. Subscript numbers are often challenging for students since the subscript is placed *after* the symbol it describes, which is an unexpected location. Using the Molecule Set students can observe both the model and its chemical formula to learn how subscripts work. Teaching this notational method takes time and practice for young students. However, this tactile approach of building molecules makes this intellectual concept visually clear for students.

Word definitions in chemistry often have students falling back on memorization to pass a test, which can be challenging for many students. Tactile approaches can help students understand technical definitions and apply them through sorting exercises with the brick models. Bruner, Goodnow, and Austin first described a technique for concept attainment in 1956, which is often called, "Describe that rule!" This learning approach has students decipher the rules for a game by watching the game (Bruner et al. 1986). In the context of LEGO bricks, molecules are built and then sorted in two categories – an element group and a compound group. Looking at a brick model of $N_2$ (see Fig. 12.1) a student can see why $N_2$ is not a compound. While two atoms are joined together, the molecule is all the same color. Therefore, this molecule is all the same element. A representation of a compound, on the other hand, would need to clearly show the two or more different elements that comprise it. The models help make the concept easier to understand by engaging multiple senses and applying the definition, rather than just requiring memorization of a written definition that students may not understand.

The definition of what is a mixture and what is "not a mixture" can be demonstrated as well with the models. A mixture of water and carbon dioxide can be modeled by building two molecules of $H_2O$ and two molecules of $CO_2$ and placing them nearby each other on the desktop. The molecules remain separate, and they could be easily separated by physical (versus chemical) means. This is the definition of a mixture. However, if a student mistakenly picks up and joins all bricks together into one molecule, she or he would no longer have an example of a molecular mixture. He or she has created a new compound. This becomes a very teachable moment where the student can be directed to write out the molecular formula for this new compound. It's no longer a mixture of separate $CO_2$ and $H_2O$ molecules moving freely, but one compound with the formula of $H_4C_2O_6$.

Given the tactile and visual nature of these activities combined with the activity mats, the models have been shown to benefit English language learners. Students learn definitions by building concrete objects. Seeing the objects at the same time as hearing the name of the object helps cement the learning (Vandiver 2013).

Educational research studies found that many middle school students associate atoms and chemicals only with experiments in test tubes, because students' only educational experience with atoms involves experiments in laboratory settings (AAAS 2018). These lab-based educational experiences lead to major misconceptions about atoms and can leave students ill-prepared for understanding the role of atoms and molecules in their environment and in their bodies. Additionally, there is a common misconception that all "chemicals" are man-made and unhealthy products. Using the various guides, teachers can facilitate the understanding that atoms exist in all kinds of matter when they provide examples of chemical reactions with the same models in multiple fields of science.

## Connecting Chemistry to Real-World Issues

The lesson "Understanding Air" reinforces STEM learning goals about atoms and molecules in the context of climate change, air pollution, and energy. When energy was included in many new state science standards, it was noted that the lesson mats showing the combustion of fossil fuels illustrate a very important exothermic chemical reaction. Employing the instructional placemats and the LEGO bricks, students build molecules and compounds, then rearrange them after combustion. This tactile approach of actively binding atoms, breaking them apart, and reconnecting them as new molecules advances student understanding of chemical reactions and creation of final products, such as air pollution (Figs. 12.2 and 12.3).

**What does a car engine need to make it run?**

**A car needs fuel, air (oxygen), and a spark!**

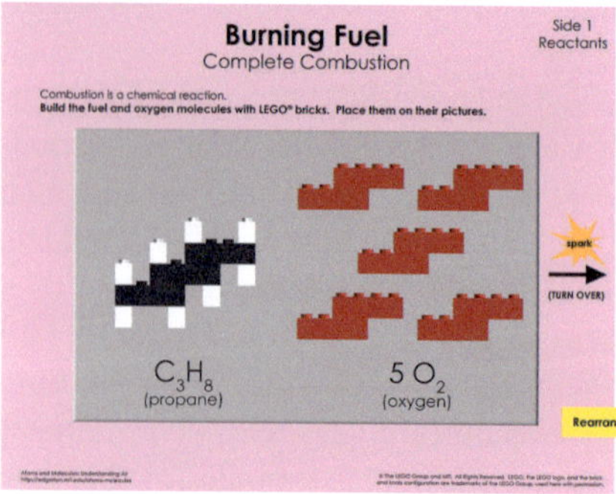


Some vehicles use propane molecules with 3 carbons in the fuel as shown in the simple LEGO reaction. However, most car engines use gasoline with longer hydrocarbon chains like octane, with 8 carbons.

**Introduce combustion (burning) as a chemical reaction.**

**Make the products from the reactants. $CO_2$ is a product.**

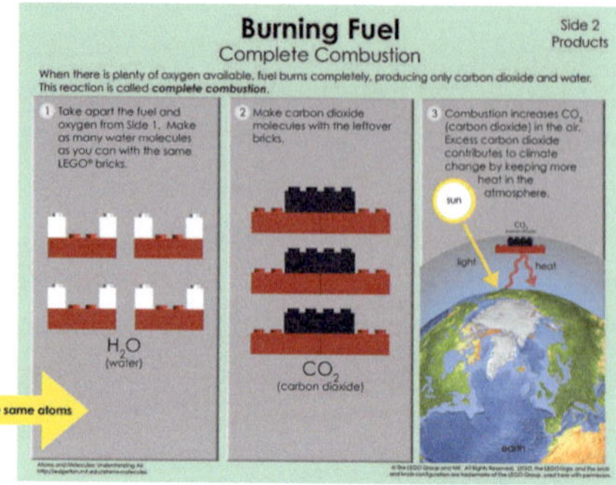


Combustion increases $CO_2$ in the air. Excess $CO_2$ contributes to climate change by keeping more heat in the atmosphere.

$C_3H_8 + 5\,O_2 \rightarrow 4\,H_2O + 3\,CO_2$

**Fig. 12.2** Complete combustion reactants are placed on Side 1. Turn over mat. Create the products with the same atoms!

**What happens if you have too few oxygen molecules?**

**This reaction has 4 instead of 5 $O_2$ molecules.**

**With too few oxygen molecules, pollutants will be produced.**

**The pollutants seen here are:**

**Carbon monoxide**

**Carbon** - Carbon makes black particles of soot.

**Discuss why soot and carbon monoxide are bad for your health.**

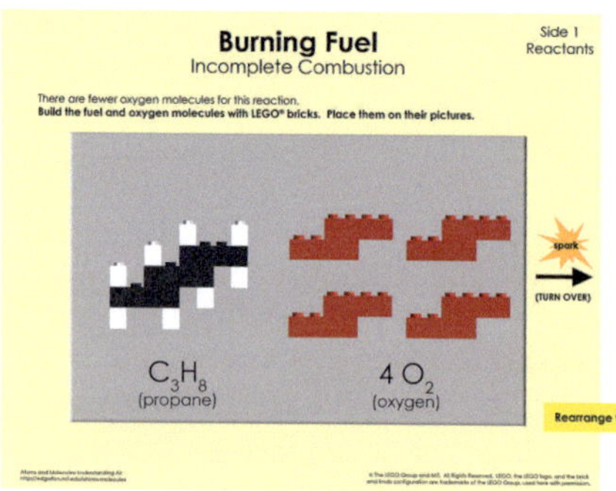


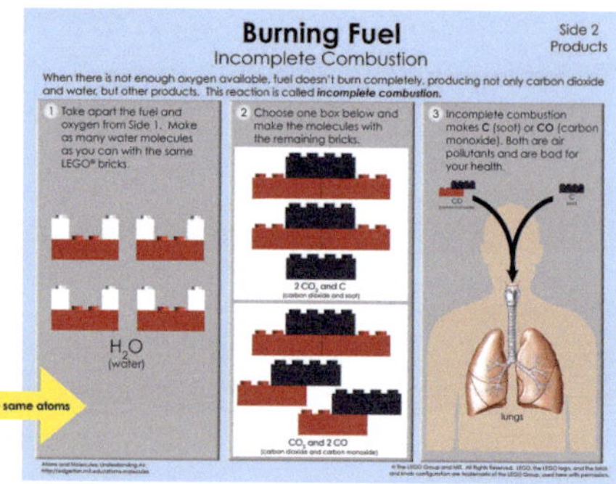


**Fig. 12.3** <u>INCOMPLETE</u> combustion reactants are placed on Side 1. Turn over mat. Create the products with the same atoms!

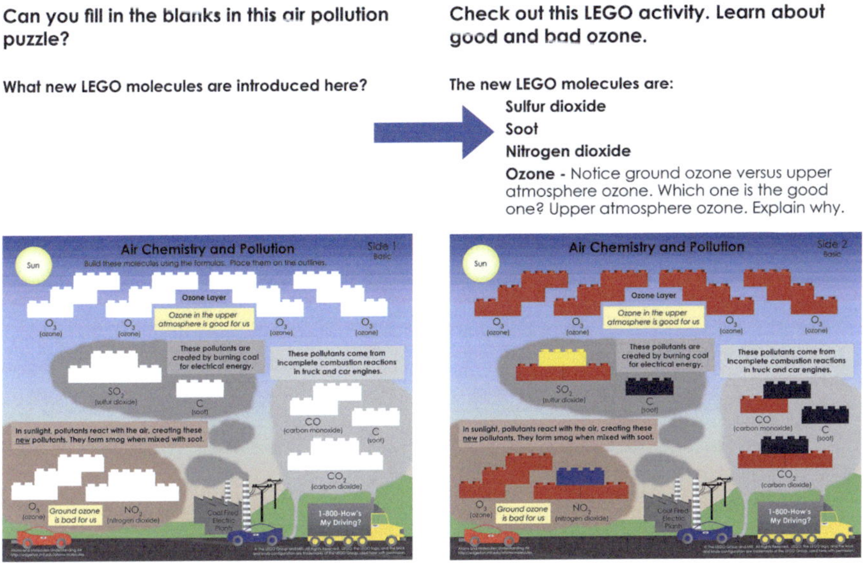


**Fig. 12.4** Air chemistry and pollution. This mat has the five EPA criteria air pollutants depicted with the LEGO brick models

Tactile models combined with evidenced-based teacher materials additionally enable students to reflect on policy issues and how they relate to human health, such as the five Environmental Protection Agency's (EPA's) criteria air pollutants. (Fig. 12.4). The EPA calculates an "Air Quality Index" based on five major air pollutants regulated by the Clean Air Act, i.e., ground ozone ($O_3$), nitrogen dioxide ($NO_2$), and sulfur dioxide ($SO_2$) carbon monoxide (CO) and particulates (also known as particulate matter) (EPA 2016). As students build these air molecules and compounds, they learn about and explore their importance to human health and the environment, and consider the rationale behind certain policies.

The manipulative models, when combined strategically with lesson mats, can be applied to systems thinking such as the inter-relationship of air, water, and the food web. The "Understanding Oceans" lesson shows students how ocean acidification is caused by increased carbon dioxide in the atmosphere. Students assemble and rearrange bricks to observe how excess $CO_2$ from the atmosphere gets absorbed into ocean waters. (See Figs. 12.5 and 12.6).

The bricks demonstrate how the pH scale is used to record the concentration of free hydrogen atoms in the water and how the ocean food webs are interdependent. The Ocean Acidification Mat builds upon the knowledge gained about pH, and prompts students to consider how these changes may affect the food web. Students use the bricks to explore how acidification can have an impact on sea creatures like corals and crabs. By manipulating the molecules, students observe how these sea creatures will no longer be able to build their shells due to the chemical changes in the water. The consequences become obvious as the students are prompted to

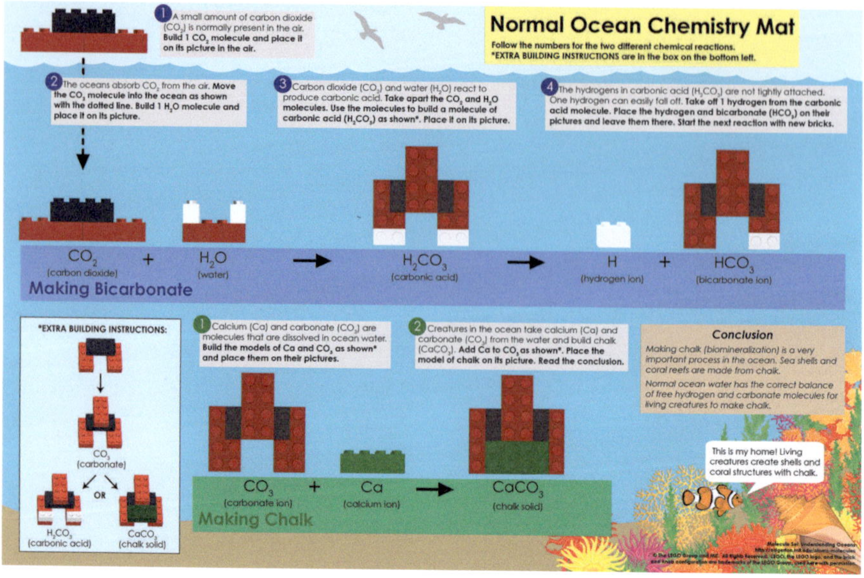


**Fig. 12.5** Normal ocean chemistry mat. This mat demonstrates how small amounts of carbon dioxide entering into the water do not interfere with the pH balance of the ocean water. Sea creatures can make chalk for their shells

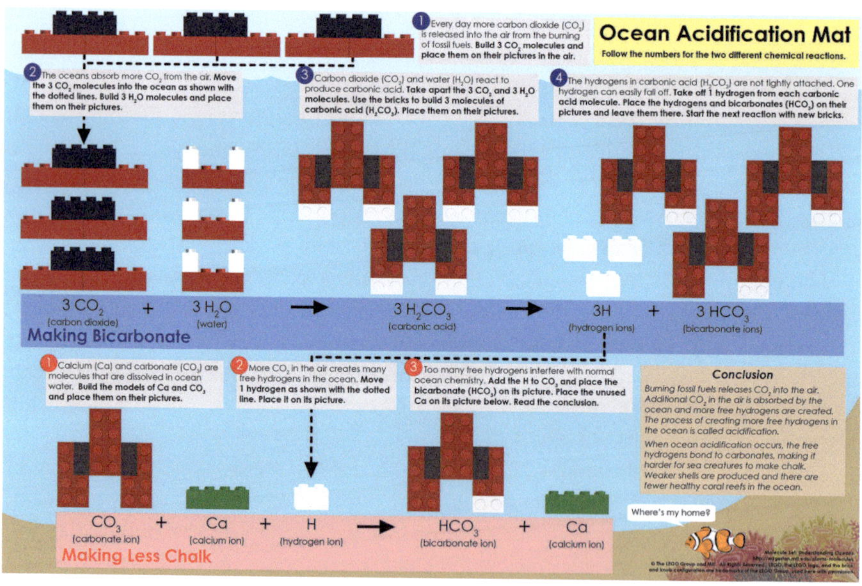


**Fig. 12.6** Ocean acidification mat. A lot of carbon dioxide entering the water causes an increase in number of free hydrogens. Too many hydrogens can interfere with a sea creatures' ability to make a chalk molecule to create their shells. The free hydrogens interfere and prevent the calcium from binding and making chalk. This reaction shown in red: Calcium can't bind. The calcium remains in the water and doesn't form chalk

consider how pH changes will affect numerous interdependent ocean species. Creatures in the ocean are dependent on each other for food, and the loss of one species often affects other species. One example are krill in the Arctic Ocean – the chemical changes that lead to their loss or reduction would affect animals that depend on them for food, including whale species.

Further extension of the mats connects pH changes in the oceans to human health. **Oceans and the pH Scale** mat builds upon the knowledge gained from the previous two lessons and prompts students to think about future implications of small changes over time for human health and strategies for taking corrective action. The mat illustrates the pH predicted for the ocean in the future (e.g., year 2100). Once again, a very abstract concept becomes obvious to the learner. Small changes in the pH number look harmless, a change of just 0.3 pH points; however, as the students work with the bricks, they observe the concentration of hydrogen in the ocean water will more than double.

Teacher materials are essential to making the connections in these activities. While the students are doing the work, and exploring the concepts with the bricks, teachers guide the inquiry process. In this lesson, for example, prompts on the mat, and in the teacher materials, help educators connect the concept of healthier air with healthier oceans, as well as get the students to think about what they can do as individuals and what we can do as a society to reduce harmful exposures. This combination of hands on activities and guided inquiry can be a powerful method for building environmental health literacy.

The "Toxic Mercury in our Environment" lesson further emphasizes the interrelationship of different components of an ecosystem and can serve as a way to challenge students to think about individual and societal solutions. Using the bricks, students learn the physical and chemical process by which mercury in the water from mining and manufacturing, and mercury in the air develop from coal-fired electric plants and volcanic sources. The combination of bricks and mats helps students see and understand that mercury in the atmosphere is the product of these specific industrial and natural sources and has become a global pollutant, since it can be deposited in lakes and streams thousands of miles away from its source.

The interactive lessons help students understand why there are advisories about eating fish. The bricks and mats walk the students through the bioaccumulation process of toxic mercury in larger predator fish, like the tuna. Students use the bricks to learn how soil bacteria take up mercury from contaminated lakes and water bodies and how the chemical nature of the mercury is altered by the addition of a methyl group ($CH_3$) to the metal. Students observe how the addition of the methyl group alters the fate of the mercury, creating a carbon compound that can enter the food chain.

Through the guided inquiry approach, students learn how fish can accumulate high levels of this mercury if they are predator fish, by eating smaller fish that earlier picked up methylmercury from eating plants that were contaminated from the soil. Teachers guide classroom discussions about solutions to these challenges based on the underlying chemical processes they have just learned about. These problem-based approaches advance the EHL of students by increasing their understanding of

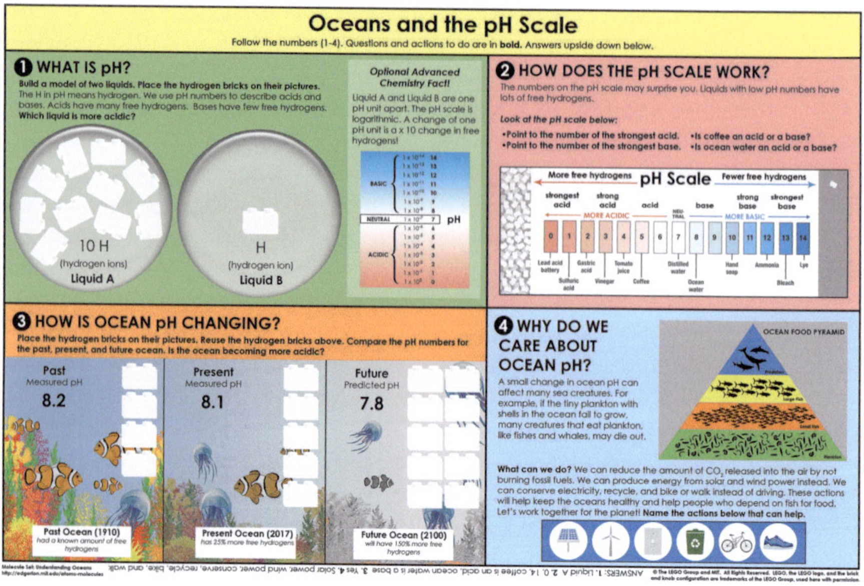


**Fig. 12.7** Oceans and the pH Scale. This mat focuses on the relationship between hydrogen concentration and the pH number scale. Some middle school students may be too young to understand the exponents needed to explain the pH scale. Therefore this mat explains the pH scale in simpler terms

the connection between environmental exposures and health, as well as building their awareness of the actions they can take to reduce those exposures (Fig. 12.7).

The Molecule Set demonstrates the potential of tactile learning to build the environmental health literacy of both students and teachers. It is one example of the value of providing engaging classroom-ready materials for teachers that enable them to integrate lessons from biology and chemistry. Such standards-based kits could lead to the inclusion of more environmental health education into mainstream science classes. Science programs in grades six through eight offer the best opportunity because curricula for these grades overlap with environmental health literacy topics almost completely. Science subjects in middle schools are also taught with greater sequencing flexibility than at the high school level. Middle schools often have a spiraling or integrated approach that touches upon several different branches of science in one year, unlike the high school "layer-cake" sequence for the sciences with a separate year for each of subject such as biology, chemistry, and physics (Education 2016; Osland 2005). Given this educational context, well-designed lessons that combine the use of models, activity placemats, and teacher guides can increase the EHL of students and teachers (Fig. 12.8).

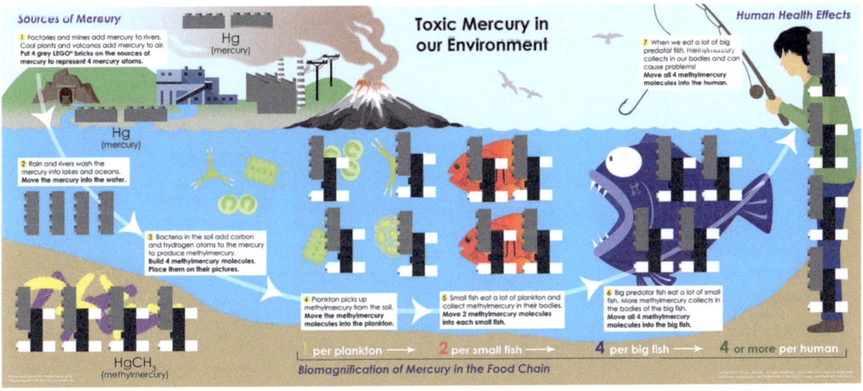


**Fig. 12.8** Toxic mercury in our environment. Four sources of mercury and the important role of microorganisms are depicted. Learners act out the steps with the bricks. The bio-magnification of mercury in the food chain helps to explain the fish consumption limits. The public environmental health messaging with this activity at informal science events with the public can be enhanced by providing information about local fish consumption advisories. Information about the dietary benefits of fish consumption to help the public find a health balance for fish as a nutritious food source can also be provided

## Manipulatives Outside the Classroom

Tactile learning can be used just as effectively outside the classroom in more informal settings. Observational research of volunteers using the "Understanding Air" lessons at informal science education events indicate that this activity can be used to build the EHL of the general public. The results suggest that "Understanding Air," which is designed for middle school audiences, is at an appropriate informational level to provide an effective tool for communicating health science information to the general public.

It is important for educators to track the efficacy of such teaching aids. However, finding a well-accepted evaluation instrument for analyzing responses independent of participant age and topic can be challenging. The "System for Assessing Mission Impact (SAMI)," which science museum professionals developed, has been validated and shared by the St. Louis Science Center. This evaluation instrument was originally designed to collect and summarize key performance indicators for educational programs at the St. Louis Science Center (Center 2014). The **SAMI Card** that they developed is a deceptively simple survey instrument, yet effective. Three simple questions assess and measure the following information: (1) a measure of self-reported learning about the topic; (2) a rating of how enjoyable the activity was; and (3) whether the activity raised their interest in learning more about the subject. Question 3 is a key question as it measures whether the educational activity has increased the person's interest in the subject area not just whether the information was understandable.

**Fig. 12.9** The American Association for the Advancement of Science (AAAS) "Family science days" event. This was our first informal tabling event. The tables had 4 different activity stations, each with a different activity mat from the "Understanding Air" curriculum. The event took place in Boston MA. The LEGO bricks attracted many people. All ages were eager to try the activities

A small study compared an 8th grade classroom presentation at an urban middle school with an informal event using the same "Understanding Air" materials. The informal education event took place at booths at "Science Family Days," sponsored by the American Association for the Advancement of Science (AAAS) in downtown Boston, MA in February 2013. Over two weekend days, about 250 people learned about climate change and air pollution at the "Understanding Air" booth, which was arranged with four stations at the tables, each with a different LEGO activity mat. These four mats were: What is Air Made Of? Guess!; Complete Combustion; Incomplete Combustion; and Air Chemistry and Pollution. The students were also shown a PM 2.5 counter demonstration. An MIT undergraduate or graduate student stood at each station to explain the concepts. He or she facilitated the activity and answered the participant's questions. The children, many who were attracted to the tables by the sight of the LEGO bricks, were accompanied by their parents, who also showed an interest in playing with the LEGO bricks. Consequently, both the parents and the youth learned about air pollution and the potential human health effects (Fig. 12.9).

The collected SAMI card results from the youth participants were compared to the classroom-led activity with the same hands-on materials at a nearby urban middle school, in Revere Massachusetts. Unfortunately, the adult responses with SAMI cards at the AAAS booth were not analyzed because the focus was on the youth data for comparison (See Fig. 12.10).

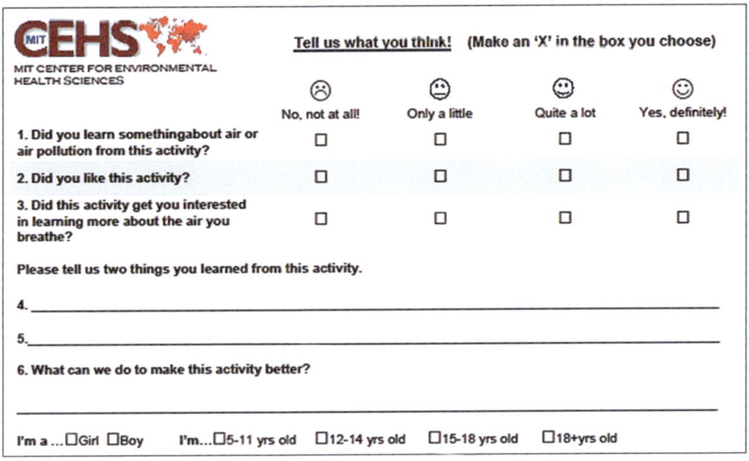
CEHS MIT
MIT CENTER FOR ENVIRONMENTAL HEALTH SCIENCES

**Tell us what you think!** **(Make an 'X' in the box you choose)**

| | No, not at all! | Only a little | Quite a lot | Yes, definitely! |
|---|---|---|---|---|
| **1. Did you learn somethingabout air or air pollution from this activity?** | ☐ | ☐ | ☐ | ☐ |
| **2. Did you like this activity?** | ☐ | ☐ | ☐ | ☐ |
| **3. Did this activity get you interested in learning more about the air you breathe?** | ☐ | ☐ | ☐ | ☐ |

**Please tell us two things you learned from this activity.**

**4.** ____________________

**5.** ____________________

**6. What can we do to make this activity better?**

____________________

**I'm a …☐Girl ☐Boy I'm…☐5-11 yrs old ☐12-14 yrs old ☐15-18 yrs old ☐18+yrs old**

**Fig. 12.10** System for assessing mission impact (SAMI) card. This half-sheet of paper was used to obtain responses to the activity "Understanding Air" at both informal and formal settings. The Likert scale questions used pictures (instead of numbers like 1–4) to improve accuracy. The answers to the two open ended questions were later catalogued by subject. Lastly, the age and gender of the participant is recorded

| Informal Setting for families: AAAS Family Science Days, Boston, MA Feb 2013 (n=88) (ages: 73%= 5-11 yr; 15 %= 12-14; 2%=18+) (gender 61%=M; 39%=F) | Eighth Grade Students in Classroom: Susan B. Anthony M.S., Revere, MA Nov 2012 (n=137) (ages: 89%=12-14 yr; 11% =15-18) (gender: 46%=M; 54%=F) |
|---|---|
| 1 Did you like this activity? Average = **3.8 ± 0.4** | 1 Did you like this activity? Average = **3.4 ± 0.6** |
| 2 Did you learn from this activity? Average = **3.3 ± 0.6** | 2 Did you learn from this activity? Average = **3.3 ± 0.6** |
| 3 Would you like to learn more? Average = **3.5 ± 0.8** | 3 Would you like to learn more? Average = **3.2 ± 0.6** |

**Fig. 12.11** Data from informal and formal learning settings for the "understanding air" curriculum is shown

Response to the "Understanding Air" activities was highly positive in settings with the general public as well as with the 8th grade students (Fig. 12.11**).** These surveys used Likert Scales; the scores to the questions for both settings averaged between 3.2 and 3.8 (out of 4.0). The results show that the informal group was a little younger and scored themselves higher on: (1) learning, (2) enjoyment and (3) interest. In the open-ended questions, the informal group subjects' top responses were that they learned about: the composition of air, how to write formulas, and the existence of particles in the air). Overall, the 8th graders' top responses were that they learned about: the composition of air and the chemical reaction of combustion. In addition, health was mentioned as a topic learned in 11% of the responses in both audiences. Taken together, these results are strong indicators that the key concepts for climate change and air pollution were acquired through this activity. These responses indicate that the "Understanding Air" curriculum material can deliver an effective environmental health outreach program to students in classrooms as well as to general audiences at public events (Fig. 12.11).

In addition to educating English speaking students, the Molecule Sets have been an effective tool for English as a Second Language students. The Molecule Sets and the "Understanding Air" curriculum can also be adapted for other languages and cultures based on their use of visual and tactile methods of communication rather than the written word. Given this emphasis on visual information, the set and curriculum were relatively easy to translate and implement at large urban Science Festivals in Asia and in Mexico. For example, preparations for the Cambodia (March 2015) event revealed that the activity mats for "Understanding Air" have minimal English text, which made it possible to easily translate the activities into Cambodian (Khmer). Using a train-the-trainer approach, local volunteers received training to host the table activities at these events and to educate the public. Bilingual university volunteers were selected for their strong background in chemistry and received thorough training before teaching the lessons at the festival booths. This approach enabled culturally and linguistically appropriate communication by engaging with participants in their own language and presenting materials that had been translated into their local language(s). The Cambodian Science and Engineering Festival was the first one ever held – it brought in a crowd of over 10,000 participants in 3 days! The organizers expected only 3000. (See Fig.12.12).

The Cambodian Science and Engineering Festival was specifically designed to encourage more Cambodian youth to enter science and engineering fields. The Molecule Set was well-received by all ages with continued popularity and attendance for the two successive years it was used at the environmental health booth. The government wanted more young people to become engineers and scientists to sustainably grow the country's infrastructure and to manage the growing population's environmental health issues. These issues include serious urban air pollution problems, created by the huge numbers of unregulated motorbikes and the high number of unpaved roads. In this context, the "Understanding Air" activities helped to fill an identified need and address a lived experience for Cambodian residents.

The training approach for the Beijing Science Festival was slightly modified from that used in Cambodia. In Beijing, approximately 30 bilingual middle school volunteers, selected from top middle schools in the region, received daily training for 4 days. These students worked at the eight activity tables and demonstrated the molecule sets. These students felt honored to be able to participate along with the university partners from MIT at the Beijing Science Festival. Similar to the experience in Cambodia, the Beijing residents found the "Understanding Air" lessons particularly relevant in the context of the poor air quality in their city. Results show that participants liked the activity and came away from the experience more knowledgeable about air pollution and its effect on their health.

In November 2016, the MIT team partnered with an outreach program at the Ecological Institute of Mexico, INECOL, in Veracruz, Mexico. The teams worked together to translate the entire Molecule Set curriculum and teacher materials into

**Fig. 12.12** The First National Cambodian Science and Engineering Festival took place in Phenom Penh, March 2015. The college volunteers who are the explainers are wearing yellow T-shirts. Kathleen Vandiver is in the middle of the line-up with the yellow shirts. All the materials were translated into Khmer

Spanish. Unlike the other two examples, this collaboration sought a more sustainable outcome; INECOL wanted to continue using the curriculum after the science days. The team trained local teachers as well as staff members at the INECOL Ecological Institute. The training was a success, as the INECOL Institute now hosts events with these materials for students on school trips during the academic year.

The Molecule Set is gaining international exposure from the MIT Blossoms website, an online repository of instructional videos with voice-overs versions in different languages. Through this portal, educators from around the world can access classroom activities and teacher guides such as the Photosynthesis and Chemical Reactions lessons, which have voice-over sound tracks in Arabic and Japanese, among other languages (Vandiver 2018a, b). These videos are specially designed to be paused from time to time to have the classroom teacher work with the students in his or her class according to the video teacher's instructions. This approach provides a scaffold for teachers using the video to help them learn new subjects and new teaching techniques.

## Conclusions and Future Directions

Much like the Montessori approach of using manipulatives to teach math concepts, the Molecular Set demonstrates the many benefits of tactile approaches to teach and communicate abstract environmental health concepts in concrete ways. The set provides educators in the United Stated with easy to use, standards-based, classroom-ready lessons. As such, teachers and school systems feel comfortable employing them in the classroom because they fill important educational gaps without requiring much additional work. From a learners' perspective, the set allows students to explore integrated concepts of biology and chemistry in a hands-on fashion and receive immediate feedback on their understanding. As described in this chapter, the sets are just as effective outside the classroom. More importantly, this tactile approach has shown that it can transcend language and culture. These successes have been documented with the SAMI feedback forms, as well as by observing the interest and enthusiasm of both the volunteers and participating members of the public.

The teachers in classrooms and the volunteers at informal events found that the visual, tangible, and kinesthetic methodologies all work to reinforce the environmental health learning. This approach can improve environmental health literacy because the methodology is learner-centered. Overall, hands-on activities with LEGO models help to express the abstract ideas about atoms in a concrete way. People can experience the concepts with their hands as well as visualize them with their eyes.

As noted in the STEM education chapter, the success of teacher training relies on quality materials (Chap. 7). The LEGO bricks represent just one part of the set. The well-designed activity mats and the teacher training guides are the two other essential parts. Material development, especially in the context of environmental health, requires transdisciplinary partnerships. The Molecule Set was developed in partnership among education experts, science communicators, and environmental health researchers. It was the combination of the skill sets of these disciplines that has contributed to the positive environmental health literacy outcomes.

The Molecule Set is just one example of how manipulatives can be used effectively in classrooms. More research is needed to better understand how similar standards-based kits could lead to the inclusion of more environmental health messages into mainstream science classes. While the MIT Blossoms webpage shows some potential for scale up of the Molecule Set lessons domestically and internationally, more work will be required to understand other sustainable approaches.

Manipulatives have demonstrated efficacy for increasing a learner's understanding and awareness of topics. In this case the connection between environment and health. However, the question remains can these hands-on approaches be used to advance EHL to higher stages? This question requires further consideration.

Tactile approaches to learning have shown the ability to transcend language and cultural barriers. Model representations of abstract concepts such as numbers, atoms, molecules, or DNA enable broader utilization because they emphasize imagery over text. As noted in the Molecule Set, it was easy to convert the activity mats because there was minimal text. Further examination into this approach in the context of EHL would be helpful to see how it relates to other communities and cultural contexts.

The Montessori approach to math using manipulatives has been successful in large part to its simplicity, its sustained application over time, and its connection to numbers. Because, EHL relies heavily on several different scientific disciplines such as chemistry, biology, as well as math, it is vital to make a connection to real-world issues to overcome this greater complexity. For example, the "Understanding Air" lessons worked well in Cambodia and Beijing in part because of the concerns related to air pollution. Boston, MA also has a strong connection to air pollution and the ocean which made those lessons particularly relevant to the local students and teachers. It would be appropriate to test these lessons in other parts of the country with different environmental health issues to validate the efficacy of these tactile approaches.

Finally, as educators embrace more digital approaches for education, there are opportunities to consider the efficacy of manipulatives in the virtual space for advancing EHL.

**Acknowledgments**

1. This work was supported by an NIEHS Grant Funded by NIH-NIEHS Core Center Grant P30-ES002109.
2. Many thanks go to the MIT Edgerton Center Director, Prof. J. Kim Vandiver for his continuous support in the production of the MIT Edgerton Center Molecule Sets.
3. Kathleen M. Vandiver would like to acknowledge the talents of Dr. Amanda Gruhl Mayer in the graphic productions of the MIT Edgerton Center Molecule Sets.
4. Special thanks to Liam O'Fallon and Symma Finn for their contributions to the review and editing of the chapter.

**Notes** All MIT Molecule Set lesson plans and supporting materials are available for free download from http://edgerton.mit.edu/molecule-set.

## References

AAAS. (2018). *Project 2061*. Available: https://www.aaas.org/program/project2061. Accessed Feb 2018.

Asokanthan, S. F. (1997). Active learning methods for teaching dynamics-development and implementation. In *Proceedings of the IEEE frontiers in education conference* (pp. 1349–1353). Pittsburgh, PA.

Bruner, J. S., Goodnow, J. J., & Austin, G. A. (1986). *A study of thinking*. New Brunswick: Transaction Books.

Center SLS. (2014). *Opening minds to science*. Available: https://www.slsc.org/2018

Education MDoEaS. (2016). *2016 massachusetts science and technology/engineering curriculum framework.* https://www.doe.mass.edu/frameworks/scitech/2016-04.pdf:Massachusetts. Departmen tof Elementary and Secondary Education.

EPA U. (2016). *Air quality index (aqi) basics.* Available: https://airnow.gov/index.cfm?action=aqibasics.aqi

Hermann-Abell, C. F., Koppal, M., & Joseman, J. E. (2016). Toward high school biology: Helping middle school students understand chemical reactions and conservation of mass in nonliving and living systems. *CBE Life Sciences Education, 15.*

Ingmire, J. (2015). *Learning by doing helps students perform better in science.* Available: https://news.uchicago.edu/article/2015/04/29/learning-doing-helps-students-perform-better-science. Accessed 19 Feb 2018.

Laski, E. V., Jor'dan, J. R., Daoust, C., & Murray, A. K. (2015). What makes mathematics manipulatives effective? Lessons from cognitive science and montessori education. *Sage Open, 5.*

Lipson, H. (2007). *Printable 3d models for customized hands-on education.*

Ormrod, J. E. (2008). *Educational psychology developing learners.*

Osland, A. (2005). *An analysis of urbanization and hazard mitigation practices near hazardous liquid transmission pipelines.* Available: https://cfpub.epa.gov/ncer_abstracts/index.cfm/fuseaction/display.abstractDetail/abstract/7599 2018

Stice, J. (1987). Using kolb's learning cycle to improve student learning. *Engineering Education, 77*, 196–291.

Vandiver, K. M. (2013). In School TsBH (Ed.), *English as a second language (esl) teacher.* Brighton.

Vandiver, K. M. (2018a). *Roots, shoots, and wood.* Available: https://blossoms.mit.edu/videos/lessons/roots_shoots_and_wood 2018

Vandiver, K. M. (2018b). *Recognizing chemical reactions.* Available: https://blossoms.mit.edu/videos/lessons/recognizing_chemical_reactions 2018

# Conclusion: Summary and Next Steps

**Abstract** This book provides the reader with a glimpse of the future for environmental health literacy (EHL) by examining its emergence, implementation and current practice. The authors discuss the connection of EHL to different theories, frameworks, and social movements. Although the exact term EHL and its conceptual framing has only been used in the last 10 years, the case studies throughout the book demonstrate it has existed in practice for much longer. While the authors touch on different aspects of EHL, there are several common themes and recommendations across the chapters. The purpose of this Conclusion is to summarize and outline the recommendations and next steps for EHL.

## Purpose of the Book

A central goal of this book is to prompt discussion about EHL and encourage testing of EHL-specific concepts in environmental health research and communication projects. While the Society of Public Health Education (SOPHE) first defined EHL in 2008, there has existed a need to solidify this concept of what it means to be environmentally health literate and how it can be applied to environmental health research and communication.

The book was organized strategically into three key sections to show the different perspectives of EHL from the vantage point of the authors. The sections flow from foundational aspects and theoretical models (Part I), to the importance and application of EHL in the context of different audiences (Part II), and finally to the different communication approaches to promote and raise EHL (Part III). Although



S. Finn, L. R. O'Fallon (eds.), *Environmental Health Literacy*,
https://doi.org/10.1007/978-3-319-94108-0

each chapter can stand alone as an exploration of a discrete topic, the chapters complement one another and, where appropriate, cross-reference common topics across chapters.

> Fundamentally, environmental health literacy begins with an understanding of the link between environmental exposures and health. (Finn & O'Fallon, 2017)

## Key Themes

### *EHL Builds Upon Extant Concepts and Theoretical Frameworks*

Although this concept is new and emerging, all chapters show how it has strong ties to existing frameworks and tested concepts. The first three chapters delineate in detail the connection to broader frameworks and relevant literacies. Chapters 4, 5, 6, 7, 8, 9, 10, 11 and 12 connect projects to specific theoretical frameworks to make the point that they are grounded in accepted practice. It is this confluence of work and understanding that EHL has gained traction and requires greater attention.

### *Knowledge to Action*

While the foundation of EHL may be the understanding and awareness of the connection between environmental exposures and health, the authors emphasize the importance of translating that knowledge into individual and community-level action. All the authors note many different actions that may be taken by the intended audience, and how some can be indicators of higher levels of EHL. For example, the Chap. 8 authors suggest that one level of EHL for healthcare professionals is simply talking to their patients about environmental health concepts, whereas a higher level of EHL may be reflected by a healthcare practitioner speaking to decision makers about the adverse effects of certain environmental exposures to human health. However, as Chap. 1 cautions, we must value the basic levels of EHL and take care not to stigmatize individuals and communities who are not able or willing to act. Chap. 2 makes the important point that action requires certain skill sets, which when lacking may prevent people from acting on environmental health information. Identifying and considering these skills when developing messages and materials can help to improve EHL. Chap. 3 and other chapters also call out the importance of understanding motivation and self-efficacy and how these attributes may promote or inhibit learning and action.

## *EHL Is About Prevention*

In the context of environmental health, preventing adverse health outcomes is central. All authors note that by raising awareness and enabling action, there is a greater likelihood of promoting health and preventing or mitigating harmful environmental exposures. Thus, being able to quantify and measure EHL is vital for developing better materials and messages (Chap. 2). By understanding the EHL of an intended audience, communicators are better able to construct effective and cultural appropriate messages for their audiences (Chap. 10). Furthermore, the application of dissemination and implementation methods to assess and measure levels of EHL will help investigators and other stakeholders to understand the full impact of materials and messages, and if they helped to achieve the goal of prevention.

## *Diverse Audiences*

Preventing exposures that may lead to adverse health outcomes or contribute to environmental health disparities requires more than just behavior change or action at the individual level. In the context of social determinants of health, EHL can be useful to individuals as well as communities. However, "communities" are not monolithic in their demographics and this often means that there are different audiences to engage with to strengthen EHL, and in some cases, the need for different languages. Several chapters touch upon the importance of knowledge brokers and trusted individuals within communities to help raise EHL. Such individuals include doctors (Chaps. 8 and 10), teachers (Chaps. 7 and 12), and community health workers or *promotoras* (Chaps. 2, 4, and 5). As such, communicators must consider not just the EHL of affected individual or community, but also the EHL of these knowledge brokers who can help to advance EHL. In addition, culture, and historical context play an important role in EHL. Chapters. 2, 4 and 9 emphasize how communication efforts and strategies for addressing EHL, can be strengthened through careful and authentic consideration of the audience's culture, language(s) and context.

## *Role of Community-Engaged Research, Citizen Science, and Popular Education*

Learning about culture and context requires honest engagement with the intended audience. Chapters. 2, 4, 5, 6, 9 and 11 emphasize the importance of dialogue with members of the targeted audience to learn from them to tailor the health risk messages and materials to their needs. These authors highlight several successful approaches that can be applied to initiatives to raise EHL, such as community-engaged research, citizen science, and popular education. The authors note that community-engaged research and citizen science have been useful for co-development of environmental health science and communication projects, and suggest the potential for such approaches in EHL efforts by democratizing the research process, building skills, and increasing knowledge of all partners. Chapter 7 posits that these approaches are also applicable and powerful for STEM education, noting how authentic, problem-based learning opportunities both raises the EHL of students and teachers, as well as inspires them to become more involved in addressing environmental health issues in their community. Chapters 5 and 11 highlight the value of popular education for promoting dialogue and co-learning; however, each takes a different perspective. Chapter 5 focuses on the spoken word between researchers and community members, whereas Chap. 11 draws attention to the value of physical and dramatic expression to communicate complex ideas. The common idea in these approaches is that they promote trust, confirm the value of every stakeholder's knowledge and skills, and encourage co-development of solutions.

## *Visualization of Exposure and Environmental Health Concepts*

Using visual modalities for communicating risk is a hallmark of EHL. GIS maps showing air pollution data (Chaps. 4 and 5) and graphics to communicate personal biomonitoring results to community members (Chap. 6) are examples. These chapters highlight how data helps to visualize the "Lived experience" and "Exposure experience." Chapter 6 provides quotes from community partners, noting how data has helped them to "put a face on pollution."

Chapters 4 and 9 emphasize the importance of using culturally appropriate imagery when communicating environmental health concepts. They note that poorly designed graphics can have unintended consequences. Chapter 10 also emphasizes that in the development of mobile apps, there must be an aesthetic element to the design and display of information to engage with the intended audience.

## *Diverse Partners*

The opening chapter makes the important observation that EHL can unite practitioners, researchers, advocates, industries, and communities in coordinated efforts to raise understanding as a first step in addressing environmental health disparities. The following chapters describe myriad examples of partnerships involving diverse stakeholders in the research process and in communication efforts. More importantly, they all call out the need for new partnerships to advance EHL. Chapter 3 suggests an opportunity to engage neuroscientists to visualize how people respond to messages and Chap. 10 notes the need for new partners in the design of effective digital communication tools.

Chapter 4 reminds us that acknowledging and working with culturally diverse partners is essential when addressing environmental justice issues.

## Recommendations

The authors present an array of recommendations for consideration, ranging from content organization and communication to research-specific ideas. Following are a few key examples:

- **Collect, classify, and share best practices and findings from these and other EHL projects.**
  Disseminating these results and best practices will help to stimulate new questions and validate on-going efforts.
- **Create tools to identify, measure, and assess EHL skills and competencies.**
- **Apply the EHL framework for different exposures and in different communities.**
- **Prioritize formative research and evaluation in communication research addressing EHL.**
  Support and encourage evaluation best practices, such as focus groups, message testing, and collaborative projects across geographic regions and cultures.
- **Support community-engaged projects.**
  Understanding local culture and context helps ensure sustainable and successful efforts to communicate and prevent exposure. Leveraging the capacity of community organizations can facilitate increased EHL. Working with culturally-diverse community groups will help strengthen EHL across all populations.
- **Support environmental health science education programs.**
  STEM learning experiences strengthen EHL of students and teachers. There is great potential in engaging youth. There is also a need to integrate environmental health science into the medical and nursing school curriculum.
- **Promote innovative and diverse partnerships.**

## Next Steps

This book marks the beginning of a new phase for EHL. In this section we call out the next steps and opportunities that the authors have touched upon in their chapters and the common themes and overarching recommendations for this field. It is our hope that this information will motivate readers to consider how they may employ and test many of the concepts and approaches described in this book.

# Index




S. Finn, L. R. O'Fallon (eds.), *Environmental Health Literacy*,
https://doi.org/10.1007/978-3-319-94108-0

## N



## O



## P



## R



## S

MIX
Papier aus verantwortungsvollen Quellen
Paper from responsible sources
FSC® C105338

If you have any concerns about our products,
you can contact us on
**ProductSafety@springernature.com**

In case Publisher is established outside the EU,
the EU authorized representative is:
**Springer Nature Customer Service Center GmbH**
**Europaplatz 3, 69115 Heidelberg, Germany**

Printed by Libri Plureos GmbH
in Hamburg, Germany